Rapid methods

Rapid methods

for biological and chemical contaminants in food and feed

Edited by:
A. van Amerongen
D. Barug
M. Lauwaars

Wageningen Academic Publishers

Subject headings:
detection
testing
residues

ISBN 9076998531

First published, 2005

Wageningen Academic Publishers
The Netherlands, 2005

Foreword

The rapid and reliable detection of biological and chemical contaminants is extremely important in managing the safety of food and feed. "Rapid Methods" is a comprehensive reference resource for anyone interested in this subject.

Developments in analytical techniques have led to the emergence of a wide range of rapid methods to complement the traditional methods. At the same time, the importance of method validation, proficiency testing, quality management, sampling and legislation have all become more widely recognised.

"Rapid Methods" presents a firm base and structured framework for considering rapid analysis of biological and chemical contaminants in food and feed. The various chapters concentrate on the state of the art in rapid methods as regards to:

- legislation;
- sampling;
- method validation;
- microbial pathogens;
- biological materials like GMOs and allergens;
- toxins like bacterial food poisoning toxins, marine toxins and biogenic amines;
- chemicals like veterinary drugs, pesticides and dioxins.

This book has arisen from a very successful international conference on rapid methods held 25-26 March 2004 in Noordwijk, the Netherlands.

The editors firmly believe that the very nature of the theme, the excellence of the peer-reviewed papers and the holistic approach chosen in this book will draw an audience from both the food and feed industry as well as from the scientific community.

The editors

A. van Amerongen
D. Barug
M. Lauwaars

Contents

Why rapid testing?

F.P. Scanlan
Nestlé Research Center Lausanne, Quality & Safety Department, P.O. Box 44, 1000 Lausanne, Switzerland

Abstract

There are several reasons justifying the importance of rapid testing. Rapid and reliable means of contaminant detection are extremely important to managing food, feed and environmental safety. Ensuring "safety" is by far the first and foremost "result" of rapid testing in these industries and takes precedence over any economic factors. The focus of this contribution is to explain some of the many reasons for the importance of rapid testing in microbiological and chemical applications from an industrial food perspective.

Keywords: process control, microbiological rapid tests, chemical rapid tests, food safety

1. Introduction

Over the last decades, safety and economical concerns have been the main driving forces behind the development of rapid test kits, instruments, screening methods as well as the birth of a multitude of companies providing these technologies to industry and government. It should also be noted that rapid intervention for managing product tampering, bio-terrorism and food contamination outbreaks have also led to the development of faster test methods.

Rapid tests help industries in determining the effectiveness of food safety measures, legal compliance as well as achieving logistical and operational goals while saving time and investments in complex instruments and staff qualifications. In some cases they also reduce costs. The main drivers for the development of rapid tests are:

- Regulatory compliance.
  Within the EU there are new rules to comply with on labelling of finished products, traceability from farm to fork and even more emphasis on the safety of food.
- Pressures of modern manufacturing practices.
  The competitive market-place forces industry to reduce costs, employ cheaper labour, and apply such logistic practices as just-in-time and zero-stock.
- Analysis complexity.
  The food industry is faced by an enormous task of controlling for a variety of chemical, biological and microbiological analytes. To do this the analyst will refer to official reference methods e.g. ISO, AOAC, CEN, IDF, USDA AFNOR etc.; in-house developed research methods; rapid or alternative methods for use in the factory or in the laboratory.

2. From lab to line

In keeping with modern QA practices and the need for real-time factory process control, more analytical methods have been gradually transferred from the factory or external lab to the factory floor. Also, where direct on/in-line analysis is not possible, near-line analysis is a solution using either a line- or close to line-lab.

Here it is considered that a rapid method is "any" method replacing an established reference method. Rapid or alternative methods are faster, less expensive, easier to use and can be moved to a near-line or to the line environment. As an example, consider the case of moisture determination for milk powders.

reference method:	drying oven at 102 °C/2 hours; or Karl-Fischer 15-20 minutes.
on-line method:	NIR spectroscopy giving results in seconds. to minutes.
at line method:	halogen balance with results in 5-10 minutes.
off-line / lab method:	NIR spectroscopy, results in seconds to minutes.

What about the cost of implementing a rapid method in the production environment? Another example is given of the measurement of total solids and fat content in an ice cream premix. Table 1 shows the numbers of samples and costs per analyte using the reference method. From this, the daily cost using the reference method would amount to 700 CHF for 10 samples. When using near infrared spectroscopy as an alternative method one would still have to run a single sample using the reference method for checking the NIR method. Therefore, the cost would be 10 x 10 CHF (cost of an NIR measurement) plus 1 x 70 CHF for the reference method. The total amounts to 170 CHF for the same 10 samples. In this example, the NIR instrument costs 100,000 CHF and the annual calibration / validation costs were 7,000 CHF. For the high number of samples tested in an ice cream factory this meant that the return on investment was less than 1 year and the discounted cash flow is zero making life easier for the company accountants.

Cost "savings" are to be had in terms of analysis time, analytical costs, shorter product release times and lower production costs. Likewise, it is also possible to cite some "earnings" in terms of process learning, process control, first time quality and product consistency.

The in-line monitoring of the quality of products at critical points in the production process is part of the industry wide application of HACCP (Hazard Analysis Critical Control Points). Knowing where a problem arises in the factory environment allows a solution to be found quickly ensuring hygiene deviation to be detected earlier and controlled.

Finally, product safety is non-negotiable while achieving product consistency leads to consumer satisfaction!

Table 1. Analytical costs of total solids and fat using reference methods for controlling ice cream premixes.

Parameters	Calibration samples	Validation samples	Total samples	Cost per analyte	Total analytical costs / CHF
Total solids	60	40	100	20	2,000
Fat	60	40	100	50	5,000

3. Off-line rapid methods

When samples are taken from raw materials arriving at a factory gate or from intermediate or finished products off the factory floor, the notion of time is critical. Consider antibiotic detection in fresh milk. For reasons of maintaining freshness the analyst has in general 5 to 15 minutes to perform the test and make a decision. This contrasts to mycotoxin or pesticide testing in cereals and fruits where several hours or days can be used to carry out the analysis. Intermediate to this is antibiotic testing in meat and farmed fish where the analyst has just a few hours to perform the tests. Microbiological and chemical rapid methods will be explored in this context.

4. Microbiological testing

The food industry is confronted with a need to identify and in some cases quantify a variety of microbial pathogens and indicator organisms in food products and production environments. A partial list of pathogens and indicator organisms is included in Table 2.

Classical microbiological methods rely on culturing microorganisms onto agar plates. These methods are often laborious and time consuming. As an indication, approximately 85% of all tests are based on counting viable cells.

Rapid methods may have a positive impact on manufacturing where analytical turn around time is critical for product release and where supply chains are tight. This may have an impact on the cost of warehousing, logistics and distribution. The use of rapid methods may also impact the availability of raw materials or enable a more rapid response to data from environmental monitoring programs. In many cases, the use of such methods reduces the need for confirmatory testing.

An "ideal" rapid method would enable a "real-time" interpretation of the presence or level of an organism or group of organisms in a sample. However, this goal is hindered by several challenges. Firstly, food represents a wide variety of complex matrices, some of which may interfere with the performance of an assay or the outgrowth of organisms in the sample. Target organisms are often present in low numbers and unevenly distributed. Other bacteria present in the foodstuff may interfere with identification and processing of foods may injure bacterial cells.

Table 2. Microbiological pathogens and indicator organisms of significance to the food industry. Organisms or groups indicated in bold text are commonly analysed in food product samples and samples taken from production environments.

Infectious pathogens	Toxigenic pathogens
• ***Listeria monocytogenes***	• ***Bacillus cereus***
• ***Salmonella spp.***	• ***Clostridium botulinum***
• ***Escherichia coli O157:H7 or other Enterohemorrhagic strains***	• ***Clostridium perfringens***
	• ***Staphylococcus aureus***
• *Campylobacter*	
• *Pseudomonas aeruginosa*	Hygienic indicator organisms/groups
• *Legionella pneumophila*	• ***Coliforms***
• *Aeromonas hydrophila*	• *Faecal coliforms*
• *Yersinia enterocolitica*	• ***Enterobacteriaceae***
• *Vibrio cholerae*	
• *Vibrio parahemolyticus*	General or spoilage indicators
• *Vibrio vulnificus*	• **Aerobic mesophilic count**
• *Shigella*	• **Mould and yeast count**
• Opportunistic pathogens	• Anaerobic plate count
	• Thermophilic sporeformers
	• Mesophilic sporeformers
	• Lactic acid bacteria

Rapid detection technologies can be divided into two classes, either Phenotypic (such as biochemical, conductance, ATP bioluminescence, immunoassays, or bacteriophage assays), which rely on expressed characteristics, or Genotypic (such as nucleic acid hybridisation, or DNA amplification), which evaluate the presence of unique genetic sequences. Today, there exists a number of commercially available biochemical and molecular assays for rapid microbial identification. Available molecular assays include ELISA, DNA hybridisation and PCR. Most of these technologies allow simultaneous detection of multiple samples and have a faster analysis time than classical methods. They also reduce the number of labour intensive steps and can be semi- or fully automated. Microarray technologies are currently under development, which incorporate glass or nylon chips and have the potential for multi-target detection. However, lengthy incubation steps remain a significant hurdle in the development of practical alternative methods. Work is ongoing at some regulatory agencies to validate alternative methods, however in many cases the reference methods are still used for official control.

5. Chemical testing

A look at the food supply chain will reveal several areas where chemical hazards can enter, cf. Figure 1. These concern pesticides and agrochemicals, mycotoxins, veterinary drug residues, processing contaminants, packaging contaminants, formation of carcinogens from the cooking process, not to mention cross-contaminations, adulteration and food tampering.

An area where rapid tests are widely used in the food industry is in the identification of veterinary drug residues in milk, meat and eggs. The discovery of penicillin G in 1929 and sulfanilamide in 1935 led to the use of antibiotics in veterinary medicine. Antibiotics were then later used as feed additives in 1951. Their use gave rise to the development of rapid screening tests, the first being

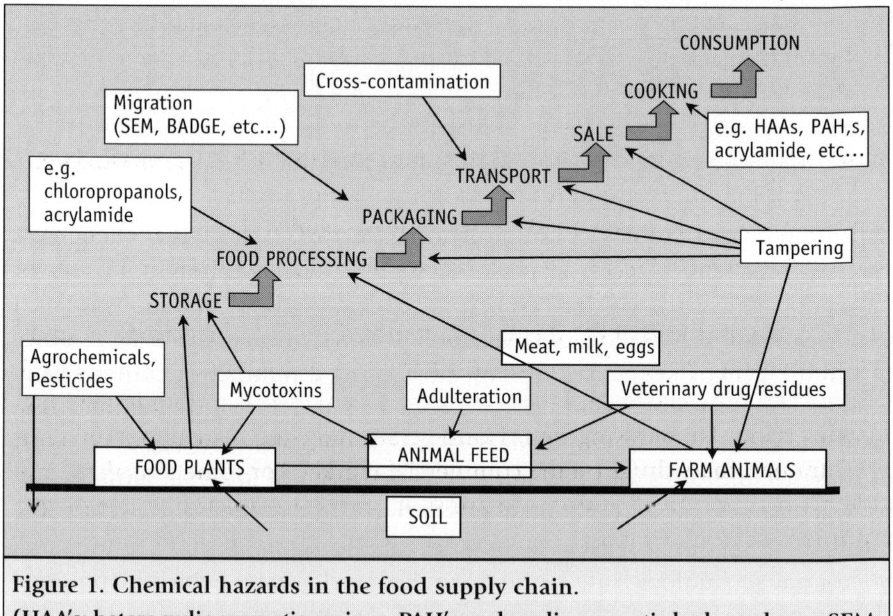

Figure 1. Chemical hazards in the food supply chain.
(HAA's: heterocyclic aromatic amines; PAH's: polycyclic aromatic hydrocarbons; SEM: semicarbazide; BADGE: bisphenol A diglycidyl ether)

the Microbial Inhibition Test (MIT) for milk in 1951. Today, there are many tests on the market and are applicable to the entire animal production chain from farmer to food producer.

Rapid tests for antibiotics have proved extremely useful in dealing with food contaminations. In the last two years alone there have been several antibiotic food contamination scares in the news headlines. Most recent examples are the chloramphenicol and nitrofuran crises in South-East Asia in spring 2002. This led to the ban of imported shrimp products into the EU resulting in the loss of hundreds of millions of US dollars to the Thai and Vietnamese economies. Again in 2003 and 2004 there have been nitrofuran contaminations of fish, poultry and honey from South-East Asia, China and Argentina.

Intensive farming practices and the mis- or illegal use of veterinary drugs are often linked to these contaminations. To illustrate intensive farming practices, the average milk production from one cow in France in 1960 was 1,360 kg/yr. In 1997, this grew to a maximum 5,400 kg/yr. Today, traditional farmers are obtaining approximately 4,000 kg/yr., while industrial scale farmers are obtaining between 8 to 9,000 kg/yr. To our knowledge the world record for milk production is 30,805 kg/yr. or 83 L/day from a cow in Wisconsin USA. Mastitits treatment using antibiotics is extremely important for milk production. As an example of the multitude of drugs on the market, Table 3 presents a list of drugs used for mastitis treatment.

Many mastitis drugs are used as intramammary injectors. Costs of injectors used in different countries are indicated in the following Table 4 (2003) clearly showing a correlation between the use of drugs and milk yields. This again highlights the use of antibiotics in intensive farming practices.

Concerning rapid tests there has been a strong growth in kit based methods having ease and speed of use. A recent survey in our laboratory (end 2003) identified commercial kits for the analysis of a total of 198 antibiotic residues. In Switzerland, United Kingdom, France, USA, Italy, there are approximately 179 intramammary drugs on the commercial market. For some tests there is a trade-off between the number of antibiotics detected and the time or cost. Example is shown in Table 5.

Table 3. Antibiotics for mastitis treatment.

Aminoglycosides	Aminosidin	**Cephalosporins**	Cefacetril
	Dihydrostreptomycin		Cefalexin
	Gentamicin		Cefalonium
	Kanamycin		Ceffazol
	Neomycin		Cefoperazon
	Streptomycin		Cefuroxim
Bacitracins	Bacitracin		Cephapirin
Beta-lactams	Amoxycillin	**Macrolides**	Erytromycin
	Ampicillin		Oleandomycin
	Benzylpenicillin		Spiramycin
	Cloxacillin	**Novobiocins**	Novobiocin
	Dicloxacillin	**Polymixins**	Colistin
	Hetacillin		Polymixin
	Nafcilin	**Rifomycins**	Rifampicin
	Oxacillin		Rifamycin
	Penethamate		Rifaxymin
Chloramphenicols	Chloramphenicol	**Sulfonamides**	Sulfadimidine
Diaminopyrimidines	Trimetoprim	**Tetracyclines**	Oxytetracycline
Lincosamides	Lincomycin		Tetracycline

Table 4. Intramammary mastitis treatment and milk production in India, EU, Brazil and USA.

Country	Cow numbers	Milk/yr. tonnes	Milk/yr./head kg	Intrammary drug cost (USD/head/yr.)
India	34,000,000	69,100,000	2,000	?
EU	21,600,000	121,000,000	5,800	5.4
Brasil	17,200,000	19,800,000	1,200	0.4
USA	9,200,000	72,400,000	8,000	8.3

Table 5. Comparison between Delvotest and Beta Star for antibiotic residue testing.			
Name of test	Antibiotics detected	Time/sample	Price/sample (CH)
Delvotest SP	Wide range	3 hours	1.66 CHF
Beta Star	Beta-lactams and Cephalosporins	5 minutes	4 CHF

6. Market for rapid tests

It is interesting to mention some figures about the market place for rapid tests. The Austrian Bureau of International Research and Technology has estimated the total food analysis global market valued at € 1.1 billion. This is shared amongst about 50 major companies. The USA and Europe represent about 2/3 of this market. € 565 million is the value of global sales for food microbiological tests. Of this, rapid methods are valued at € 115 million or approx. 20% of the total market share. Tables 6 and 7 indicate the world market for microbiological tests and the market of tests by class. This is expected to grow for both microbiological and chemical testing.

Table 6. World market for microbiological tests.		
Technology	Sales (US $ millions)	Market share (%)
ELISA/EIA	70	61
DNA	25	22
Impedance	10	9
Flow Cytometry	5	4
ATP	5	4
Total	115	100

Table 7. World market - tests by class.			
	Food Analysis (€ million)	Food Diagnostics (€ million)	% Food Diagnostics (€ million)
Food pathogens	97.5	13	15
Spoilage	220	37.5	17
Chemical contaminant	53	19.5	40
Process chemicals	120	16	10
Total	490.5	86	

7. Conclusions

The importance of rapid methods to the food industry has been described in terms of their benefits for the supply chain and manufacturing; best practice in quality assurance and for public health.

The old proverb that "time is money" is very appropriate to the use of rapid methods. Naturally when deciding on whether to use the reference or rapid method a consideration of the total analytical cost in proportion to the benefits must be made. Rapid tests have an impact on the turn-around time on manufacturing and can be a cost efficient way to screen for positives, which would then be analysed by a suitable confirmatory method. They are known for their ease of use by non-skilled operators and have the advantage to be accessible worldwide. They generally require small sample sizes and are often portable for use outside of the laboratory environment.

An important remark is that rapid methods allow the effectiveness of food safety measures to be determined. A proactive approach using preventative testing by robust real-time process control instead of reactive testing is the backbone of modern QA systems in factories. Rapid tests are a very useful tool in the HACCP plans of factories.

Finally, rapid methods can be used in conditions where urgent intervention is required to manage food contaminations by either preventing or minimising the effects of these and in cases of product tampering and potential bio-terrorist acts.

Acknowledgements

Gratitude to my colleagues who have contributed to this paper: Dr. Jacques Prodolliet (Factory process control); Dr. Tim Jackson, Ms. Frédérique Cantergiani and Dr. Joey Marugg (Microbiology); Mr. Jean-Marc Diserens (Antibiotic testing); Dr. Rupert Rorschnik (Quality Management); Dr. Yasmine Motarjemi (Food Safety Management).

A European approach to validation and certification of rapid microbiological methods

R. Betts
Campden & Chorleywood Food Research Association, Chipping Campden, Glos., GL55 6LD, United Kingdom

Abstract

Over the past twenty years, there has been increased research and development into rapid test procedures for the food industry. Such procedures can offer significant benefits to users when compared with conventional reference methods, however, in order to give the user confidence in their results, these methods must be properly validated. This paper discusses the reasons for validation, the different types of validation and certification that can be used for food microbiological testing and their positive and negative attributes.

Keywords: Rapid microbiological methods, validation, third party validation schemes, certification

1. Method choice

When any laboratory is considering the use of a new test method, it is likely to have a number of methods from which to choose. This makes the choice far from easy, but by following simple rules, the laboratory will get the method that best fits its purposes.

The list of questions that any prospective user should ask themselves includes:

1. What types or levels of microorganisms or microbial groups should my test measure?
2. How many tests per day or week do I need to do? (Future requirements must also be considered.)
3. What will the new test cost in terms of capital equipment and consumables?
4. What sample types will the new test have to work with?
5. How fast do I want to get a result?

6. What level of laboratory experience does the new test require and does this match my current staff?
7. Do I have confidence that the method manufacturer/distributor will give me back up service and will be able to solve any problems that I have?
8. Do I have confidence that the results I will get from the new method will be valid?

This list is a good general start to most laboratories considering using a new method, but perhaps the most important answer will be that from point VIII. Any method that is not valid, does not give 'correct' results or is not fit for the purpose the laboratory wishes to use it for, should not be considered for use. This, of course, brings up the question of how does any laboratory know that a new method is valid and fit for purpose and the only way to answer this question is to undertake method validation.

2. What is validation?

Validation of a method concerns its vigorous testing against previously agreed parameters in order to measure its ability to give an acceptable result. In most cases with microbiological methods, the parameters against which a new method is tested will be a relevant reference method (e.g. an international standard) and the 'acceptable result' will be some form of correlation with the results of the reference method. Therefore users are looking at the ability of the new method to give results that are comparable to those that would be obtained with a reference method, if both test methods were used on the same samples.

3. Why is validation required?

As noted earlier, it is important for any laboratory to know that any method they use gives a 'correct' result. For most microbiology laboratories wishing to use a rapid method, a 'correct' result will mean a method that gives an equivalent result to a reference method. This information will give the laboratory the confidence that the rapid method works to their requirements giving a 'correct result'. It also gives the user of the results the confidence to make decisions based on those results. Such decisions are highly significant and may be the release of a food product onto the market, or in case of poor results, the recall of products from the market. In either case, an incorrect result could compromise a food producer's reputation and potentially result in a large financial penalty to that company.

The additional benefit of rigorous method validation is that it is a requirement for any laboratory wishing to put a method into the scope of a third party laboratory accreditation programme. Many laboratories now seek accreditation to a standard such as ISO 17025, and to gain accreditation, laboratories must show that their methods are valid for use. This will mean they must validate their methods.

4. Types of validation

Generally there are two types of method validation:

1. In-house method validation
2. External third party validation

In-house validation is done by the laboratory intending to use the method. It is primarily to assure the user laboratory that the method is fit for its intended purpose in that laboratory. It will also satisfy the laboratory's clients that the method gives 'correct' results, and it is required by any third party laboratory accreditation body to show that the method can be included in the laboratory's scope of accreditation. It should always be noted, however, that in-house validation of this type relates only to the laboratory in which the work has been done. All laboratories should undertake some form of in-house validation for any method that they intend to use and suggested ways in which this type of validation can be done are given in Baylis *et al.* (2001).

External third party validation of methods is done or organised by a third party, independent of the method manufacturer and the method's end users. The role of third party validation is to satisfy and assure a wide range of potential users that a new method is generally fit for the purpose intended. This type of validation will satisfy third party laboratory accreditation bodies that a method will give equivalent results to appropriate reference methods. However, such accreditation bodies will still require user laboratories to do their own additional in-house validation to show that a method is fit for the purpose in their laboratory.

5. Third party validation schemes

Unfortunately for both method manufacturers and users, there is not a single third party method validation scheme that satisfies all users throughout the world. A number of well-founded and recognised schemes exist, and these are noted in Table 1.

The situation of having a number of different validation bodies leads method manufacturers having to potentially invest a considerable level of funding and time into multiple validations around the world and causes confusion amongst method users who often simply wish to know 'which validation gives me greatest confidence that the method works?'

In the 1990's, a European EUREKA project set out to develop a European validation system for microbiological test methods. The project was called MicroVal and was made up of twenty-one partner organisations from seven different EU member states. The MicroVal project had two aims:

1. To develop a comprehensive, scientifically valid way of validating microbiological rapid methods.
2. To develop a certification scheme that would allow methods that had been successfully taken through I., to be certified by an appropriate certifying body.

Table 1. Third party validation schemes.

Name	Geographical area of recognition	Single laboratory study required	Collaborative study required
• AOAC Research Institute Performance Tested Methods Programme	USA, parts of Europe, other parts of the world	Yes	No
• AOAC Peer Verified Methods Programme	USA	Yes	One or two laboratories only
• AOAC Official Methods Programme	USA, parts of Europe, other parts of the world	Yes	Yes
• AFNOR	France	Yes	Yes
• NordVal	Nordic countries	Yes	Yes
• MicroVal	Europe	Yes	Yes

The comprehensive validation procedure developed by the MicroVal project was published in 2003 as CEN and ISO standard 16140. This standard has the following scope:

- To provide the principle and a technical protocol for the validation of alternative methods.
- It is appropriate for methods directed at foods, animal feeds, environmental and veterinary samples.
- It is appropriate for the validation of methods for official control purposes.
- It allows an international acceptance of the results obtained from the alternative method.

The standard does however, note that a less stringent validation may be appropriate for methods used on a routine basis for 'internal laboratory use'.

6. EN ISO 16140 - What does it require?

The standard details validation procedures for both qualitative and quantitative microbiological methods, but not for any form of identification test or any method for which a recognised reference method does not exist.

In order to gain an impression of the vigorous nature of the validation testing detailed by EN ISO 16140, the requirements are given in brief in Tables 2 and 3, for Qualitative and Qualitative Methods respectively.

As well as the requirements noted in Tables 2 and 3, EN ISO 16140 also requires verification that the production of the alternative method is done in such a way as to ensure consistent quality and therefore consistent results. This is important as the practical testing of the method will utilise kits from a limited production run. It is important to assure that, as far as possible, every production batch of kits will give the same results if used on similar food samples. To this end, the standard requires a manufacturing quality system to a level of ISO 9000 to be in place at the method manufacturers.

7. Certification of alternative methods

Certification of alternative methods, that have successfully undertaken the requirements of EN ISO 16140, can only be done by an appropriate certification body.

Table 2. EN ISO 16140 - Qualitative method validation requirements.		
Stage of validation	Test parameters to be determined	Basic testing requirements
Methods comparison study	1. Relative accuracy 2. Positive deviation 3. Negative deviation 4. Relative sensitivity 5. Relative specificity 6. Relative detection level 7. Inclusivity 8. Exclusivity	1. 5 food categories to be used 2. 60 samples from each category 3. Approximately 50% of samples positive for target organism/group 4. 1 food from each category used for relative detection level determination 5. Inclusivity requires 50 target organisms/strains to be tested (*Salmonella* requires only 30) 6. Exclusivity requires 30 non-target organisms to be used
Inter-laboratory study	1. Specificity 2. Sensitivity 3. Relative accuracy	1. Results from 10 laboratories 2. 1 food type used 3. 3 contamination levels 4. 8 replicates

8. Importance of EN ISO 16140

As there is a greater requirement for a worldwide trade in food, there is an increasing requirement for the international acceptance of both laboratories and methods used for testing. Such acceptance will reduce the requirement for multiple testing of food consignments as they leave and enter different countries, but will ensure that the overall confidence in results from tests increases.

The Laboratory Quality Standard ISO 17025 allows for the acceptance of laboratories, whilst EN ISO 16140 will allow this for alternative microbiological methods. Already the EN ISO 16140 standard is being mentioned in EU documentation, such as the Draft EU Microbiological Criteria Document (SANCO/4198/2001) and this will undoubtedly continue. There is, however,

Table 3. EN ISO 16140 Quantitative method validation requirements.

Stage of validation	Test parameters to be determined	Basic testing requirements
Methods Comparison Study	1. Linearity 2. Relative Accuracy 3. Detection Limit 4. Quantification Limit 5. Sensitivity 6. Inclusivity 7. Exclusivity	1. 5 food categories 2. 5 levels of organism (minimum) per category 3. 2 to 10 replicates 4. Inclusivity requires 30 target organisms/strains to be tested 5. Exclusivity requires 20 non-target strains to be tested
Inter laboratory study	1. Relative Accuracy 2. Repeatability 3. Reproducibility	1. Results from 8 laboratories required 2. 1 food type used 3. 3 contamination levels 4. 2 replicates (minimum

still some understandable confusion amongst both method producers and users concerning which method validation scheme they should rely upon.

It is difficult to see how national or regional schemes (e.g. AFNOR and NordVal respectively) will gain acceptance outside their own geographical area. The obvious way forward is the more widespread use and acceptance of true international method validation. EN ISO 16140 fulfils these requirements, as it is both a European and International Standard, and there are indications that certifying organisations, such as AOAC, AFNOR and NordVal, are moving towards the use of EN ISO 16140 as a technical solution to method validation (MicroVal will use only this standard for its method validation). So perhaps the future of the international recognition of alternative microbiological method validation lies with different certification organisations mutually accepting each others work, all done to a common standard (EN ISO 16140). This may be a little way in the future, but there is some evidence that method validation is moving in this direction, making it easier for method producers and users to gain the credibility and confidence they require to use these methods for food testing.

References

EN ISO/IEC 17025, 1999. General requirements for the competence of testing and calibration laboratories.

EN ISO 16140, 2003. Microbiology of food and animal feeding stuffs - protocol for the validation of alternative methods. ISBN 0 580 41919 3.

Baylis, C.L., Jewell, K., Oscroft, C.A. and Brookes, F.L., 2001. Guidelines for establishing the suitability of food microbiology methods. Campden & Chorleywood Food Research Association, Guideline No. 29.

SANCO/4198/2001. Rev 9, 2004. Draft Commission Regulation on Microbiological Criteria for Foodstuffs. European Commission Health and Consumer Protection Directorate-General. Brussels.

Sampling and sample preparation: An underestimated issue?

M.C. Spanjer
Dutch Food and Consumer Product Safety Authority, Primary Agricultural & Horticultural Products Department, Hoogte Kadijk 401, 1018 BK Amsterdam, the Netherlands

Abstract

Sampling and sample preparation are issues that are generally underestimated in any analytical procedure. Tests can be as rapid as desired nowadays, whereas sampling and sample preparation are still time consuming, but nevertheless crucial steps, due to the fact that the judgement of a lot is based on the total procedure. In this contribution details of these aspects will be outlined in the field of primary agricultural and horticultural products, as to be precisely: mycotoxins, pesticides and genetically modified organisms (GMOs). From these diverse examples it will be clarified that these basic concepts can be applied to any type of other analytical problem.

Keywords: sampling, sample preparation, mycotoxins, pesticides, GMOs

1. Introduction

Whatever the type of analysis, nobody in the world bothers about the analytical result of the sample itself, since no consumer will ever eat it. Therefore the crucial point in any analytical procedure is the judgement of the lot from which the sample originates. Acceptance or rejection of a lot is based on the corresponding operating characteristics curve. In these curves the probability on false positive (producer's risk) or negative (consumer's risk) results are plotted against the contamination level in a lot as determined by the applied sampling plan. This principle holds for chemical as microbiological contamination as well, as illustrated by Figure 1. In the latter figure the influence of increasing the number of sample units from n=5 to n=20 can also be seen. This leads to a specific sampling plan for different contaminant-commodity combinations (Gilbert and Vargas, 2003).

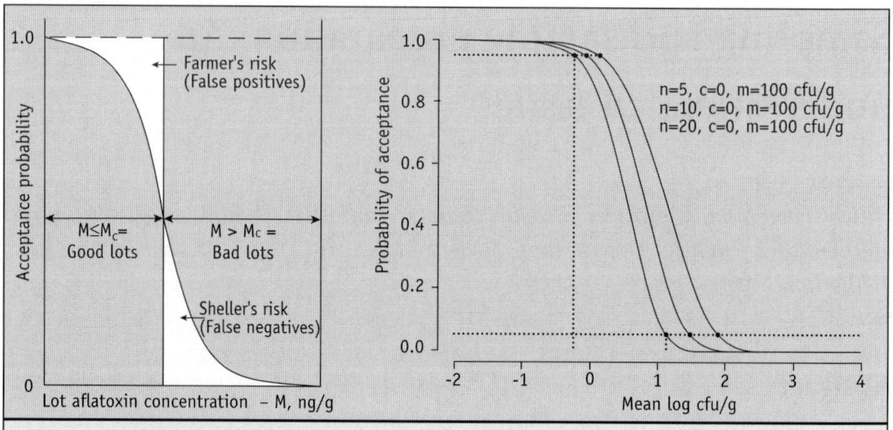

Figure 1. Typical operating characteristic curves for evaluating farmer's stock sampling plans for aflatoxin in peanuts (Whitaker *et al.*, 1994) and for mean microbial concentrations (Dahms, 2004).

2. Mycotoxins

The sampling plans for aflatoxin in peanuts are based on extensive scientific mathematical statistical calculations of analytical results of a large amount of subsamples of a lot, as to investigate the number of necessary incremental samples what is needed for a representative sample. Sampling plans and limits for the USA (21.8 kg at 20 ng total aflatoxin/kg), UK (10 kg at 10 ng total aflatoxin/kg) and the Netherlands (4x 7.5 subsamples at 5 ng/kg for aflatoxin B1 only) have been evaluated as to investigate the differences in exporting and importing aspects (Whitaker *et al.*, 1995). This is illustrated in Figure 2. Since 1998 the EU plans are harmonised to 3 subsamples of 10 kg each per lot (European Commission, 1998). To complete the overall view it has to be kept in mind that the total error consists of a sampling, sample preparation and an analytical error (Campbell *et al.*, 1986) as illustrated in Figure 3.

To judge about the scope of these items, let us consider the sampling Directive (European Commission, 1998) of import control on, for example nuts packed in sacks. The container of an incoming ship is put ashore. Before inspection this container must be completely unloaded. A 21-ton container (i.c. truck load) may hold 350 sacks of 60 kg each. All sacks must be removed from the container. Food inspectors must take 100 incremental samples of 300 g each, requiring

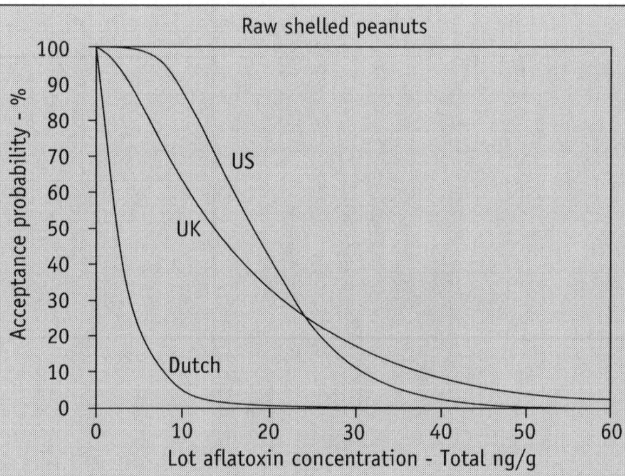

Figure 2. Probability of accepting lots of raw shelled peanuts by using the US, UK and NL procedures (Whitaker *et al.*, 1995).

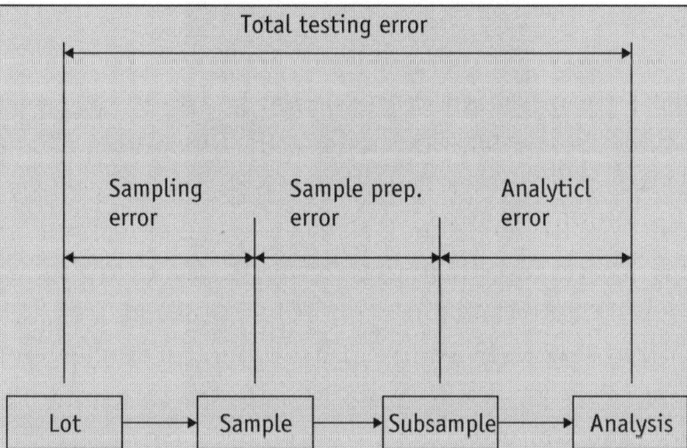

Figure 3. Total variance in test procedures (Whitaker *et al.*, 2000).

them to open ca. 3 out of every 10 sacks by means of sampling spears (ISO 13690:1999, see Figures 4 and 5). These 100 sacks then must be closed and the container must be reloaded again to create a proper truckload. Two food inspectors need half a working day to sample this one container, because they have to move all these heavy bags and they must do it together because the weight exceeds the maximum one person can lift. The owner of the lot loses

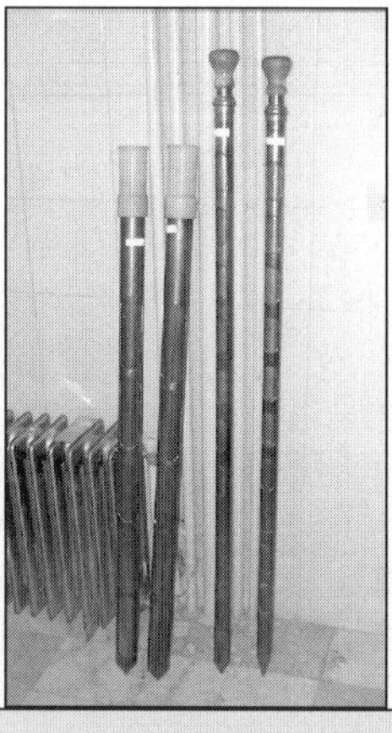

Figure 4. Sampling spears.

Figure 5. Sampling bags.

time waiting for sampling and analysis. Sometimes the result must be delivered within a certain period, as regulated by directives (i.e. customs regulations), which puts time constraint on all interested parties and also on preparing the affected cargo for further trading. And all this work is required to sample only one container of just one ship! The result of this inspection is a 30 kg sample, which equals half a bag. Compared to the load we achieved a reduction of the lot by a factor 700.

Since this is not a portion one can just analyse, the next step in the procedure will be to reduce sample size. This can only be done under the strict condition that the sample remains unaffected by a comminution step. Milling and mixing will therefore be the next step in sample preparation. On the milling process (dry grind) and on slurry mixing (grind with water added) data on the quality of these processes are available from the literature (i.e. Dorner and Cole, 1993). Recently, a comparison was made (Spanjer *et al.*, 2003) between dry milling and slurry mixing as a comminuting step preceding mycotoxin analysis. Green coffee, almonds and pistachio samples of 10 kg were milled by a Romer Analytical Sampling (RAS) mill (Figure 6) and all three subsamples were analysed for aflatoxin B1 or ochratoxin A content. Normally only subsample A is analysed, but this experiment was carried out by the following procedure:

1. Sampling according to the EC directive, resulting in 10 kg sample.
2. Milling the 10 kg sample by a Romer mill with a split ratio of 10%.
3. Taking a dry sample out of the 10% part as usual for Romer mill users (subsample A).
4. Slurry mixing of the remaining part of the 10% part of the sample (subsample B).
5. Slurry preparation of the 90% part by Silverson mixing (subsample C).
6. Analysing the three subsamples A, B and C by HPLC methods.

By doing so the mycotoxins content in the sample can be reconstructed afterwards by calculation.

Figure 6. Romer Analytical Sampling mill.

Consideration of the results has to be done from the starting point of this experiment: milling the 10 kg sample by a Romer mill, which creates a division of the original sample in 2 subsamples of different weight. When RAS milling is used in daily routine analysis this step is followed by taking an incremental sample out of the smallest subsample for further clean up and chemical analysis. This situation is comparable with the results for subsample A in this experiment with the crucial difference that data as presented for subsamples B and C are never measured in daily practice. In case of sample preparation by means of slurry, the whole sample is dealt with. A portion of the slurry is taken for further analysis. Regarding the methods in detail reveals that it will never be possible to do an experiment by applying both preparations towards one sample. Therefore the best estimate of a measurement of these samples, as if they were handled by preparing a slurry, can only be made by calculating the amount of mycotoxins from the individual A, B and C subsample values. This calculated value is presented as "sample value" in Table 1.

Table 1. Analytical results (µg/kg) from all subsamples and the reconstructed sample value.

	sample	sub A	sub B	sub C
Ochratoxin A				
Green Coffee	1.5	8.1	0.4	1.6
Green Coffee	1.9	1.8	2.3	1.8
Green Coffee	2.0	2.7	2.6	2.0
Green Coffee	2.0	1.5	2.0	2.0
Aflatoxin B1				
Almonds	2.0	1.0	0.2	2.2
Almonds	2.4	1.0	4.2	2.2
Almonds	3.1	0	0	3.4
Almonds	4.1	0.5	6.7	3.8
Pistachio in shell	33.8	88.2	38	33
Pistachio in shell	44.1	51.4	42.4	44.2
Pistachio kernels	114.1	250	108	114
Pistachio kernels	126.0	204	122	126

For the enforcement of a directive the analytical results are important at the point of accepting or rejecting a lot. Aflatoxin B1 is regulated in EC directives: 2 μg/kg for nuts and 3 μg/kg is drafted for ochratoxin A (OTA) in coffee beans. With these figures, without adding measurement uncertainties, the differences between judgements of a lot based on dry milling (subsample A data) are compared with the data that would have been obtained after slurry preparation of the sample as a whole (sample value data). Doing so it can be concluded from Table 1 that 3 from 4 almond samples would pass, whereas the actual contamination is higher then 2 μg/kg. In 1 coffee bean sample the dry milling procedure would reject a lot, which would be accepted by the slurry preparation method. This is an interesting detail, since both commodities are rather expensive, so from this point of view even false positive results are not desirable. The aflatoxin levels in the pistachios are so high that the measurements lead to rejection in any case.

The homogenisation process can be evaluated in view of coefficients of variation for different types of milling and from particle size distributions. CV values of 20-33% for dry milling are reported by Dorner and Cole (1993) while applying Hobart and Stephan mills. CV values that can be obtained by slurry mixing with a Silverson machine are typically lower then 10% (Table 2).

This difference was explained on the basis of measured particle size distributions for both milling types, which were carried out by Schatzki in 2001 and published in 2003 (Schatzki and Toyofuku, 2003). Particle sizes achieved by Hobart milling are significantly larger then by Silverson milling (Figure 7).

Table 2. Homogeneity data of 7.5 to 30 kg nut sample slurries from Member States (MS).

Matrix	n	mean	STD	CV	MS
Pistachio	4	3.9	0.2	4.4	F
Pistachio	20	5.7	0.2	3.9	NL
Peanut	10	7	0.3	3.8	NL
Pistachio	4	7.5	0.6	8.2	F
Pistachio	10	10.4	1.0	9.7	D
Pistachio	20	10.9	0.5	4.5	NL
Pistachio	20	14.5	1.1	7.3	A

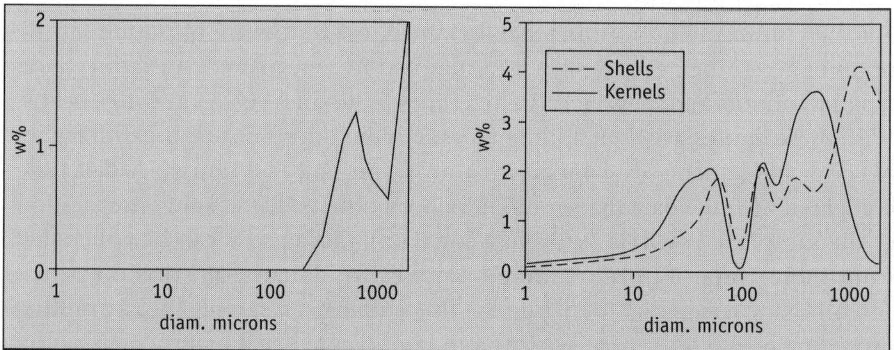

Figure 7. Particle size distribution after (dry) Hobart milling and (slurry) Silverson mixing.

An interesting aspect of the difference between dry milling and slurry mixing was found for the determination of quantitation of *Aspergilli* species in peanuts. As can be seen from Table 3 the coefficient of variance (n=5) increases enormously in case of dry milling. This is due to the fact that the sample is heated up to 90 °C, which killed the fungi at higher temperature (Dorner, 2002). In Figure 8 pictures of several mills are given.

In most EU laboratories the slurry mixing method is preferred due to the fact that it can be applied at sample sizes of 10-30 kg, is easy to perform and cleaning after milling takes only 5 minutes of washing by means of a water shower head. Typically from the slurry a portion is taken for immediate analysis. A second portion is stored in the freezer in case of dispute. The amount of nut in those

Table 3. Quantity and variability of *Aspergilli* species in peanut subsamples at different mills.

Milling time (minutes)	dry grind CFU/g	slurry CV %	CFU/g	CV %
1	471600	9.3	548400	16.5
3	332400	10.3	572000	9.1
5	5400	46.5	589000	5.1
7	7200	63.2	599000	2.5

Figure 8. Pictures of Hobart, Stephan and Silverson mill; first 2 for dry, the latter for slurry grind.

portions is 63.5 g. Regarding the starting point of 30 kg this implies another reduction by a factor 472. Analysis by the appropriate CEN method causes another reduction due to extraction, redissolve in 1 ml and finally injection of 20 µl into an automated HPLC system. This makes up for another reduction by a factor of 50 times. In total the cumulative reduction factor is 700x472x50= 16,520,000. Compared to the Netherlands' population this is comparable by analysing just 1 person and conclude that you know the characteristics of all inhabitants of the Netherlands. With regard to the container this means that the size of the sample that is actually analysed is only 1.27 g, which equals just one single peanut kernel! These data illustrate clearly why sample handling is critical. This 1.27 g has to contain particles from all peanuts in the 30 kg sample. So, the finer the milling the better. And once again, the judgment of the whole container depends on the analytical result of this part of it. In case of pistachios it has been calculated what the potential cost can be for a false positive result leading to the rejection of an entire batch of pistachio nuts, at retail level (Ramsey *et al.*, 2001). Including the loss of corporate reputation this is estimated to be to be 80,000 Euro. In case of almonds or coffee beans this might even be a higher amount.

Thus far the above demonstrated the knowledge on analysis and sample preparation errors. The error of sampling as such was not taken into account. In case of sampling 5 kg peanuts, sample preparation by USDA subsampling mill, 250 g subsample analysed by TLC, the respective sampling, subsampling and analytical errors were calculated to be 83, 9 and 8% (Whitaker, 2003). For DON in wheat the problem appeared to be less severe. Using a 454 g wheat sample, a Romer mill, analysing 25 g by the fluoroquant method, these data were 6.3, 10 and 6.3% (Whitaker *et al.*, 2000). Whether this is due to a less heterogeneous spread of DON in wheat kernels is not exactly known yet. On the other hand, contamination of ochratoxin A in raisins seems more comparable to the spread of aflatoxin in peanut. This can be concluded from a Swedish marketing survey on consumer packages of raisins (Möller and Nyberg, 2003). They investigated duplicate samples, from which could be expected that the analytical results should be comparable. In 37 of 54 samples the ochratoxin A content was above the detection limit of 0.3 μg/kg. As can be seen in Figure 9 the scatter between the duplicates is significant. Further investigations will be inevitable, since measurement uncertainty raises more and more interest. Worldwide sampling is considered long since to be the largest source of analytical error, especially at heterogeneous distribution of contamination in a lot, which is raised as an interesting topic again (Thompson, 2001).

3. Pesticides

At first glance the situation looks much better for pesticide sampling. In general pesticides were considered to be spread more homogeneous, until the moment that in the UK unexplainable results on pesticide residues were found in carrots

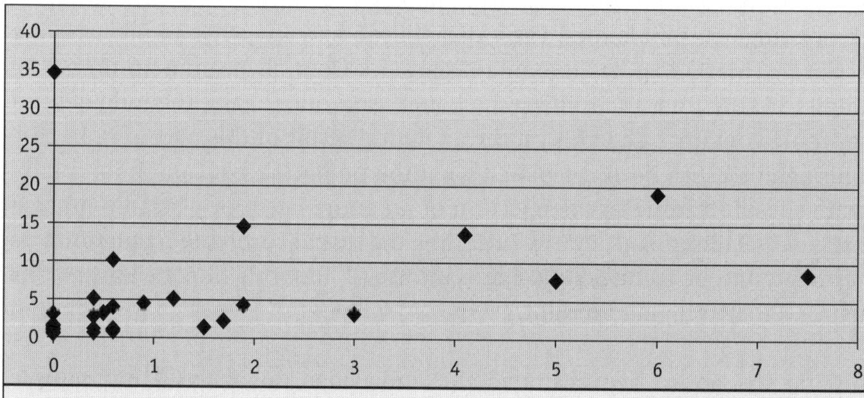

Figure 9. Ochratoxin A in 37 duplicate consumer packages (Möller and Nyberg, 2003).

during routine sampling in the early 1990s (Harris, 2000). A significant proportion of the carrot samples exceeded the Maximum Residue Levels (MRLs), whereas the farmers applied Good Agriculture Practice (GAP) at crop control. A field test was performed with a blue dye to investigate whether the water flow, drain and tillage caused this variability (Carter *et al.*, 2000). This is illustrated in Figure 10. Table 4 gives corresponding data (Earl *et al.*, 2000).

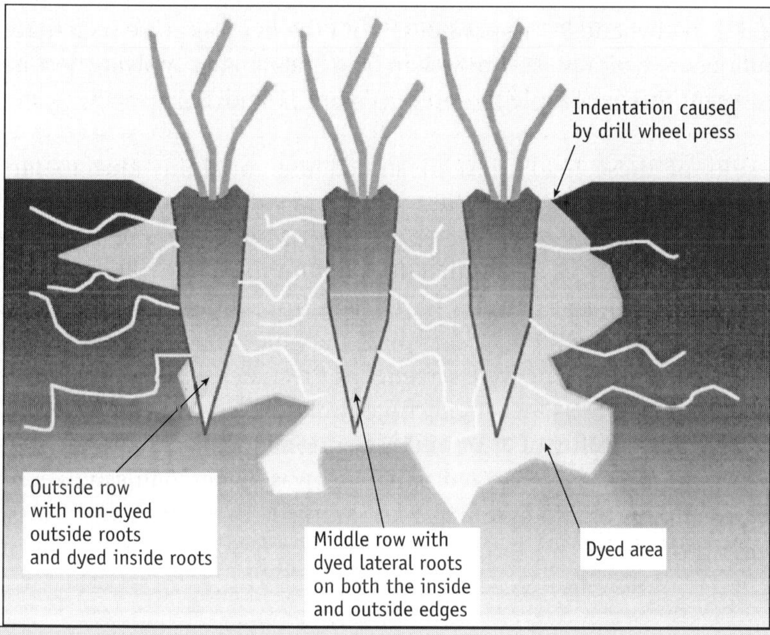

Figure 10. Staining of roots in carrot rows.

Table 4. OP (organophosphate) residue levels in individual carrots.

Trial	Range of residue levels in individual carrots (mg/kg)	Mean residue level (mg/kg)
Trial 1	0.01-0.16	0.06
Trial 2	0.01-0.45	0.18
Trial 3	0.05-0.80	0.22
Trial 4	0.07-1.4	0.50

Earl *et al.* mentioned these data in their publication on their experiments from supervised industry field trials. They reported ranges of residue levels in individual oranges of <0.03-0.34 at a mean of 0.17 mg/kg, and in peaches of <0.03-0.14 at a mean of 0.05 mg/kg. From these data variability factors of 2-3 can be calculated, as these are defined by the ratio of the highest level and the mean of the composite sample from which individual units originate. Ambrus (2000) generated more data on field trials in 1996, which were reported in 2000. From these data he calculated maximum variability factors of 3 in apples, 7.2 in kiwi and 9.1 in potatoes. From the figures can be seen that these distributions are typically skew, as well in field trials as after applying post-harvest applied agent thiabendazole in apples (Figure 11 and 12).

The results of these field trials illustrated the variability due to farming practice. But they also raised interest in the individual contribution of a fruit item. Because for carrots and potatoes one can assume that more than one item is eaten per consumer meal. But that does not hold for most fruits. Due to the carrot variability a lot of samples were taken from different fruits on the UK market: 120-140 crop items of apples, banana, oranges, peaches, pears, nectarines, kiwi and plums were screened for residues, individually as well as composite samples of 10 units. A long list of fruit and residue levels was reported, with ranges of variability of 2-34 in the mentioned matrices for a total of 104 batches (Harris, 2000). As expected, variability was higher in this market survey, because retail sources and import was sampled, thus revealing units from

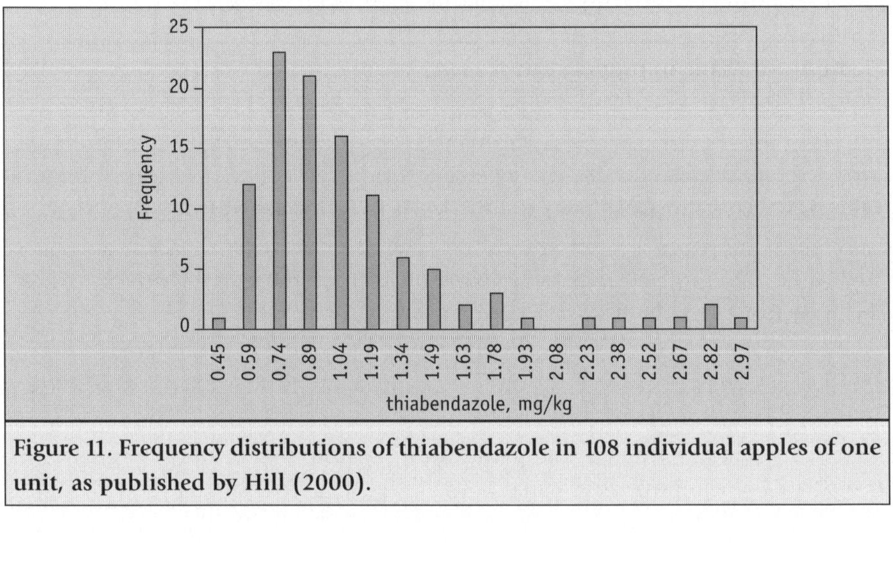

Figure 11. Frequency distributions of thiabendazole in 108 individual apples of one unit, as published by Hill (2000).

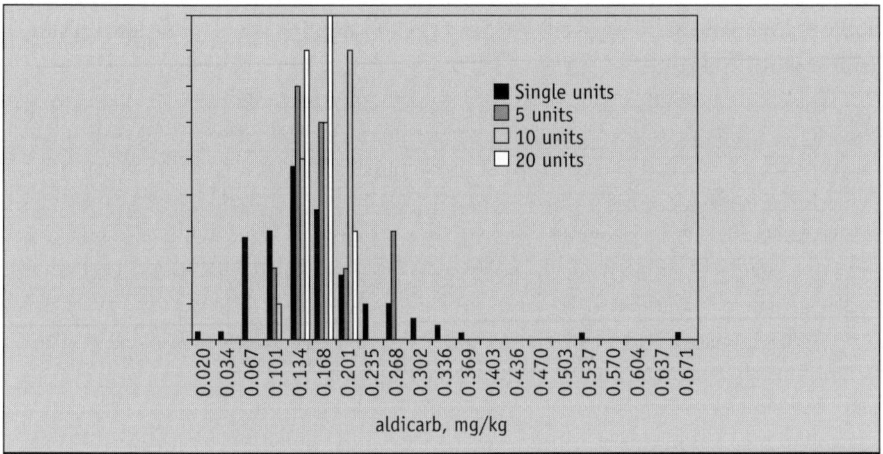

Figure 12. Frequency distributions of aldicarb in 1, 5, 10 and 20 potato samples, as published by Hill (2000).

different orchards/farmers. This is due to the way of production in intensive agriculture in Southern and Eastern European regions, where typically small farm units, e.g. greenhouses of 1-2 hectares, are cultivated by different farmers. As a consequence the treatment in terms of pesticide usage and levels can vary greatly, even in the same region and the same crop. The production mix of all these products results in a very different distribution of the values, and in turn makes it a critical factor for an adequate sampling process.

The EC Directive for sampling fruit and vegetables for pesticides control is simple: a sample of 10 units suffices (European Commission, 2002). But ten years ago this principle was already questioned by the Swedish National Food Administration (NFA), who set their own sampling plan: 3 subsamples of 1-2 kg and at least 8-10 units each. Codex requires 15 units to be sampled. In a study NFA compared EC, Codex and Swedish sampling methods. From these few results in Table 5 it can be concluded that these data do not differ significantly.

As these were only 2 sets, Hill calculated variation on residues in apples, bananas, nectarines, oranges, peaches and pears in samples sizes of 1, 5, 10 and 20 units from the available UK data sets (Hill, 2000). Results were compared with the 10 units EC sample size (indicated as 'real' in Table 6). As can be learned from

Table 5. Comparison between NFA, EC and Codex sampling plans for residues testing (Andersson, 2000).

Pesticide (mg/kg)	commodity	units	min	max	mean	max/ mean	NFA A	NFA B	NFA C	EC	Codex
Thiabendazole											
	pear	30	2,1	8	4,86	1,6	5,6	5,6	4,8	5,8	5,7
Methamidophos											
	cucumber	15	0,02	0,05	0,03	2,7	0,05	0,05	0,06	0,05	0,06

Table 6. Sample size and variation for some residue-fruit combinations. Data from Hill (2000).

	Sample size (units)				'Real'[b] 10-units
	1	5[a]	10[a]	20[a]	
Mean level, ratio to 100-unit mean	1.0	1.0	1.0	1.0	1.2
% CV	104	48	34	23	59
Highest level, ratio to 100-unit mean	11.1	3.5	2.4	2.5	4.0
Lowest level, ratio to 100-unit mean	<0.01	<0.01	0.3	0.4	0.4
Highest max./min. result ratio for any batch	432	110	8	6	-
Number of data points	4090	800	800	800	40

[a] Data from 20 random selections, made with replacement, taken from each of forty 100-unit data sets.
[b] Data from screening, i.e. real 10-unit composite samples.
All values are relative to the weighted mean level (mg/kg) calculated for each 100-unit batch

Table 6, CV values are only reaching reasonable levels at 10 or 20 unit samples. When sampling so many units the variability decreased drastically from 100-400 to a level of 6-8. This is comparable with the internationally agreed default variability of 7, which is applied by regulatory authorities to the assessment of consumer risks when they evaluate data for acute toxicity of pesticide residues (CAC/GL33 and FAO, 1999). This figure was also calculated to be the upper limit for 95% probability for the variability factor in fruit and vegetable sampling of 10 units, whereas that factor theoretically would be 10 (Hill and Reynolds, 2002). In the same article they mention the generally accepted view that analytical variance contributes not more than 11% to the total variance.

4. Genetically modified food

The above paragraph on pesticides shows that apparently sampling of contaminants that occur or are widely spread in nature is more complex than has been thought of on forehand. This becomes even more evident when sampling for GMO analysis is considered (Gilbert, 1999). In this case sampling is complicated by the fact that GMO free material is shipped and handled by the same food chain elements as GMO food. The remaining GMO material after cleaning a ship or truck is therefore contaminating the GMO free commodity in a way, which is likely also to be heterogeneous. Lischer (2001) has described this aspect in more detail. All together this was reason to develop a simulation study on kernel lot sampling to investigate the effect of heterogeneity on the detection of GMO contamination (Paoletti *et al.*, 2003). They created a mathematical model for kernel lots at any stratification level, which can be sampled to produce bulk samples. A wide range of possible sampling rates (from 5 to 250 units per population) is presented in Figure 13. In Table 7 the associated CVs are shown for the impurity levels 0.1% and 1% at 10, 20, 50 and 100 units per population. From this table can be seen how difficult it will be in daily practice to sample a load with CV values of less than 10%. This model is presently on going tested in the pan-EU project KeLDA (Kernel Lot Distribution Assessment) within the ENGL (European Network of GMOs Laboratories). The KeSTE, Kernel Sampling Technique Evaluation, software is downloadable at http://biotech.jrc.it/sampling_software.htm. Further information on the project is to be found on the website http://biotech.jrc.it/kelda/login.asp.

It has to lead to full understanding the spread of GMO in a load, which in turn will form the basis for a final EU sampling scheme (European Commission, 2004).

Table 7. CV values at 2 GMO rates and 5 stratification levels.

%GM kernels	Number of increments drown from the seed lot	Stratification				
		73%	36%	15%	7%	4%
0.1	10	21%	40%	76%	111%	150%
	20	13%	30%	60%	78%	111%
	50	10%	19%	33%	45%	72%
	100	7.8%	15%	22%	35%	72%
1	10	19%	40%	76%	108%	151%
	20	15%	29%	49%	87%	118%
	50	9.4%	18%	27%	47%	70%
	100	6.4%	15%	18%	31%	57%

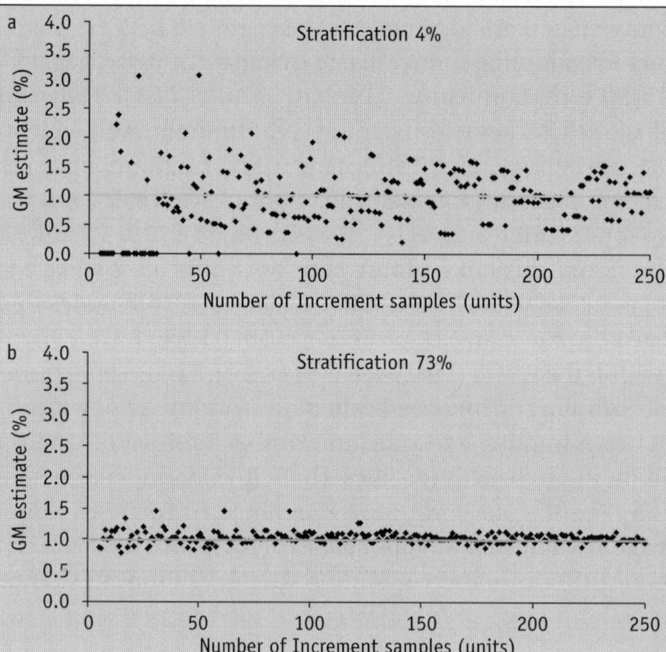

Figure 13. GM impurity estimate based on (a) 4% and (b) 73% stratification (Paoletti *et al.* 2003).

5. Future developments

The examples on sampling problems given thus far, all have one crucial point in common: they all plead for analysing larger samples to reduce sampling uncertainty. But in daily routine analysis it is far from practical to take larger samples. Therefore recently some ideas have been worked out at a prototype scale for nuts, coffee beans, corn and wheat.

5.1 Sampling lance

Instead of collecting incremental samples while using a sample lance it would be worthwhile to put it in as a probe to collect dust of the material for analysis by means of a vacuum pump and thus avoiding any damage to the goods tested (Stroka *et al.*, 2003). A prototype-sampling lance was constructed (Figure 14). This lance consisted out of a sampling head of 20 cm diameter with solid walls on the bottom. The upper walls of this head were made out of a grid with 2.5 mm holes and an inner tube was placed in the sample head on a metal tube. The inner tube was connected to a SM 16274 filter chamber (Sartorius, Germany). The filters tested were GF/A glass fibre filters (Whatman, UK) and various ash-less cellulose filters (Schleicher and Schuell, Germany). Vacuum was applied with a SV 1040 pump (Busch, Switzerland) that was connected to the filter chamber. With this set-up, 3-10 kg sample amounts in plastic buckets were tested. The lance was manually moved to penetrate the goods for 3 minutes by rapid up and down movements, under vacuum. After this treatment the filters were analysed.

Results from the experiments carried out for ochratoxin A (OTA) in coffee beans are shown in Figure 15. Despite the little number of samples that could be tested, a clear correlation between the two sets of results can be seen. In the case of green coffee beans the samples as well as the material on the filters were analysed according to CEN methods for ochratoxin A.

In case of sampling experiments for DON in corn, the spectra of the filter materials were obtained by Fourier-Transform-Infrared Attenuated-Total-Reflection (FTIR-ATR) and evaluated by Principle-Component-Analysis (Kos *et al.*, 2003), to obtain comparable correlation. For both matrices results look more promising when compared to collaborative static trials where variation between samplers contributed significantly to sampling reproducibility (Thompson *et al.*, 2002). Other matrices such as cocoa beans, peanuts, cashew nuts and pistachios caused problems when sampled with the lance, due to high fat content.

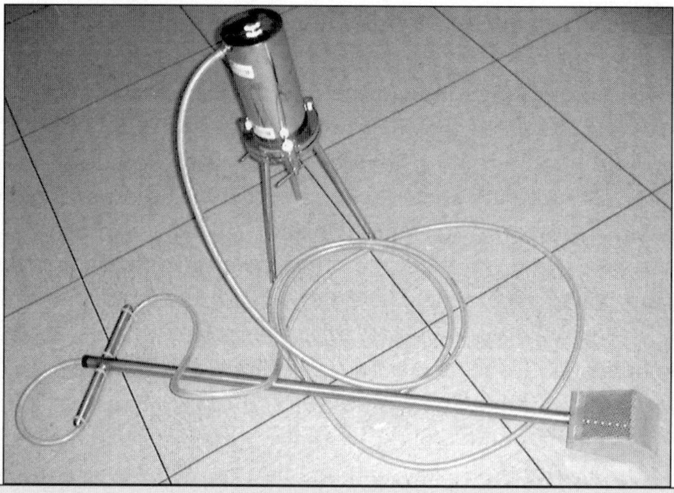

Figure 14. Prototype of sampling lance.

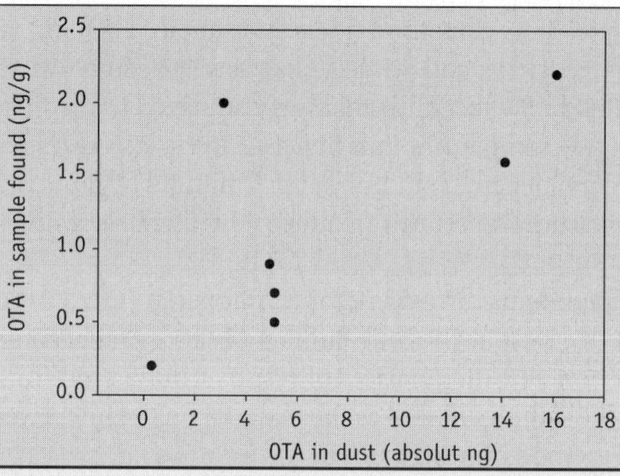

Figure 15. Dust vs. sample OTA results.

5.2 Inspection tunnel

Another idea could be to have a device in which a complete pallet load can be placed, after which air can be circulated and contaminants can be captured by membranes, which can be analysed by an appropriate method, aimed to specific compounds. Such technology is available as it was originally developed and tested for detection of ultra-trace levels of explosives in cargo shipments at airports, seaports and border crossings (Figure 16). The system uses an inspection chamber in which pressure cycling, temperature, vibration, air jetting, air blasts, sonic shock waves and other conditions are controlled for physical agitation of items to knock loose vapour and particles from the interior and outer surfaces of inspected items. The system also encompasses a selective collection option for transporting, directing, and harvesting the released and suspended analyte. Following collection on glass fibre filters, the samples are analysed and identified using standard equipment. Nevertheless these physical processes, the content of this inspection tunnel remains unharmed, unaffected and unchanged.

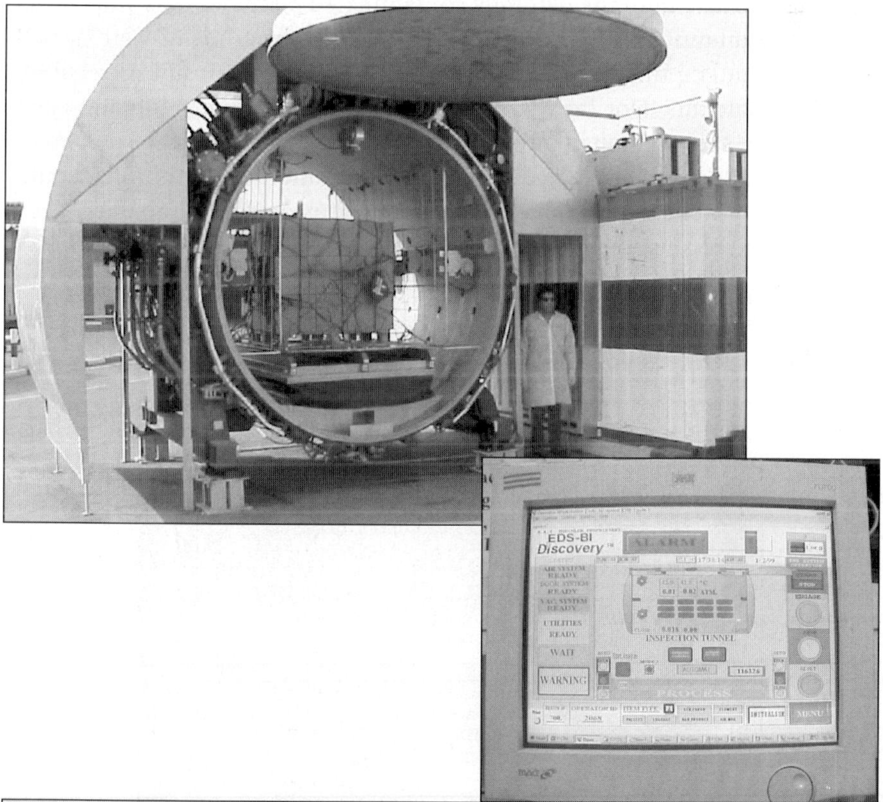

Figure 16. Discovery system as in use for aviation security purposes (Spanjer *et al.* 2001).

In order to assess the potential of this technology for detecting contamination in bulk food shipments, a preliminary test on extraction and detection was performed using existing equipment dedicated to routine inspections of cargo on explosives, to detect aflatoxins in pistachios. Naturally contaminated pistachio nuts were placed into the Discovery system that can accept for inspection, in one cycle, up to two fully loaded ISO or Euro pallets (Figure 17) and the automated sampling process was carried out for 10 minutes. Filters were placed at collection ports. In a blank experiment, filters were collected after processing of an empty Discovery system. After all these sampling processes the membranes were collected for subsequent analysis as appropriate for the contaminant of interest.

The experiments as carried out, demonstrated that this novel technology could extract aflatoxins out of the naturally contaminated pistachio nuts and collect them on membranes prior to chemical analysis. The experiments were carried out without attempting to optimise detection conditions, but still successfully extracted aflatoxins. Not from a spiked, but from a naturally contaminated pistachio nut sample. Apparently there was no problem due to the fat content of nuts. It is likely that changes to, and optimisation of conditions, could considerably improve detection. The experiment as described (Spanjer *et al.*, 2001) showed correlation with classical sampling and analysis, which provides

Figure 17. Loading a pallet of fruit into the inspection chamber of the Discovery system.

a firm basis for development of further research of this technology and its adaptation to contamination detection in food and feed.

This new approach, allowing the sampling of a fraction up to fully loaded shipment pallets of the commodity material as a whole, in one cycle, showed promising results for mycotoxins analysis. As a matter of fact non-destructive automated bulk sampling as it has been applied here offers several advantages for all kind of food and feed applications. In a first step the Discovery CERT FQS^{TM} system was tested for general feasibility and applicability with fortified coffee beans (OTA analysis) and wheat (3-acetyl-DON analysis) as well as the naturally contaminated pistachios (aflatoxin analysis) and naturally contaminated green coffee (OTA analysis). Fortification was used to allow a well-known defined contamination level. Additional experiments clearly showed significant detection of pesticides in fruit and salad, GMO in maize, quarantine pests and pathogens in salads, potatoes and eggplants and bacteria. These applications have been presented at several symposia in the past 3 years. In 2004 a FP6 proposal has been sent to the EU Cordis services as to start a large project to test this novel method of non-destructive bulk sampling technique in the field of food and feed. This could especially become a topic of significance importance when import control gets concentrated at Border Inspection Posts (BIPs), as indicated in the latest food and feed control draft regulation from EC.

6. Conclusions

The cases as presented here are based on daily routine controls as performed by any EU food authority. They are given as an example of practical problems one has to encounter when applying analytical tests, whatever its kind. The scope is limited to these areas, because the author works in the Primary Agricultural & Horticultural Products Department of the Dutch Food and Consumer Product Safety Authority, where these topics belong to the daily work program. From the given examples it will be clear that similar problems have to be faced in other fields of food and feed control. So sampling and sample preparation aspects are items that have to be investigated thoroughly before starting any analysis at all, slow or rapid the same. The crucial topic that has to be evaluated for every correct application of sampling is the fitness of the analytical procedure in relation to the aim of the final analytical test:

- Does the sample result give the correct information of the lot?

- Every contaminant/commodity combination needs its specific sampling strategy.
- And always remember: nobody ever eats the sample, so be sure about your analysis!

Regarding the given examples, some references are given as a quick reference to enter the crucial aspects in the illustrated fields:

1. Mycotoxins: Journal of AOAC INTERNATIONAL, September/October 2000 Vol. 83, No. 5 pp. 1027-1292, a special guest editor section on sampling; the book of the second World Mycotoxin Forum (as referred to by listing Adams and Whitaker, 2004); and by the EMAN website http://www.lfra.co.uk/eman2/index.asp
2. Pesticides: Food Additives and Contaminants, 2000, Vol. 17, No. 7, is a special volume reporting the International Conference on Pesticide Residues Variability and Acute Dietary Risk Assessment, edited by Caroline A. Harris *et al.*; and the website http://www.pesticides.gov.uk/home.asp
3. GMO: Simulating kernel lot sampling - the effect of heterogeneity on the detection of GM contamination (https:\\engl.jrc.it\crl\oj\simulation.pdf); modelling sampling of non-randomly distributed contaminants in kernel lots (https:\\engl.jrc.it\crl\oj\modelling.pdf); sampling for the detection of GM grains in grain lots - guidelines for decision making (https:\\engl.jrc.it\crl\oj\grain sampling.pdf); and the websites as mentioned above in the paragraph on genetically modified food.

Acknowledgements

Colleagues from the Primary Agricultural & Horticultural Products Department of the Dutch Food and Consumer Product Safety Authority in Amsterdam are acknowledged for the supply of literature references and documents on the pesticides and GMO topics. Colleagues from several EU Member States are acknowledged for their delivery of data of pistachio nut sample preparations by slurry mixing.

References

Adams, J. and Whitaker, T.B., 2004. Peanuts, aflatoxin, and the U.S. origin certification program. In: Meeting the mycotoxin menace, edited by D. Barug, H.P. van Egmond, R. López-García, W.A. van Osenbruggen and A. Visconti, Wageningen Academic Publishers, Wageningen, the Netherlands, p. 183-196.

Ambrus, A. 2000. Within and between field variability of residue data and sampling implications. Food Additives and Contaminants 17, p. 519-537.

Andersson, A. 2000. Comparison of pesticide residues in composite samples and in individual units: the Swedish approach to sampling. Food Additives and Contaminants 17, p. 547-550.

CAC/GL 33. 1999. Recommended methods of sampling for the determination of pesticide residues for compliance with MRLs. Codex Alimentarius. FAO, Rome, Italy

Campbell, A.D., Whitaker, T.B., Pohland, A.E., Dickens, J.W. and Park, D.L. 1986. Sampling, sample preparation, and sampling plans for foodstuffs for mycotoxin analysis. Pure and Appl. Chem. 58, p. 305-314.

Carter, A.D., Fogg, P. and Beard, G.R. 2000. Investigations into the causes of residue variability on carrots in the UK. Food Additives and Contaminants 17, p. 503-509.

Dahms, S. 2004. Microbiological sampling plans - Statistical aspects. Mitt. Lebensm. Hyg. 95 32-44. Excel sheet on site: http://www.foodscience.afisc.csiro.au/icmsf/excel.htm

Dorner, J.W. 2002. Simultaneous quantitation of *Aspergillus flavus*/*A. parasiticus* and aflatoxins in peanuts. Journal of AOAC INTERNATIONAL 85, p. 911-916.

Dorner, J.W. and Cole, R.J. 1993. Variability among peanut subsamples prepared for aflatoxin analysis with four mills. Journal of AOAC INTERNATIONAL 76, p. 983-987.

Earl, M., Kaethner, M. and Uihlein, M. 2000. Unit to unit variation of pesticide residues options for acute dietary risk assessment. Food Additives and Contaminants 17,p. 583-589.

European Commission, 1998. Commission Directive (98/53/EC) of 16 July 1998 laying down the sampling methods and the methods of analysis for the official control of the levels for certain contaminants in foodstuffs. Official Journal of the European Communities, L 201, p. 93-101.

European Commission, 2002. Commission Directive (2002/63/EC) of 11 July 2002 establishing Community methods of sampling for the official control of pesticide residues in and on products of plant and animal origin and repealing Directive 79/700/EEC. Official Journal of the European Communities, L 187, p. 30-43.

European Commission, 2004. draft Commission Recommendation on technical guidance for sampling and detection of genetically modified organisms and material produced from genetically modified organisms as or in products in the context of Regulation (EC) No 1830/2003, Brussels, ENV/04/04.

FAO, 1999. Pesticide residues in food 1999. Report of the Joint Meeting of the FAO panel of experts on pesticide residues in food and the environment and the WHO Core Assessment Group on pesticide residues. FAO plant production and protection paper 153. FAO, Rome, Italy.

Gilbert, J., 1999. Sampling of raw materials and processed foods for the presence of GMOs. Food Control 10, p. 363-365.

Gilbert, J. and Vargas, E.A. 2003. Advances in sampling and analysis for aflatoxins in food and animal feed. Journal of Toxicology, Toxin Reviews 22, p. 381-422.

Harris, C.A., 2000. How the variability issue was uncovered: the history of the UK residue variability findings. Food Additives and Contaminants 17, p. 491-495.

Hill, A.R.C., 2000. Residue variability and sampling-practical problems and consequences for residues monitoring. Food Additives and Contaminants 17, p. 539-546.

Hill, A.R.C. and Reynolds, S.L., 2002. Unit-to-unit variability of pesticide residues in fruit and vegetables. Food Additives and Contaminants 19, p. 733-747.

ISO 13690:1999. Cereals, pulses and milled products - Sampling of static batches. International Organization for Standardization. Geneva, Switzerland.

Kos, G., Lohninger, H. and Krska, R., 2003. Development of a method for the determination of *Fusarium* fungi on corn using Mid-Infrared Spectroscopy with Attenuated Total Reflection (FTIR-ATR) and Chemometrics. Analytical Chemistry 75, p. 1211-1217.

Lischer, P., 2001. Sampling procedures to determine the proportion of genetically modified organisms in raw materials. Part I: Correct sampling, good sampling practice. Part II: Sampling from batches of grain. Mitt. Lebensm. Hyg. 92, p. 290-304 and p. 305-312.

Möller, T.E., and Nyberg, M., 2003. Ochratoxin A in raisins and currents: basic extraction procedure used in two small marketing surveys of the occurrence and control of the heterogeneity of the toxins in samples. Food Additives and Contaminants 20, p. 1072-1076.

Paoletti, C., Donatelli, M., Kay, S. and Van den Eede, G. 2003. Simulating kernel lot sampling: the effect of heterogeneity on the detection of GMO contaminants. Seed Science and Technology 31, p. 629-638.

Ramsey, M.H., Lyn, J. and Wood, R. 2001. Optimised uncertainty at minimum overall cost to achieve fitness-for-purpose in food analysis. The Analyst 126: 1777-1783.

Schatzki, T.F. and Toyofuku, N. 2003. Sample preparation and pre-sampling of pistachios. J. Agric. Food Chem. 51, p. 6068-6072.

Spanjer, M., Stroka, J., Patel, S., Buechler, S., Pittet, A. and Barel, S. 2001. Non-destructive automated sampling of mycotoxins in bulk food and feed - a new tool for required harmonization. Mycotoxin Research. 17 A, p. 198-202.

Spanjer, M.C., Kastrup, S. and Schatzki, T.F., 2003, Sample preparation: slurry or dry milling? Book of abstracts of the second World Mycotoxin Forum, p. 94. 17-18 February 2003, Noordwijk aan Zee, the Netherlands.

Stroka, J., Spanjer, M., Buechler, S., Barel, S. and Anklam, E., 2003. Sampling methods for the analysis of deoxynivalenol. Proceedings of the ILSI Workshop on DON. Dublin, Northern Ireland.

Thompson, M., 2001. Sampling: the uncertainty that dares not speak its name. J. Environ. Monit. 1, 19N-21N.

Thompson, M., Willetts, P., Anderson, S., Brereton, P. and Wood, R., 2002. Collaborative trials of the sampling of two foodstuffs, wheat and green coffee. The Analyst 127, p. 689-691.

Whitaker, T.B. 2003. Standardisation of mycotoxin sampling procedures: an urgent necessity. Food Control 14, p. 233-237.

Whitaker, T.B., Hagler, Jr., W.M., Giesbrecht, F.G. and Johansson, A.S., 2000. Sampling, sample preparation, and analytical variability associated with testing wheat for deoxynivalenol. Journal of AOAC INTERNATIONAL 83, p. 1285-1292.

Whitaker, T.B., Springer, J., Defize, P.R., De Koe, W.J. and Coker, R., 1995. Evaluation of sampling plans used in the United States, United Kingdom and The Netherlands to test raw shelled peanuts for aflatoxin. Journal of AOAC INTERNATIONAL 78: 1010-1018.

Whitaker, T.B., Wu, J., Dowell, F.E., Hagler, Jr., W.M. and Giesbrecht, F.G., 1994. Effects of sample size and acceptance level on the number of aflatoxin-contaminated farmer's stock lots accepted and rejected at the buying point. Journal of AOAC INTERNATIONAL 77, p. 1672-1680.

Rapid methods and EU legislation

F. Verstraete
European Commission, Health and Consumer Protection Directorate-General, B-1049 Brussels, Belgium

"This contribution expresses exclusively the author's personal opinions and does not, in any case, bind the European Commission"

Abstract

To maintain the quality and safety of the food supply in the European Union (EU), feed and food legislation contains standards and specifications detailing compositional requirements as well as maximum levels of additives, residues and contaminants in food. An essential element in determining whether there is compliance with legislation is the need for authorities, food and feed business operators and laboratories to have available appropriate and validated methods of analysis. This review will focus on the provisions in EU legislation on methods of analysis related to control on contaminants in feed and food.

Keywords: Rapid methods, EU legislation, food, feed

1. Current legislation as regards methods of analysis and feed legislation

1.1 Framework legislation

Council Directive 70/373/EEC of 20 July 1970 on the introduction of Community methods of sampling and analysis for the official control of feedingstuffs (OJEC, 1970) provides that competent authorities of all Member States have to take all measures necessary to ensure that official controls of feedingstuffs, for the purpose of checking compliance with requirements arising under provisions laid down by law, regulation or administrative action concerning the quality and composition of feedingstuffs, are carried out using the Community methods of sampling and analysis. These Community methods of sampling and analysis are established by specific Commission Directives.

1.2 Disadvantages of approach

While the approach of establishing Community methods of analysis and the requirement to apply these uniform methods of sampling and analysis for the official control of feedingstuffs ensures a harmonised enforcement of EU legislation, this approach does have some disadvantages. Some of these disadvantages are summed up hereafter:

- methods of analysis are steadily improved through scientific and technological developments and it is not always possible to include in a timely manner these improvements into the methods established by EU legislation;
- laboratories performing analysis for official control are obliged to use the method of analysis established in EU-legislation although the laboratory can have more experience, expertise and the appropriate equipment to perform the same analysis with another valid and reliable method of analysis;
- it is not possible to establish by EU legislation a complete set of official methods of analysis to control compositional requirements as well as maximum levels of additives, residues and contaminants in feed and food.

1.3 Provisions in legislation in the absence of Community method of analysis

Council Directive 95/53/EC of 25 October 1995 fixing the principles governing the organisation of official inspections in the field of animal nutrition (OJEC, 1995) provides that competent authorities of the Member States shall ensure that sampling and analysis for official control are carried out in accordance with the Community rules, including the use of methods of analysis, established by EU legislation. However, as mentioned above, it is not possible to have a complete set of official methods of analysis in EU legislation and therefore it is furthermore foreseen that in the absence of Community rules and methods, competent authorities of the Member States shall take all the necessary steps to ensure that inspections are carried out:

- in accordance with standards recognised by international bodies;
- in the absence of such standards, in accordance with scientifically recognised national rules.

Important to note is that all the abovementioned provisions under II.2 and II.3 are applicable for the official control performed by competent authorities. Controls performed in the frame of internal control, quality assurance done by the food business operators do not have to follow these provisions necessarily and other sampling procedures and methods of analysis, including rapid screening methods may be used.

1.4 Implementation into EU-legislation

Methods of analysis for ingredients in animal feed, feed additives, compositional requirements and contaminants have been established in about 20 Commission Directives and a sampling procedure has also been established in a specific Directive.

1.5 Mandate to the European Committee for Standardisation (CEN)

A closer co-operation between EC Expert Committee "Methods of Analysis in animal feed", competent for the elaboration of Community methods of analysis pursuant Directive 70/373/EEC and CEN/TC 327 "Animal Feeding Stuffs - Methods of Analysis" for the standardisation of methods of analysis for feedingstuffs has been agreed to avoid overlapping activities and efforts. The objective of the co-operation was to have available a more completed and updated set of standardised methods of analysis and sampling and this to the benefit of the competent authorities and the operators. Therefore a mandate for standardisation has been addressed to CEN in 2001 ordering the standardisation of a method of analysis for about 10 items concerning compositional requirements and contaminants. An extension of this mandate containing about 16 items for standardisation is in preparation covering feed additives and contaminants. The extension of the mandate contains also a specific item on sample preparation.

In practice the EC Expert Committee "Methods of Analysis in animal feed" is since 2000 no longer active as regards elaboration of Community methods of analysis and sampling, but more general items as regards sampling and analysis, as well interpretation of analytical results are discussed. All standardisation work is currently referred to the CEN/TC 327.

The following requirements have been established as regards methods of analysis to be standardised in the framework of the mandate from the European Commission addressed to CEN:

- methods of analysis must be applicable to a wide range of feed materials and feedingstuffs;
- methods of analysis must have been successfully validated at the level of interest;
- standards must be delivered within a strict time schedule.

2. Current legislation as regards methods of analysis and food contaminant legislation

2.1 Framework legislation

Council Regulation (EEC) No 315/93 of 8 February 1993 laying down procedures for contaminants in food (OJEC, 1993 p.1) provides that maximum tolerances for specific contaminants shall be established where necessary in order to protect pubic health. These tolerances may include:
- limits for the same contaminant in different foods;
- analytical detection limits;
- a reference to the sampling and analysis methods to be used.

Council Directive 85/591/EEC of 20 December 1985 concerning the introduction of Community methods of sampling and analysis for the monitoring of foodstuffs intended for human consumption (OJEC, 1985) foresees that Community methods of sampling or analysis can be adopted where it is necessary. For the purpose of determining whether it is necessary to introduce Community methods of sampling and analysis, the following criteria has to be taken into consideration:
- the need to ensure that Community law is uniformly applied;
- the existence of barriers to intra-Community trade.

It is furthermore foreseen that Community methods of sampling must take account of the state of scientific and technical knowledge, in particular of proven methods of sampling and analysis.
The Community methods of analysis shall comply with the following criteria:
a. Methods of analysis must be examined with respect to following criteria:
    - specificity;
    - accuracy;

- precision: repeatability intra-laboratory (within laboratory) and reproducibility inter-laboratory (within and between laboratories) variabilities;
- limit of detection;
- sensitivity;
- practicability and applicability;
- other criteria which may be selected as required.

b. The precision values must be obtained from a collaborative trial, which has been conducted in accordance with an internationally recognised protocol on collaborative trials. The repeatability and reproducibility values shall be expressed in an internationally recognised form (e.g. the 95 % confidence intervals as defined by ISO 5725/1981). The results from the collaborative trial shall be published or freely available.

c. Methods of analysis which are applicable uniformly to various groups of commodities should be given preference over methods which apply only to individual commodities.

There is an important difference with the legislation as regards methods of analysis in relation with feedingstuffs (see 1). In Directive 85/591/EEC it is explicitly foreseen that the establishment of Community methods of sampling and analysis does not preclude Member States from using other tested and scientifically valid methods provided that this does not hinder the free movement of products recognised as complying with the rules by virtue of Community methods. However, in the event of differences in the interpretation of results, those obtained by the use of Community methods shall be determinant.

2.2 Official control and methods of analysis

Council Directive 89/397/EEC of 14 June 1989 on the official control of foodstuffs (OJEC, 1989) provides that the analysis in the framework of official control has to be carried out by official laboratories and other laboratories, empowered by the competent authorities of the Member States to carry out such analyses.

Council Directive 93/99/EEC of 29 October 1993 on the subject of additional measures concerning the official control of foodstuffs (OJEC, 1993 p.14) provides that the aforementioned official laboratories must comply with the general criteria for the operation of testing laboratories laid down in European

Standard EN 45001, now replaced by ISO/IEC/17025:1999. The Directive furthermore provides that the validation of the methods of analysis used within the context of official control must comply with the criteria laid down in Council Directive 85/591/EEC (see 2.1)

2.3 Implementation into EU legislation on contaminants in food

Besides the establishment of maximum levels for contaminants in Commission Regulation (EC) No 466/2001 of 8 March 2001 setting maximum levels for certain contaminants in foodstuffs (OJEC, 2001 p.1), pursuant to the provisions of Council Regulation (EEC) 315/93, several Commission Directives have been adopted providing for provisions as regards sampling procedure, sample preparation and performance criteria for the method of analysis. These Directives are:

- Commission Directive 1998/53/EC of 16 July March 2001 laying down the sampling methods and the methods of analysis for the official control of the levels of certain contaminants in foodstuffs (OJEC, 1998) (aflatoxin)
- Commission Directive 2001/22/EC of 8 March 2001 laying down the sampling methods and the methods of analysis for the official control of the levels of lead, cadmium, mercury and 3-MPCD in foodstuffs (OJEC, 2001 p.14)
- Commission Directive 2002/26/EC of 13 March 2002 laying down the sampling methods and the methods of analysis for the official control of the levels of ochratoxin A in foodstuffs (OJEC, 2002 p.38).
- Commission Directive 2002/69/EC of 30 July 2002 laying down the sampling methods and the methods of analysis for the official control of dioxins and the determination of dioxin-like PCBs in foodstuffs (OJEC, 2002 p.5).
- Commission Directive 2003/78/EC of 11 August 2003 laying down the sampling methods and the methods of analysis for the official control of the levels of patulin in foodstuffs (OJEC, 2002 p.40).
- Commission Directive 2004/16/EC of 12 February 2004 laying down the sampling methods and the methods of analysis for the official control of the levels of tin in canned foods (OJEC, 2004 p.16).

These Directives determine that the methods of analysis used for food control purposes must comply with the criteria laid down in Council Directive 85/591/EC and contains specific requirements as regards detection limit, limit of quantification, recovery and precision.

In case there are a limited number of fully validated methods of analysis, alternatively, a "fitness-for-purpose" approach, defining a single parameter - a fitness function - to evaluate the acceptability of methods of analysis for official control can be used. A fitness function is an uncertainty function that specifies maximum levels of uncertainty regarded as fit for purpose. The maximum standard uncertainty can be calculated using the following formula:

$$Uf = \sqrt{(LOD / 2)^2 + (\alpha \times C)^2}$$

where:
- Uf is the maximum standard uncertainty (μg/kg);
- LOD is the limit of detection of the method (μg/kg);
- α is a constant, numeric factor to be used depending on the value of C (the values to be used are given in Table 1);
- C is the concentration of interest (μg/kg).

3. Reporting and interpretation of analytical results

3.1 Introduction

Besides the establishment of harmonised specifications (e.g. maximum level) at EU level, and of a harmonised sampling procedure and harmonised requirements as regards the methods of analysis, associated the harmonised specification it is also of major importance that the reporting and interpretation of analytical results is harmonised.

Table 1. Numeric values to be used for α as constant in the above mentioned formula, depending on the concentration of interest.

C (μg/kg)	α
≤ 50	0.2
51-500	0.18
501-1000	0.15
1001-10000	0.12
> 10000	0.1

3.2 Background

The results of a Scientific Co-operation task on "Preparation of a working document in support of the uniform interpretation of legislative standards and the laboratory quality standards prescribed under Directive 93/99/EEC" indicated that different approaches existed between the Member States with regard to the taking into account of the analytical variability when making the assessment if a consignment/lot complies with the legislation. In particular in the case the analytical result exceeds the maximum level by less than the expanded uncertainty, the evaluation of the acceptability of the consignment was varying within the EU.

3.3 Harmonised reporting and interpretation of results

In order to ensure that the reporting and interpretation of analytical results are harmonised across the EU, specific provisions have been introduced in EU legislation concerning contaminants in food and feed. These provisions consist of the following:

- The analytical result has to be reported as x +/- U whereby x is the analytical result and U is the expanded measurement uncertainty, using a coverage factor of 2 which gives a level of confidence of approximately 95%. An alternative approach to the determination of the expanded measurement uncertainty is the establishment of the decision limit (CCα) according to the provisions of Commission Decision 2002/657/EC of 14 August 2002 implementing Council Directive 96/23/EC concerning the performance of analytical methods and the interpretation of results (OJEC, 2002 p.8).
- The analytical result is to be reported corrected or uncorrected for recovery. The manner of reporting and the level of recovery must be reported.
- A batch, a consignment is considered as being not in conformity with the EU requirements as regards the contaminants in feed and food in case the analytical result of the representative sample, corrected for recovery, exceeds beyond any reasonable doubt the EU maximum level, taking into account the expanded measurement uncertainty.

More extensive information on this issue can be found in the document "Report on the relationship between analytical results, the measurement uncertainty, recovery factors and the provisions in EU Food and Feed legislation with particular focus on the community legislation concerning:

- contaminants in food (Council Regulation (EEC) No 315/93 of 8 February 1993 laying down community procedures for contaminants in food) (OJEC, 1993 p.1);
- undesirable substances in feed (Directive 2002/32/EC of the European Parliament and of the Council of 7 May 2002 on undesirable substances in animal feed) (OJEC, 2002 p.10).

The document is available at the SANCO Food Safety website: http://europa.eu.int/comm/food/food/chemicalsafety/contaminants/report-sampling_analysis_2004_en.pdf.

4. Specific reference in current EU legislation to the use of rapid screening methods in the frame of official control

4.1 Introduction

In the following two headings more details are given for two pieces of EU-legislation how the use of rapid screening methods can be integrated into the official control.

- Commission Decision 2002/657/EC of 14 August 2002 implementing Council Directive 96/23/EC concerning the performance of analytical methods and the interpretation of results (OJEC, 2002 p.8).
- Commission Directive 2002/69/EC of 30 July 2002 laying down the sampling methods and the methods of analysis for the official control of dioxins and the determination of dioxin-like PCBs in foodstuffs (OJEC, 2002 p.5).and Commission Directive 2002/70/EC of 26 July 2002 laying down the sampling methods and the methods of analysis for the official control of dioxins and the determination of dioxin-like PCBs in feedingstuffs (OJEC, 2002 p.5)

4.2 Control of certain substances and residues thereof in live animals and animal products

Commission Decision 2002/657/EC establishes criteria and procedures for the validation of analytical methods to ensure the quality and comparability of analytical results generated by official laboratories. Moreover, the Decision establishes common criteria for the interpretation of test results and introduces a procedure to progressively establish minimum required performance limits

(MRPL) for analytical methods employed to detect substances for which no permitted limit (maximum limit) has been established. This is in particular important for substances whose use is not authorised or is specifically prohibited in the EU.

In the Decision it is mentioned that as a result of advances in analytical chemistry, the concept of routine methods and reference methods has been superseded by the criteria approach, in which performance criteria and procedures for the validation of screening and confirmatory methods are established. A screening method is defined as "methods that are used to detect the presence of a substance or class of substances at the level of interest. These methods have the capability for a high sample throughput and are used to sift large numbers of samples for potential non-compliant results. They are specifically designed to avoid false compliant results."

The Decision furthermore specifies that only those analytical techniques, for which it can be demonstrated in a documented traceable manner that they are validated and have a false compliant rate of < 5 % (β-error) at the level of interest shall be used for screening purposes. In the case of a suspected non-compliant result, this result shall be confirmed by a confirmatory method. β-error means the probability that the tested sample is truly non-compliant, even though a compliant measurement has been obtained (false compliant decision). A confirmatory method means methods that provide full or complementary information enabling the substance to be unequivocally identified and if necessary quantified at the level of interest.

It is also of major importance that the screening method is sensitive enough to enable detection at levels lower than the defined minimum required performance limits (MRPL). MRPL is defined as the minimum content of an analyte in a sample, which at least has to be detected and confirmed. It is also intended to harmonise the analytical performance of methods for substances for which no permitted limit has been established.

4.3 Control of dioxins and dioxin-like PCBs in feed and food

Commission Directive 2002/69/EC of 30 July 2002 laying down the sampling methods and the methods of analysis for the official control of dioxins and the determination of dioxin-like PCBs in foodstuffs and Commission Directive 2002/70/EC of 26 July 2002 laying down the sampling methods and the methods of analysis for the official control of dioxins and the determination of dioxin-like PCBs in feedingstuffs establish requirements and performance criteria for the methods of analysis for the official control of the levels of dioxins and dioxin-like PCBs in feed and food.

Both directives explicitly provide that official control for the presence of dioxins in feed and food can be performed by a strategy involving a screening method with proven, widely acceptable validation and high throughput in order to select those samples with levels of dioxins and dioxin-like PCBs that are less than 30-40% below or exceed the level of interest. The concentration of dioxins in those samples with significant levels needs to be determined/ confirmed by a confirmatory method. Both Directives establish therefore similar requirements for the confirmatory methods of analysis and for the screening method.

4.3.1 Basic requirements for acceptance of analytical procedures for the analysis of dioxins

High sensitivity and low limits of detection

For polychlorinated dibenzo-p-dioxins (PCDD) and polychlorinated dibenzofurans (PCDF), detectable quantities have to be in the picogram TEQ (10^{-12} g) range because of extreme toxicity of some of these compounds. Polychlorinated biphenyls (PCBs) are known to occur at higher levels than the PCDDs and PCDFs. For most PCB congeners sensitivity in nanogram (10^{-9} g) range is already sufficient. However, for the measurement of the more toxic dioxin-like PCB congeners (in particular non-ortho substituted congeners), the same sensitivity must be reached as for the PCDDs and PCDFs.

High selectivity (specificity)

A distinction is required for PCDDs, PCDFs and dioxin-like PCBs from a multitude of other, co-extracted and possibly interfering compounds present at concentrations up to several orders of magnitude higher than those of the analytes of interest. For gas chromatography/mass spectrometry (GC/MS) methods, a differentiation among various congeners is necessary, such as between toxic (e.g. the seventeen 2,3,7,8-substituted PCDDs and PCDFs and dioxin-like PCBs) and other congeners. Bioassays should be able to determine TEQ values selectively as the sum of PCDDs, PCDFs and dioxin-like PCBs.

High accuracy (trueness and precision)
The determination should provide a valid and reliable estimate of the true concentration in a sample. High accuracy (accuracy of the measurement: the closeness of agreement between the result of a measurement with the true or assigned value of the measurand) is necessary to avoid the rejection of a sample analysis result on the basis of poor reliability of the estimate of toxic equivalents (TEQs). Accuracy is expressed as trueness (difference between the mean value measured for an analyte in a certified material and its certified value, expressed as percentage of this value) and precision (precision is usually calculated as a standard deviation including repeatability and reproducibility, and indicates the closeness of agreement between the results obtained by applying the experimental procedure several times under prescribed conditions).

Screening methods can comprise bioassays and GC/MS methods; confirmatory methods are high-resolution gas-chromatography/high-resolution mass-spectrometry (HRGC/HRMS) methods. The following criteria have to be complied with on total TEQ value:

	Screening methods	Confirmatory methods
False negative rate	< 1%	
Trueness		- 20% to + 20%
CV (coefficient of variation)	< 30%	< 15%

Important to note is the very strict requirement for screening methods to have a false negative rate of less than 1%.

4.3.2 Different analytical approaches using a screening method

Different analytical approaches can be performed using a screening method: a pure screening approach and a quantitative approach.

Screening approach
The response of samples is compared to that of a reference sample at the level of interest. Samples with a response less than the reference are declared negative, those with a higher response are suspected positives. Requirements are:

- A blank and a reference sample(s) have to be included in each test series, which is extracted and tested at the same time under identical conditions.

The reference sample must show a clearly elevated response in comparison to a blank.
- Additional reference samples at 0.5x and 2x the level of interest should be included to demonstrate the proper performance of the test in the range of interest for the control of the level of interest.
- When testing other matrices, the suitability of the reference sample(s) has to be demonstrated, preferentially by including samples shown by HRGC/HRMS to contain a TEQ level around that of the reference sample or else a blank spiked at this level.
- Since no internal standards can be used in bioassays, tests on repeatability are very important to obtain information on the standard deviation within one test series. The coefficient of variation should be below 30%.
- For bioassays, the target compounds, possible interferences, and maximum tolerable blank levels should be defined.

Quantitative approach
The quantitative approach requires standard dilution series, duplicate or triplicate clean up and measuring as well as blank and recovery controls. The result may be expressed as TEQ, thereby assuming that the compounds responsible for the signal correspond to the TEQ principle. This can be performed by using TCDD (or a dioxin/furan standard mixture) to produce a calibration curve to calculate the TEQ level in the extract and thus in the sample. This is subsequently corrected for the TEQ level calculated for a blank sample (to account for impurities from solvents and chemicals used), and a recovery (calculated from the TEQ level in a quality control sample around the limit of interest). It is essential to note that part of the apparent recovery loss may be due to matrix effects and/or differences between the TEF values in the bioassays and the official Toxic Equivalency Factors (TEF) values set by the World Health Organization (WHO).

4.3.3 Requirements for methods of analysis used for screening

Requirements for methods of analysis used for screening are:
- GC/MS methods of analysis and bioassays may be used for screening.
- For cell-based bioassays and kit-based bioassays specific requirements are provided for (see 4.3.4 and 4.3.5).
- Information on the number of false-positive and false-negative results of a large set of samples below and above the maximum level or action level is necessary, in comparison to the TEQ content as determined by a confirmatory method of analysis. Actual false negative rates should be under 1%. The rate

of false positive samples should be low enough to make the use of a screening tool advantageous.

- Positive results have always to be confirmed by a confirmatory method of analysis (HRGC/HRMS). In addition, samples from a wide TEQ-range should be confirmed by HRGC/HRMS (approximately 2-10% of the negative samples). Information on correspondence between bioassay and HRGC/HRMS results should be made available

4.3.4 Specific requirements for cell based bioassays

Specific requirements for cell based bioassays are:

- When performing a bioassay, every test run requires a series of reference concentration of TCDD or a dioxin/furan mixture (full dose-response curve with a $R^2 > 0.95$). However, for screening purposes, an expanded low level curve for analysing low-level samples could be used.
- A TCDD reference concentration (about 3x limit of quantification) on a quality control sheet should be used for the outcome of the bioassay over a constant time period. An alternative could be the relative response of a reference sample in comparison to the TCDD calibration line since the response of the cells may depend on many factors.
- Quality control (QC) charts for each type of reference material should be recorded and checked to make sure the outcome is in accordance with the stated guidelines.
- In particular for quantitative calculations, the induction of the sample dilution used must be within the linear portion of the response curve. Samples above the linear portion of the response curve must be diluted and re-tested. Therefore, at least 3 or more dilutions at one time are recommended to be tested.
- The percent standard deviation should not be above 15% in a triplicate determination for each sample dilution and not above 30% between three independent experiments.
- The limit of detection may be set as 3x the standard deviation of the solvent blank or of the background response. Another approach is to apply a response that is above the background (induction factor 5x the solvent blank) calculated from the calibration curve of the day. The limit of quantification may be set as 5 to 6x the standard deviation of the solvent blank or of the background response or to apply a response that is clearly above the background (induction factor 10x the solvent blank) calculated from the calibration curve of the day.

4.3.5 Specific requirements for kit based bioassays

Specific requirements for kit based bioassays are:

- Manufacturer's instructions for sample preparation and analyses have to be followed.
- Test kits should not be used after the expiration date.
- Materials or components designed for use with other kits should not be used.
- Test kits should be kept within the specified range of storage temperature and used at the specified operating temperature.
- The limit of detection for immunoassays is determined as the sum of the mean and 3x the standard deviation, based on 10 replicate analysis of the blank, to be divided by the slope value of the linear regression equation.
- Reference standards should be used for tests at the laboratory to make sure that the responsiveness to the standard is within an acceptable range.

Important to note is that for the time being no evidence has yet been submitted of commercially available kit-based bioassays having sufficient sensitivity and reliability to be used for screening for the presence of dioxins at the required levels in samples of foodstuffs and feedingstuffs.

5. The new Official Feed and Food Control Regulation

Regulation (EC) No 882/2004 of the European Parliament and of the Council of 29 April 2004 on official controls performed to ensure the verification of compliance with feed and food law, animal health and animal welfare rules (OJEC, 2004 p.83). This Regulation establishes at Community level a harmonised framework of general rules for the organisation of such controls. The Regulation is applicable from 1 January 2006 onwards and replaces the provisions contained in Directives 70/373/EEC, 85/591/EEC, 89/397/EEC, 93/99/EEC and 95/53/EC. However, the implementing rules adopted on the basis of those acts, shall remain in force in so far as they are not in contradiction with this Regulation, pending the adoption of the necessary provisions on the basis of this Regulation.

5.1 Methods of sampling and analysis for official control

As regards the sampling and analysis this Regulation does provide that sampling and analysis methods used in the context of official controls shall comply with relevant Community rules or

a. if no such rules exist, with internationally recognised rules or protocols, for example those that the European Committee for Standardisation (CEN) has accepted or those agreed in national legislation; or
b. in the absence of the above, with other methods fit for the intended purpose or developed in accordance with scientific protocols.

In case the abovementioned does not apply, validation of methods of analysis may take place within a single laboratory according to an internationally accepted protocol.

5.2 Characterisation of methods of analysis

Wherever possible, methods of analysis shall be characterised by the following appropriate criteria:

- accuracy;
- applicability (matrix and concentration range);
- limit of detection;
- limit of determination;
- precision;
- repeatability;
- reproducibility;
- recovery;
- selectivity;
- sensitivity;
- linearity;
- measurement uncertainty;
- other criteria that may be selected as required.

The precision values shall either be obtained from a collaborative trial which has been conducted in accordance with an internationally recognised protocol on collaborative trials (e.g. ISO 5725:1994 or the IUPAC International Harmonised Protocol) or, where performance criteria for analytical methods have been established, be based on criteria compliance tests. The repeatability and reproducibility values shall be expressed in an internationally recognised form (e.g. the 95 % confidence intervals as defined by ISO 5725:1994 or IUPAC). The results from the collaborative trial shall be published or freely available.

Methods of analysis which are applicable uniformly to various groups of commodities should be given preference over methods which apply only to individual commodities.

In situations where methods of analysis can only be validated within a single laboratory then they should be validated in accordance with e.g. IUPAC Harmonised Guidelines, or where performance criteria for analytical methods have been established, be based on criteria compliance tests.

5.3 Further possible measures as regards methods of sampling and analysis

The Regulation further provides for the possibility to establish:

- methods of sampling and analysis, including the confirmatory or reference methods to be used in the event of a dispute;
- performance criteria, analysis parameters, measurement uncertainty and procedures for the validation of the methods mentioned in the previous indent;
- rules on the interpretation of results.

The Regulation furthermore foresees in the right of feed and food business operators whose products are subject to sampling and analysis to apply for a supplementary expert opinion

5.4 Requirements as regards official laboratories

The Regulation provides also that the competent authority shall designate laboratories that may carry out the analysis of samples taken during official controls. However, competent authorities may only designate laboratories that operate and are assessed and accredited in accordance with the following European standards:

- EN ISO/IEC 17025 on 'General requirements for the competence of testing and calibration laboratories';
- EN 45002 on 'General criteria for the assessment of testing laboratories';
- EN 45003 on 'Calibration and testing laboratory accreditation system. General requirements for operation and recognition'.

The accreditation and assessment of testing laboratories may relate to individual tests or groups of tests.

6 The new Hygiene Regulations on Food and Feed

6.1 General Hygiene Regulation on Food

Regulation (EC) No 852/2004 of the European Parliament and of the Council of 29 April 2004 on the hygiene of foodstuffs (OJEC, 2004 p.3) lays down general rules for food business operators on the hygiene of foodstuffs with the objective to ensure a high level of consumer protection with regard to food safety. The provisions of this Regulation become applicable as from 1 January 2006. As regards sampling and analysis it provides that food business operators shall as appropriate adopt specific hygienic requirements as regards procedures necessary to meet targets set to achieve the objectives of this Regulation and associated sampling and analysis. The Regulation furthermore foresees that the targets and the associated sampling and analysis method can be established by the Commission. However, when the sampling or analysis method is not specified, food business operators may use appropriate methods laid down in Community legislation or national legislation, or in the absence of such methods, methods that offer equivalent results to those obtained using the reference method, if they are scientifically validated in accordance with internationally recognised rules or protocols.

The Regulation furthermore foresees that the food business operators shall put in place, implement and maintain a permanent procedure or procedures based on hazard analysis and critical control points (HACCP) principles. The HACCP principles consist amongst others of the establishment and implementation of an effective monitoring procedure at critical control points.

The availability and use of rapid reliable screening methods is of major importance for the application of an effective monitoring procedure.

6.2 Hygiene Regulation on Feed

The Regulation of the European Parliament and of the Council laying down requirements for feed hygiene is expected to be adopted by the end of 2004. It will apply also as from 1 January 2006. The objective of the hygiene rules set out in this Regulation is to ensure a high level of consumer protection with regard to feed and food safety.

In line with the hygiene rules for food business operators also feed business operators shall put in place, implement and maintain a permanent procedure or procedures HACCP principles. It furthermore provides that feed business operators, with the exception of feed businesses at the level of primary production, must draw up a quality control plan, including in particular checks on the critical control points in the manufacturing process, sampling procedures and frequencies, methods of analysis and their frequency, compliance with the specifications - and the destination in the event of non-compliance - from processed materials to final products.

It goes without saying that the availability and use of rapid reliable screening methods is of major importance for the application of the provisions of this Regulation.

7. Concluding remarks as regards the use of rapid screening methods

Although specific references to the use of rapid screening methods are not so frequent in EU legislation, the availability and application of reliable and rapid screening methods are without any doubt of major importance to achieve the objectives of EU food safety legislation, i.e. a high level of consumer protection with regard to food safety.

However when using reliable rapid screening methods of analysis, special attention should be paid to following aspects (non exhaustive list):

- A reliable rapid screening method should be applied on a sample that is representative for the batch to control.
- The applied extraction and clean-up procedures are of major importance for the reliability of the analytical outcome.
- The sensitivity and specificity of the rapid screening method should be sufficient and fit for the purpose.
- The rapid screening method can only be used on matrices and at levels of interest for which it has been validated.
- Although the use of a rapid screening method might require less experience and expertise from the analyst than the application of a high technological

confirmatory method such as HRGC-HRMS, a certain level of experience and expertise is always necessary in order to ensure that reliable results are obtained by using the screening method.
- Certified reference materials are an absolute necessity enabling the verification of the reliability of rapid screening methods at regular time periods.

References

Official Journal of the European Communities, L 170, 3.8.1970, p. 2

Official Journal of the European Communities, L 372, 31.12.1985, p.50

Official Journal of the European Communities , L186, 30.6.1989, p.23.

Official Journal of the European Communities, L 37, 13.2.1993, p.1

Official Journal of the European Communities , L290, 24.11.1993, p.14

Official Journal of the European Communities, L 265, 8.11.1995, p.17

Official Journal of the European Communities, L 201, 17.7.1998, p.93. As last amended by Commission Directive 2004/43/EC.

Official Journal of the European Communities, L 77, 16.3.2001, p.1. As last amended by Commission Regulation (EC) 684/2004

Official Journal of the European Communities, OJ L 77, 16.3.2001, p.14

Official Journal of the European Communities, OJ L 75, 16.3.2002, p.38. As last amended by Commission Directive 2004/43/EC

Official Journal of the European Communities, L 140, 30.5.2002, p. 10

Official Journal of the European Communities, OJ L 209, 6.8.2002, p.5. As last amended by Commission Directive 2004/44/EC

Official Journal of the European Communities, OJ L 203, 12.8.2002, p.40

Official Journal of the European Communities L 221, 17.8.2002, p. 8. Last amended by Commission Decision 2004/25/EC of 22 December 2003

Official Journal of the European Communities, OJ L 42, 13.2.2004, p.16

Official Journal of the European Union L 139, 30.4.2004, p. 1. Corrigendum published in OJ L 226, 25.6.2004, p. 3

Official Journal of the European Union L 165, 30.4.2004, p. 1. Corrigendum published in OJ L 226, 25.6.2004, p. 83

Advances in biosensors for detection of pathogens in food and water

R. O'Kennedy[1], P. Leonard[1], S. Hearty[1], S. Daly[1], P. Dillon[1], J. Brennan[1], L. Dunne[1], A. Darmaninsheehan[1], S. Stapleton[1], E. Tully[1], J. Quinn[1,2] and T. Chakraborty[3]
[1]School of Biotechnology and National Centre for Sensor Research, Dublin City University, Dublin 9, Ireland, [2]Texas Instruments Inc., 13536 N. Central Expressway, Dallas ,Texas 75243, USA, [3]Institute of Med. Microbiology, Justus-Leibig-Universität Giessen, Frankfurter Str. 107, D-35392 Giessen, Germany

Abstract

Microorganisms play many important roles in nature, but certain potentially harmful microbes can contaminate food and water, resulting in infectious and often potentially life-threatening diseases in animals and humans. Many established procedures for the detection of such contamination have required pre-enrichment steps prior to biochemical identification and this approach may require a considerable time-period. Recent research has focused on the development of sensors for the detection of microorganisms, with rapid and 'real-time' identification. Some of the most commonly used systems will be described with particular reference to surface plasmon resonance (SPR) sensors. Their applications for the detection of pathogens in food and water are reviewed. The advantages and limitations of biosensors for the detection of pathogens will be discussed. A number of case studies will be given. In addition, new approaches and potential areas for research will be outlined.

Keywords: biosensors, pathogens, food, water

1. Introduction

There is now a very major demand for the measurement of pathogens in a variety of situations and recent events related to terrorism and security have accelerated the need for and development of such systems. We have recently reviewed the use of sensors for the detection of pathogens in food and water and this paper will extend that review and examine areas and technologies that may, in future, play a key role in the analysis of pathogens (Leonard *et al.*, 2003). It is evident that a variety of pathogens pose major problems. Examples of these are given in Table 1.

At present pathogenic organisms present one of the major challenges to world health and this is often exacerbated in times of war and civil strife when water supplies become contaminated, food supplies are destroyed and individuals are forced to consume material, which would otherwise be discarded, in order to ensure their survival. However, there are also major problems encountered when food becomes contaminated due to lack of sufficient care during production, poor handling and insufficient attention to storage conditions. This can result in large numbers of people becoming infected to varying degrees with the associated health effects and may result in huge food recalls, loss of consumer confidence and major legal and financial implications for food outlets and suppliers. The supply of clean, potable, contamination-free water is also an essential requisite for health and, despite major advances worldwide, there are still many reports of contaminated water supplies often arising from problems associated with inappropriate or inadequate treatment of sewage, industrial effluents and incidents that occur following flooding.

Biosensor development over the last few decades has grown enormously, and now, biosensors are an established and reliable analytical tool.

A biosensor may be defined as any probe or transducer that incorporates a biological component as the key functional element in the overall transducer sequence (Vadgama and Crump, 1992). It consists of three essential elements:

Table 1. Examples of some pathogenic organisms.
Campylobacter jejuni
Listeria monocytogenes
Escherichia coli 0157
Staphylococcus aureus
Salmonella typhimurium
Streptococci
Clostridium botulinum
Bacillus cereus

- A biological entity that interrogates the sample and generally binds to the target analyte;
- a transducer that converts the biological binding event into a measurable signal;
- and an output that records the signal generated.

The biological element may be an antibody or antibody fragment, a nucleic acid or a derivative (DNA, RNA, peptide nucleic acid or aptamer), an enzyme, a lectin, a specific receptor, a whole cell, a membrane or cell preparation, an animal part (e.g. insect antennae) or a molecularly imprinted polymer. Indeed, many sensors have a number of these entities combined examples being the use of enzyme-labelled antibodies or DNA and a huge variety of different combinations have been recorded in the literature. In a sensor both the transducer and biological element work together to relate the concentration of analyte to a measurable electronic signal. The first biosensor reported (Clark and Lyons, 1962) incorporated an enzyme as the bio-component operating in combination with an electrochemical transducer for the detection of glucose in samples. Since then interest in the biosensor area has grown substantially, and this development is extensively reviewed elsewhere (Rogers, 2000; Patel, 2002). Future trends in biosensor development today show the generation of miniaturised, chip-based, microarrays capable of multi-analyte detection (Ekins, 1998; de Wildt *et al.*, 2000; Templin *et al.*, 2002; Darmaninsheehan *et al.*, 2003).

When the biological entity interacts specifically with the target analyte a change in one or more physio-chemical parameters occurs. These changes may produce ions, electrons, gases, heat, mass alterations, fluorescence etc., which are then converted into electrical signals by the transducers that process and display the information in a suitable form. A number of different transducer elements are available for use in biosensors (Sethi, 1994). These include electrochemical, piezo-electric and optoelectronic.

Biosensors-based on optoelectronics often incorporate the phenomenon of surface plasmon resonance (SPR) as a method of detection. A number of SPR-based biosensors are commercially available which include the Biacore range of instruments (Biacore AB), the IAsys instrument (Affinity Sensors), the Texas Instruments SPR device (TI-SPR) and the BIOS-1 system (Windsor Scientific Ltd).

Whatever biosensor format is chosen it must fulfil a number of requirements as outlined in Table 2. In relation to the detection of pathogens there may be additional elements that must be taken into consideration. These may include the need to adequately protect the operator during analysis, the ability to cleanse the unit or to use sensing elements that do no attract fouling by the sample matrix or can be readily discarded, problems associated with the relative size, distribution of antigens and extraction of the pathogen, and, finally the necessity for a pre-concentration step. The latter may require the use of an enrichment growth medium or the need for a concentration step to collect sufficient material for analysis from a sample of large volume.

It should be noted that these requirements represent the ideal scenario while, in reality, it has proved very difficult to achieve these demands. However, with new developments in technology and, recent advances, particularly in relation to miniaturisation, the feasibility of exploiting the potential of sensors for many

Table 2. Requirements for biosensor-based assay systems.

Accuracy	need to eliminate false positives/negatives
Sensitivity	compatible with relevant standards and this may be a major problem for some systems but new approaches and increased sensitivity are being developed
Specificity	essential for detection of pathogenic species
Need for pre-enrichment	ideally no such step should be required, but in practice it is often necessary
Speed of analysis	ideally this should be in 'real-time' or at least very rapid
Robust	the system should be able to function in the field or often in relatively harsh conditions
'User-friendly'	increasingly the demand is for instrumentation that can be used with minimum manipulation by relatively unskilled personnel
Capable	of working directly in complex matrices or with minimum extraction required.
Portable	necessary in some applications e.g. 'on-site' analysis for environmental or military applications
Re-useable	sensing elements may be capable of multiple use following regeneration or may be very cheap enabling replacement after each analysis

practical applications for the detection of pathogens in food and water now seems achievable within a reasonable time-frame. Hence, there is great interest commercially in the development of such systems.

Examples of a range of sensor systems that are used for the detection of pathogens are given in Table 3.

2. Surface plasmon resonance

Surface plasmon resonance is of significance in many well-developed sensor-based applications. The principle of SPR will be described in the context of Biacore instruments. They incorporate the SPR phenomenon to monitor biomolecular interactions in 'real-time'. At an interface between two transparent media of different refractive index such as glass and water, light coming from

Table 3. Examples of the use of sensors for the detection of pathogens.

Sensor System	Sensing element	Target	Sensitivity	Reference
• Quartz crystal microbalance	Antibody	*E. coli*	1.7×10^5- 8.7×10^7	Kim and Park (2002)
• Potentiometric	Antibody	*E. coli*	10 cells/ml	Ercole *et al.* (2003)
• Flexural plate wave	Antibody	*E. coli* K12 and J5	3×10^5-6.2×10^7	Pyun *et al.* (1998)
• SPR/MS	Antibody	Staphylococcal toxin	1ng/ml	Nedelkov *et al.* (2000)
• Spectral SPR	Antibody	Staphyloccal enterotoxin B	0.5 ng/ml	Homola *et al.* (2002)
• Electrochemical screen-printed array	Lectins	*E. coli*	-	Ertl *et al.* (2003)
• Fluorescence resonance energy transfer	Antibody	*Listeria* or *Salmonella* antigens	2.0 µg/ml	Ko and Grant (2003)
• Quartz crystal microbalance	Antibody	*Listeria*	10^7 cells/ml	Vaughan *et al.* (2001)
• Silicon chip	Antibody	*S. typhimurium*	119 CFUs	Dill *et al.* (1999)

the side of higher refractive index is partly reflected and partly refracted. Above a certain critical angle of incidence no light is refracted across the interface and total internal reflection (TIR) occurs at the metal film-liquid interface. This is where light travels through an optically dense medium such as glass, and is reflected back through that medium at the interface with a less optically dense medium such as buffer. Although the incident light is totally reflected, the electromagnetic field component, termed the evanescent wave, penetrates a distance on the order of one wavelength into the less optically dense medium. The evanescent wave is generated at the interface between a glass prism (high refractive index) and a layer of buffer (lower refractive index). If the interface between the media of higher and lower refractive indices is coated with a thin metal film (a fraction of the light wavelength), then the propagation of the evanescent wave will interact with the electrons on the metal layer. Metals contain electron clouds at their surface, which can couple with incident light at certain angles. These electrons are also known as plasmons, and the passage of the evanescent wave through the metal layer causes the plasmons to resonate, forming a quantum mechanical wave known as a surface plasmon. Therefore, when surface plasmon resonance occurs, energy from the incident light is lost to the metal film resulting in a decrease in the reflected light intensity. The resonance phenomenon only occurs at an acutely defined angle of the incident light. This angle is dependent on the refractive index of the medium close to the metal-film surface. Changes in the refractive index of the buffer solution (e.g. an increase in surface concentration of solutes), to a distance of about 300 nm from the metal film surface will therefore alter the resonance angle. Continuous monitoring of this resonance angle allows the quantitation of changes in refractive index of the buffer solution close to the metal-film surface.

In 'real-time' Biacore, the metal film properties, wavelength, and refractive index of the glass (denser medium) are all kept constant, and as a result SPR can be used to monitor the refractive index of the aqueous layer immediately adjacent to the metal (gold) layer. In the Biacore system the chip is composed of glass, has 4 channels and the associated gold layer is covered with a layer of dextran chemically modified to facilitate immobilisation of ligands such as antibodies or antigens. Any changes in mass that occur due to binding of the analyte with the immobilised antibody on the sensor chip will cause a change in SPR angle, which is monitored in 'real-time' and quantified as a sensorgram. A mass change of approximately 1 kRU (1,000 RU) corresponds to a mass change in surface protein concentration of 1 ng/mm^2. Typical responses for surface binding of proteins are of the order of 0.1-20 kRU. A more extensive review of SPR is reported by Liedberg and Lundstrom (1993).

3. Resonant mirror-based devices

An example of a resonant mirror-based device is the IAsys system. It is also similar to the Biacore device in that polarised light is totally internally reflected at the underside of the sensor surface. In this case, the sensor surface has the metal layer replaced by a dielectric resonant layer (Cush *et al.*, 1993). It is approximately 100 nm thick, and made of titanium which is of high refractive index. This is separated from the prism by a low index layer made of silica. Incident light focused on the prism-silica interface results in frustrated total internal reflection (FTIR). This evanescently couples light into the high index or wave guiding layer. Within this waveguiding layer, total internal reflection (TIR) occurs generating an evanescent field at the biointerface. Efficient coupling occurs only for certain incident angles where phase matching between the incident beam and the resonant modes of the high index layers is achieved. At the resonant point, light couples into the high index layer, and propagates along the sensing surface before coupling back into the prism. During resonance, a phase shift of the order of 2π in reflected light occurs and causes constructive interference that produces sharp peaks of light at resonance. These changes in angular position are monitored in 'real-time' as they are directly related to changes in refractive index at the biointerface surface. The IAsys system has also utilised a cuvette-based design as opposed to a flow cell configuration, which continually stirs to reduce mass transport effects and has a much larger surface area than the Biacore chip.

4. Miniature Texas Instruments-SPR device

In 1996 Texas Instruments Incorporated developed and released a fully integrated and miniature surface plasmon resonance sensor. The TISPR-1 contains a light emitting diode (LED), polarizer, thermistor (to allow for correction due to temperature fluctuations) and two 128 silicon photo diode arrays housed in an epoxy resin moulding which was designed in the form of the Kretschmann geometry prism (Figure 1). The device was developed for use as a non-destructive, low-cost, and small-sized smart sensor. The wedge shaped beam reflects off the SPR sensing layer onto the PDA array by means of a mirror. The SPR minimum is then measured in 'real-time' following dedicated signal processing (Kukanskis *et al.*, 1999; Chinowsky *et al.*, 2003). The system offers the capability for 'remote-sensing' and is available in a hand-held format called the Spreeta™ device. A scheme of the device is illustrated in Figure 1.

Recent developments in this system offer the potential of multi-analyte detection at relatively low cost and this can facilitate many different applications.

5. Detection of bacteria using SPR

Biosensors offer the possibility of rapid and 'real-time' detection of pathogens, and therefore a number of sensors and assay formats (Table 4) have recently been reported in the literature (Leonard *et al.*, 2003). One of the earliest attempts to use SPR for direct detection of microbes (Fratamico *et al.*, 1998) achieved a detection limit for *E. coli* O157:H7 of 5-7 x 10^7 CFU/ml. Fratamico and co-workers used a sandwich assay format in which they immobilised an anti-*E. coli* O157:H7 monoclonal antibody on the surface of the sensor chip. They then passed various concentrations of cells over the surface and amplified the signal with the injection of polyclonal anti-*E. coli* antibodies. Perkins and Squirrel (2000) described the detection of bacteria using light scattering in combination with a Biacore 2000 surface plasmon resonance and reported that they could

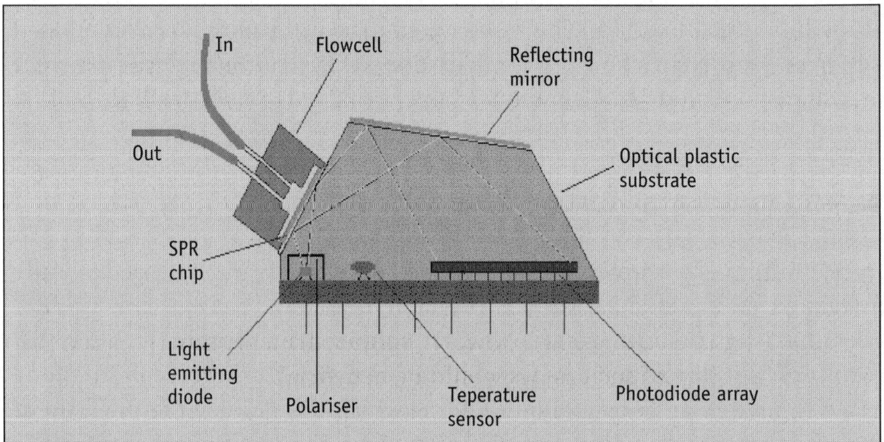

Figure 1. Cross-section of miniature TI-SPR instrument. Mass changes are related to changes in the resonant angle and position of the reflectance minima, which are detected in 'real-time'.

Table 4. List of biosensor-based immunoassays for the detection of *Listeria monocytogenes*.

Format	*Limit of detection*	*Reference*
SPR (IAsys)	-	Lathrop *et al.* (2003)
SPR (miniature prism)	10^6 cells/ml	Koubova *et al.* (2001)
SPR (Biacore)	10^4 cells/ml	Haines and Patel (1995)
Modified electrochemical	10^5 cells/ml	Minett *et al.* (2003)
Electrochemical redox probe	10^3-10^4/ml	Susmel *et al.* (2003)
Piezoelectric	10^7 cells/ml	Vaughan *et al.* (2001)
Piezoelectric	10^5/crystal	Minunni *et al.* (1996)
Electrochemical	$2.5x10^3$/ml	Benhar *et al.* (2001)

detect 10^7 bacterial spores ml^{-1}. Koubova *et al.* (2001) described the detection of *Salmonella enteritidis* and *Listeria monocytogenes* at concentrations down to 10^6 cells/ml. Koubova and co-workers used a surface plasmon resonance biosensor-based on prism excitation of surface plasmons and spectral interrogation to directly detect cells on antibody immobilised surfaces. Various cell concentrations were injected over the surface and the shift in resonant wavelength recorded. Recently, Bokken *et al.* (2003) described the detection of *Salmonella* with a Biacore surface plasmon resonance-based biosensor. Anti-*Salmonella* antibodies, immobilised on the biosensor chip surface bound injected bacteria and this was followed by a pulse with soluble anti-*Salmonella* immunoglobulins to intensify the signal. The target organism was successfully detected at 1.7 x 10^5 CFU/ml. These recently reported publications highlight the potential of SPR-based biosensors for the detection of pathogens. Other sensor formats such as electrochemical sensors (listed in Table 4 below and reviewed more extensively by Leonard *et al.*, 2003) have been reported in the literature with limits of detection as low as approximately 10^3 cells/ml (Benhar *et al.*, 2001; Susmel *et al.*, 2003). However, further characterisation to establish the reproducibility of such assays would be beneficial.

We have generated polyclonal, monoclonal and recombinant antibodies against *Listeria* proteins Internalin A and B and incorporated them into a number of assay formats (Leonard *et al.*, 2004).

Over the last fifteen years, genetic and molecular techniques have been used as a means of generating monoclonal antibodies and antibody fragments (Smith, 1985). This has resulted in the engineering of antibody fragments which include the antigen-binding (Fab) and single chain variable fragment's (scFv). This truncated Fab fragment consists of one V_L domain and one V_H domain. The Fv fragment is the smallest part of the antibody required for antigen binding. The Fv fragment can be very unstable for use in therapeutics or immunoanalysis because it lacks the inter-chain disulphide bond, which is present in the Fab fragment (Killard *et al.*, 1995). Protein engineering has facilitated the development of a more stable Fv fragment (Wörn and Plückthun, 1999). A synthetic peptide linker (or in some cases the insertion of cysteine residues to form a disulphide bridge) can be incorporated into the Fv fragment to produce a more stable single chain Fv (scFv) fragment. Smaller antibody fragments may also be engineered capable of binding to antigen. These include the Fd region which consists of one V_H, one C_H1 domain from the heavy chain and the complementary determining region (CDR), which is the smallest antibody fragment with the ability to bind to antigen (Figure 2).

The DNA of the genes required to code for antibody fragments is available from a number of sources, which include the genetic material from non-immunised or immunised animals or humans and genes from a repertoire of naïve, semi-synthetic or synthetic genes (Ditzel *et al.*, 1993). Messenger RNA may also be used from hybridomas expressing specific functional antibodies to an antigen and used to provide the raw material for cloning the antibody variable gene segments (Winter and Milstein, 1991).

Phage-display technology has become increasingly powerful, and has transformed the way in which we produce binding-ligands, such as, antibodies or peptides for a given target (Roovers, 1998). The technology has enabled the tailor-made design of repertoires of ligands *ab initio*, and their expression on the surface of filamentous phage for the selection of a ligand with the desired properties (Hoogenboom *et al.*, 1998). Phage-display has proven to be particularly successful for the isolation of scFv antibodies, either from a hybridoma, naïve spleen cells, or spleen cells that were pre-immunised with antigen *in vivo*. The enrichment technique of affinity selection (panning) is particularly advantageous as it allows for the selective amplification of phage particles carrying antibody specificities of interest from the rest of the population. In our work we have utilised the 'Krebber'-based system (Krebber *et al.*, 1997), which utilises phage-display technology for the production of

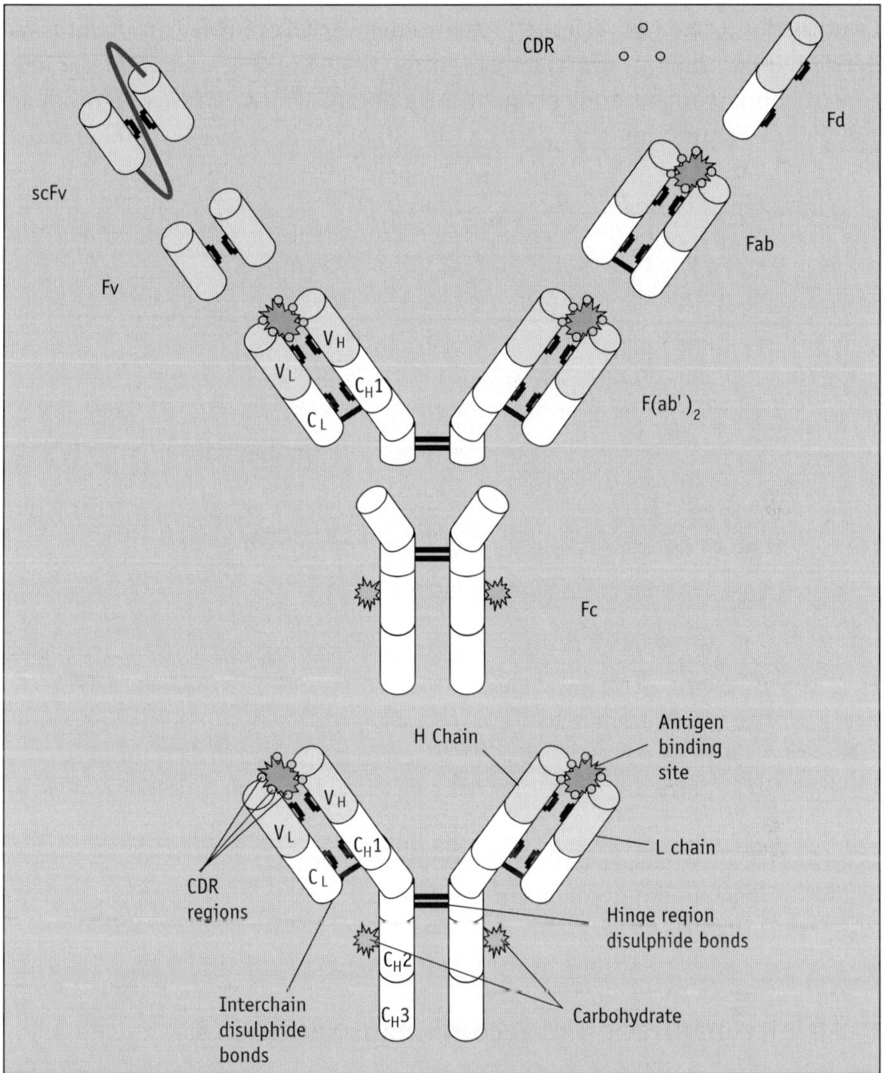

Figure 2. Illustration of an antibody and antibody fragments.
The fragments illustrated above may be produced by genetic or enzymatic and chemical manipulation. The whole antibody may be initially broken up into either Fab (antigen-binding fragment) or $F(ab')_2$ (two antigen-binding fragments linked) and Fc (crystalline fragments) regions. Fab fragments may be further broken up into Fv (variable fragments), scFv (single chain Fv) and individual CDR regions, which are all capable of antigen binding.

soluble scFv antibodies (Figure 3). Mice were immunised with various antigens derived from *Listeria* and the RNA from their spleens was used for the construction of an antibody phage-display library, which was used to produce soluble scFv antibodies.

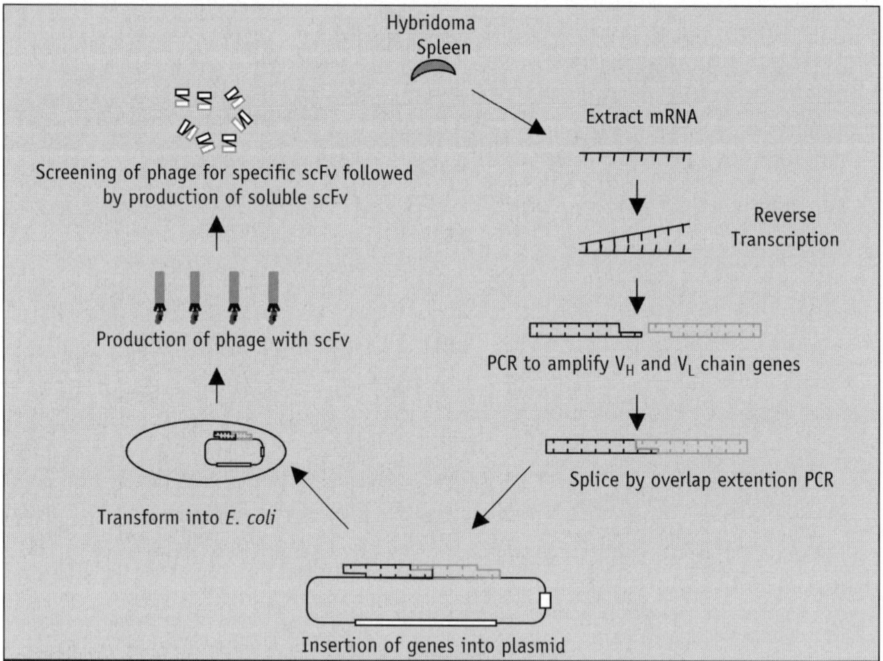

Figure 3. Generation of recombinant antibodies.

6. Affinity maturation of recombinant antibodies

In the immune system of the host, high affinity antibodies for antigen are produced by the combinatorial rearrangement of antibody germline V gene segments and by somatic hypermutation. The affinity of recombinant antibody fragments for their antigen can be affinity maturated by processes such as chain shuffling, error prone PCR and mutagenesis, resulting in the formation of new antibody fragments with enhanced sensitivity and specificity (Marks *et al.*, 1992; Adams and Schier, 1999; Nielsen and Marks, 2001). The affinity maturation of recombinant scFv antibodies reactive with *Listeria*, derived from hybridoma cells, will involve identifying important amino acids in the CDR 3 regions of both the heavy and light chains, mutating these residues using random primers, selection of specific binders and screening for improved binding characteristics. Depending on the level of affinity maturation achieved, further CDR regions can be mutated ('CDR walking'). Each CDR region will be mutated sequentially as parallel mutation of CDR regions has been report to sometimes have a negative effect on binding (Schier *et al.*, 1996).

7. Antibody utilisation in sensors

Whatever the choice of antibody fragment or source it is important that it is optimally orientated on the sensor to maximise its efficiency. This is often overlooked and can result in large percentages of the antibody preparation unable to bind the ligand of interest and, therefore, effectively taking up space on the sensor surface that cannot respond effectively and this greatly diminishes overall sensitivity. In selecting antibodies it is also necessary to consider aspects such as their affinity. For example, if the surface is to be regenerated it might be more useful to have a lower affinity antibody that would be easier to dissociate. The specificity should also be considered as high specificity may be important for the detection of a particular micro-organism whereas wider specificity might be useful for the detection of a group of closely related compounds or pathogens.

Other factors to be considered include cost, availability, stability and ease of labelling where this may aid isolation or detection. Increasingly investigators are also examining the use of a number of other binding molecules including molecularly imprinted polymers and aptamers in order to develop high stability and increased diversity in selectivity.

8. Recent approaches to the detection of *Listeria* using antibody-based SPR sensor assays

We have used a Biacore 3000 instrument to detect *Listeria monocytogenes* cells in various assay formats. Two polyclonal antibodies, an anti-*Listeria monocytogenes* antibody and an anti-InlB extract antibody were used to directly detect *Listeria monocytogenes* cells while the anti-InlB scFv antibody was used to detect *Listeria monocytogenes* InlB extract and, therefore, indirectly detecting *Listeria monocytogenes* cells.

A subtractive inhibition assay was developed to rapidly detect *Listeria monocytogenes* cells in solution using the anti-*Listeria monocytogenes* polyclonal antibody. The antibody was incubated with various concentrations of *Listeria monocytogenes* cells and free unbound antibody separated from bound antibody by centrifugation. By removing cells and bound antibody from solution with a stepwise centrifugation step (gradually increasing the centrifugation speed in small increments), free unbound antibody could be detected by an immobilised anti-Fab antibody on the sensor chip surface. A decrease in free antibody concentration was observed with increasing *Listeria monocytogenes* cell concentrations. A calibration curve was constructed by plotting the average change in response in antibody levels for each *Listeria monocytogenes* cell concentration.

Internalin B (InlB) is an antigen found in *Listeria monocytogenes.* A polyclonal antibody against InlB extract was used to develop a direct inhibition assay to accurately detect *Listeria monocytogenes* cells in solution. Protein G-purified antibody was incubated with various concentrations of *Listeria monocytogenes* cells and subsequently injected over a CM5 sensor chip with recombinant InlB immobilised on its surface. A decrease in antibody binding response was observed with increasing *Listeria monocytogenes* cell concentrations and a standard curve generated. Intra- and inter-day assay variability studies were carried out to evaluate the precision and reproducibility of the assay.

An anti-InlB single chain Fv fragment (scFv) antibody was also used to develop an inhibition assay to indirectly detect *Listeria monocytogenes.* Antibody, incubated with InlB extract from a known concentration of *Listeria monocytogenes* cells, was injected over a recombinant InlB protein immobilised sensor chip surface. The decrease in binding response observed with increasing InlB extract concentrations was recorded and used to determine equivalent *Listeria monocytogenes* cell numbers (equivalent cell numbers at each InlB extract concentration). The format of one of these assays is outlined in Figure 4.

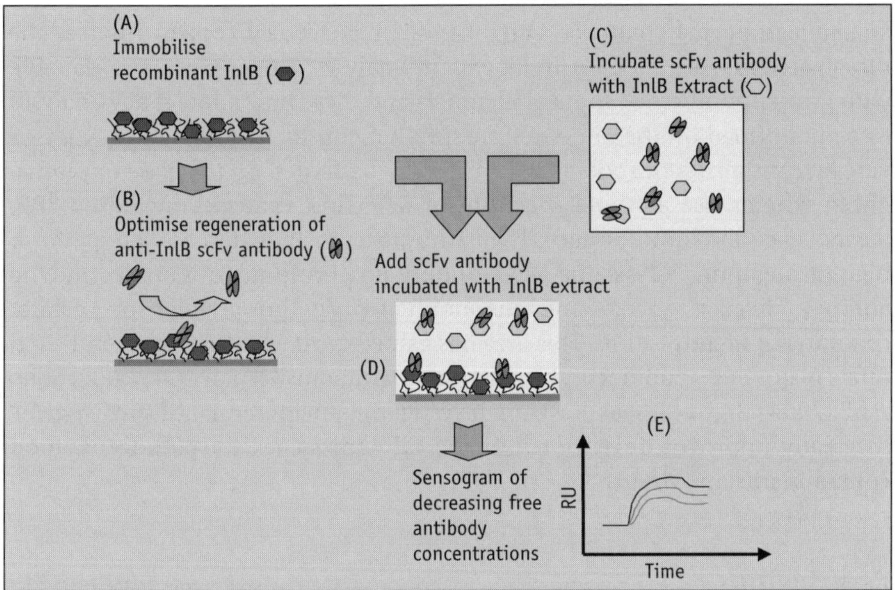

Figure 4. Illustration of the Biacore inhibition assay used to indirectly detect *Listeria monocytogenes* cells with the anti-InlB scFv antibody.

A. Purified recombinant InlB (rInlB) protein is immobilised on the surface of a CM5 chip.
B. The surface regeneration conditions are optimised and the surface binding capacity determined for a number of binding-regeneration cycles.
C. Various concentrations of *Listeria monocytogenes* InlB extracts are incubated with an anti-InlB scFv antibody.
D. Antibody and extract from (c) above are passed over the rInlB-immobilised surface and the response observed.
E. A sensorgram of increasing free antibody concentrations with decreasing *Listeria monocytogenes* cell numbers is observed and a standard curve plotted.

9. Biochips and arrays

A biochip may be defined as a small plastic or glass chip that contains minute concentrations of various proteins or DNA immobilised onto its surface. It can be used as part of a detection device for multiple and simultaneous analyses of biologically significant molecules. Antigen-specific antibodies are commonly used on the chip surface. A typical biochip, such as that found on the Biacore system, can have 4 channels. One of these may act as a control whereas the other three may be used for independent analyses. Larger chips are available with greater numbers of spots or channels and these may allow many analyses to be performed on the same sample simultaneously. In a high density arrays very large numbers of spots may be carefully patterned on the glass or plastic. These systems use a whole spectrum of detection strategies including SPR, fluorescence and luminescence. Their advantages include the ability to perform high throughput analysis, the need for only small volumes of samples and the ability to detect a range of antigens simultaneously thus giving rise to greater certainty for identification. This area has gained great momentum recently and offers major opportunities and challenges (Darmaninsheehan *et al.*, 2002; Song *et al.*, 2004). In our research we are investigating the potential of such systems to greatly improve conventional analytical approaches for pathogens, food contaminants and drugs.

Conclusions

It is evident that sensor formats offer great potential for the development of rapid methods for the detection of pathogens in food and water. The main constraint remains the need for an enrichment or a concentration step before analysis can be performed due to limitations in sensitivity. However, with improved formats and detection strategies and the exploitation of advances in antibody engineering and nucleic acid detection these problems can be overcome thus ensuring enhanced quality control during food production and handling, better and more timely environmental monitoring and greater safety for the consumer.

Acknowledgements

The support of Enterprise Ireland is gratefully acknowledged.

References

Adams, G.P. and Schier, R., 1999. Generating improved single-chain Fv molecules for tumour targeting. J. Immunol. Meth. 231 p. 249-260.

Benhar, I., Eshkenazi, I., Neufeld, T., Opatowsky, J., Shaky, S. and Rishpon, J., 2001. Recombinant single chain antibodies in bioelectrochemical sensors. Talanta 55 p. 899-907.

Bokken, G.C.A.M., Corbee, R.J., van Knapen, F. and Bergwerff, A.A., 2003. Immunochemical detection of *Salmonella* group B, D and E using an optical surface plasmon resonance biosensor. FEMS Microbiol. Lett. 222 p. 75-82.

Chinowsky, T.M., Quinn, J.G., Bartholomew, D.U., Kaiser R. and Elkind, J.L., 2003. Performance of the Spreeta 2000 integrated surface plasmon resonance affinity sensor. Sensors and Actuators B 6954 p. 1-9.

Clark, J.R. and Lyons, C., 1962. Electrode systems for continuous monitoring in cardiovascular surgery. Ann. N.Y. Acad. Sci. 102 p. 29-45.

Cush, R., Cronin, J.M., Stewart, W.J., Maule, C.H., Molloy, J., and Goddard, N.J., 1993. The resonant mirror: a novel optical biosensor for direct sensing of biomolecular interactions. Part 1: Principle of operation and associated instrumentation. Biosensors and Bioelectronics 8, p. 347-353.

Darmaninsheehan, A. Quinn, J. Daly, S., Dillon P and O'Kennedy, R., 2002. The development of novel miniaturised immuno-sensing devices. A review of a small technology with a large future Analytical Letters 36, p. 511-537.

de Wildt, R.M.T., Mundy, C.R., Gorick, B.D. and Tomlinson, I.M., 2000. Antibody arrays for high-throughput screening of antibody-antigen interactions. Nature Biotech. 18, p. 989-994.

Dill, K., Stanker L.H., and Young C.R., 1999. Detection of salmonella in poultry using a silicon chip-based biosensor. J. Biochem. Biophys. Meth. 41, p. 61-67.

Ditzel, H., 1993. In: Gosling, J.P. and Reen, D.J. (eds.), Immunotechnology, Portland Press, London, p. 27-38.

Ekins, R.P., 1998. Ligand assays: from electrophoresis to miniaturised microarrays. Clin. Chem. 44, p. 2015-2030.

Ercole, C., Del Gallo, M., Mosiello, L., Baccella, S. and Lepidi, A., 2003. *Escherichia coli* detection in vegetable food by a potentiometric biosensor Sensors and Actuators B 91, p. 163-168.

Ertl, P., Wagner, M., Corton, E. and Mikkelsen S.R., 2003. Rapid identification of viable *Escherichia coli* subspecies with an electrochemical screen-printed biosensor array. Biosensors and Bioelectronics 18, p. 907-916

Fratamico, P.M., Strobaugh, T.P., Medina, M.B. and Gehring, A.G., 1998. Detection of *Escherichia coli* O157:H7 using a surface plasmon resonance biosensor. Biotechnol. Tech. 12, p. 571-576.

Haines, J. and Patel, P.D., 1995. Detection of foodborne pathogens using BIA. Biacore AB Biotechnology note, p. 31.

Homola, J., Dostalek, J., Chen, S., Rasooly, A., Jiang, S. and Yee, S.S., 2002.

Spectral surface plasmon resonance biosensor for detection of staphyloccal enterotoxin B in milk. Int. J. Food Microbiol. 75, p. 61-69.

Hoogenboom, H.R., de Bruine, A.P., Hufton, S.E., Hoet, R.M., Arends, J.W., and Roovers R.C., 1998. Antibody phage-display technology and its applications. Immunotechnology 4, p. 1-20

Killard, A.J., Deasy, B., O'Kennedy R. and Smyth, M.R., 1995. Antibodies: Production, function and applications in biosensors. Trends in Analytical. Chemistry 14, p. 257-266.

Kim, N. and Park, In-S., 2003. Application of a flow-type antibody sensor to the detection of *Escherichia coli* in various foods. Biosensors and Bioelectronics 18, p. 1101-1107

Ko, S. and Grant, S.A., 2003. Development of a novel FRET method of detection of *Listeria* or *Salmonella.* Sensors and Actuators B 96, p. 372-378.

Koubova, V., Brynda, E., Karasova, L., Skvor, J., Homola, J., Dostalek, J., Tobiska, P. and Rosicky, J., 2001. Detection of foodborne pathogens using surface plasmon resonance biosensors. Sensors and Actuators B74, p. 100-105.

Krebber, A., Bornhauser, S., Burmester, J., Honegger, A., Willuda, J., Bosshard, H.R., and Pluckthun, A., 1997. Reliable cloning of functional antibody variable domains from hybridomas and spleen cell repertoires employing a reengineered phage- display system. J. Immunol. Meth. 201 35-55.

Kukanskis, K., Elkind, J., Melendez, J., Murphy, T., Miller, G., and Garner, H., 1999.Detection of DNA hybridisation using the TISPR-1 surface plasmon resonance biosensor. Anal. Biochem. 274, p. 7-17.

Lathrop, A. A., Jaradat, Z. W., Haley, T. and Bhunia, A. K., 2003. Characterisation and application of a *Listeria monocytogenes* reactive monoclonal antibody C11E9 in a resonant mirror biosensor. J. Immunol. Meth. 281, p. 119-128.

Leonard, P., Hearty, S., Brennan, J., Dunne, L., Quinn, J., Chakraborty, T. and O'Kennedy, R., 2003. Advances in biosensors for the detection of pathogens in food and water. Enzyme Microb. Tech. 32 3-13.

Leonard, P., Hearty, S., Quinn, J. and O'Kennedy, R., 2004. A generic approach for the detection of whole *Listeria monocytogenes* cells in contaminated samples using surface plasmon resonance. Biosensors and Bioelectronics 19, p. 1331-1335.

Liedberg, B. and Lundstrom, I., 1993. Principles of biosensing with an extended coupling matrix and surface plasmon resonance. Sensors and Actuators B11, p. 63-72.

Marks, J.D., Griffiths, A.D., Malmqvist, M., Clackson, T.P., Bye, J.M., and Winter, G., 1992. By-passing immunisation: building high affinity human antibodies by chain shuffling. Bio/tech. 10, p. 779.

Mello, L.D and Kubota, L.T., 2002. Review of the use of biosensors as analytical tools in the food and drinks industry. Food Chem. 77, p. 237-256.

Minett, A.I., Barisci, J.N. and Wallace, G.G., 2003. Coupling conducting polymers and mediated electrochemical responses for the detection of *Listeria*. Analytica Chimica Acta 475, p. 37-45.

Minunni, M., Mascini, M, Carter, R.M., Jacobs, M.B., Lubrano, G.J. and Guilbault, G.G., 1996. A quartz crystal microbalance displacement assay for *Listeria monocytogenes*. Analytica Chimica Acta 325, p. 169-174.

Nedelkov, D. Rasooly, A. and Nelson, R.W., 2000. Multitoxin biosensor-mass spectrometry analysis: a new approach for rapid real-time, sensitive analysis of staphylococcal toxins in food. Int. J. Food Microbiol. 60, p. 1-13.

Nielsen, U.B. and Marks, J. D., 2001. Affinity Maturation by Chain Shuffling and Site Directed Mutagenesis. Antibody Engineering, Kontermann, R. and Dübel, S. (eds.), Springer Lab Manual, Springer-Verlag, Berlin Heidelberg, p. 515-539.

Patel P.D., 2002 .Biosensors for measurement of analytes implicated in food safety. Trends in Chem. 21, p. 96-115.

Perkins, E.A. and Squirrell, D.J., 2000. Development of instrumentation to allow the detection of microorganisms using light scattering in combination with surface plasmon resonance. Biosensors and. Bioelectronics 14, p. 853-859.

Pyun, J.C., Beutel, H., Meyer, J.U. and Ruf, H.H., 1998. Development of a biosensor for *E. coli* based on a flexural plate wave (FPW) transducer. Biosensors and Bioelectronics 13, p. 839-845

Rogers, K.R., 2000. Principles of affinity-based biosensors. Molec. Biotech. 14, p. 109-129.

Roovers, R.C., 1998. Antibody phage-display technology and its applications. Immunotechnology 4, p. 1-20.

Schier, R., McCall, A., Adams, G.P., Marshall, K.W., Merritt, H., Yim, M., Crawford, R.S., Weiner, L.M. and Marks, J.D., 1996. Isolation of picomolar affinity anti-c-erbB-2 single chain Fv by molecular evolution of the complementarity determining regions in the centre of the antibody-binding site. J. Molecular Biol. 263, p. 551-567.

Sethi, R.S., 1994. Transducer aspects of biosensors. Biosensors and Bioelectronics 9, p. 243-264.

Smith, G.P., 1985. Filamentous fusion phage: novel expression vectors that display cloned antigens on the virion surface. Science 228, p. 1315-1317.

Song S-P., Li, B., Hu, J., and Li., M-Q., 2004. Simultaneous multi-analysis for tumour markers by antibody fragments microarray system. Anal. Chem. Acta (in press).

Susmel, S., Guilbault, G.G. and O'Sullivan, C.K., 2003. Demonstration of label-less detection of food pathogens using electrochemical redox probe and screen-printed gold electrodes. Biosensors and Bioelectronics 18, p. 881-889.

Templin, M.F., Stoll, D., Schrenk, M., Traub, P.C., Vohringer, C.F. and Joos, T.O., 2002. Protein microarray technology. Trends in Biotechnology, 20(4), p. 160-166.

Vadgama, P. and Crump, P. 1992. Biosensors: Recent trends, a review. Analyst 117, p. 1657-1670.

Vaughan, R.D., O'Sullivan, C.K. and Guilbault, G.G., 2001. Development of a quartz crystal microbalance (QCM) immunosensor for the detection of *Listeria monocytogenes*. Enzyme Microbiol. Technol. 29, p. 635-638.

Winter, G. and Milstein, C., 1991. Man-made antibodies. Nature (Lond) 349, p. 293-299.

Wörn, A. and Plückthun, A., 1999. Different equilibrium stability behaviour of scFv fragments: identification, classification, and improvement by protein engineering. Biochemistry 38, p. 8739-8750.

Simple and rapid bacterial protein and DNA diagnostic methods based on signal generation with colloidal carbon particles

A. van Amerongen and M. Koets
Agrotechnology & Food Innovations, Wageningen University and Research Centre, P.O. Box 17, 6700 AA Wageningen, the Netherlands

Abstract

To speed up bacterial detection several rapid and simple methods have been developed in which black colloidal carbon particles generate the signal that indicates the presence of the particular organism(s). Based on the lateral flow principle through nitrocellulose membranes cellular antigens can be detected (lateral flow immunoassay; LFIA), and/or amplified genetic material (nucleic acid LFIA; NALFIA). These methods only require addition of (culture) supernatant or amplification product and the signal is visible within 15 minutes. In a new Mini-array method specifically labelled carbon particles are incubated with sample on small membranes mounted on a slide. Per slide up to 16 different samples can be screened for a maximum of 30 different target compounds. The results on the membranes can be digitised by flatbed scanning followed by image analysis and data processing.

Keywords: bacterial/microbial detection, (nucleic acid) lateral flow immunoassays, LFIA, NALFIA, Mini-array, colloidal carbon particles

1. Introduction

The ideal method for detection of microorganisms offers real-time detection of low numbers of living cells. Driving forces to develop this kind of methods are particular applications in human and veterinary clinical diagnostics, bio-warfare and some life threatening conditions in food and feed safety. However, the actual situation is quite different, since in a lot of cases bacterial detection is still done by culture plate methods. These methods involve various phases, such as sampling, resuscitation of injured cells, pre-enrichment, selective enrichment, detection and identification. The total assay time may be from two to seven days. Other methods that measure phenotypic characteristics have been developed and introduced for high-throughput screening. These methods are based on fluorescence, ATP bioluminescence, pH, specific antibodies and some electrical phenomena such as conductance, impedance and capacitance (Fung, 2002). Microbial genotypic characteristics are used in PCR and NASBA based methods (Johnson, 2000). Although a large number of molecular biological methods have been described in literature the implementation of these methods in daily practice is lacking. In principle, the nucleic acid technology is much more sensitive than the phenotypic methods, however, it still suffers from some drawbacks such as a) contamination that is a potential problem, b) inhibitors of the PCR procedure that may be found in food products in various concentrations, and c) in most cases a pre-enrichment step that is still necessary to have sufficient cells for doing PCR (Johnson, 2000; Ellis and Goodacre, 2001). Furthermore, it is difficult to discriminate between DNA of dead and living cells, the viable cell count problem. This last drawback may be tackled by amplifying RNA instead of DNA. RNA is much more unstable than DNA and probably will be denatured in dead cells. However, examples of long survival times of RNA have been described in literature (Birch *et al.*, 2001). Moreover, the genotypic methods have not yet been accepted officially by governmental and regulatory bodies. These aspects have been addressed in a European project (FOOD-PCR) directed at the standardisation of diagnostic PCR for the detection of foodborne pathogens (Malorny *et al.*, 2003).

Further refinement of existing methodology is one of the strategies to improve microbial diagnostics. To the aspects addressed are robotisation of particular steps in the procedures, automation of data recording and processing, and design of rapid and simple assay protocols.

We have focused on these last aspects by developing rapid and simple procedures in which coloured colloidal particles are used as final test signal (Wichers *et al.*, 1987; Van Amerongen *et al.*, 1994). The particles are used in Lateral Flow ImmunoAssays (LFIAs). LFIAs (Figure 1) can be used for the rapid detection of proteins (e.g., allergens), microorganisms and small chemicals (e.g., antibiotics) in food products (Niessen *et al.*, 1999; Aldus *et al.*, 2003; Van Dam *et al.*, 2004). Up to four or eight lines can be accommodated on a single- or double-lane one-step device, respectively. The tests are simple and straightforward, easily interpretable and the results are visible within 10 to 15 minutes, whereas necessary equipment is very cheap. These assays are especially

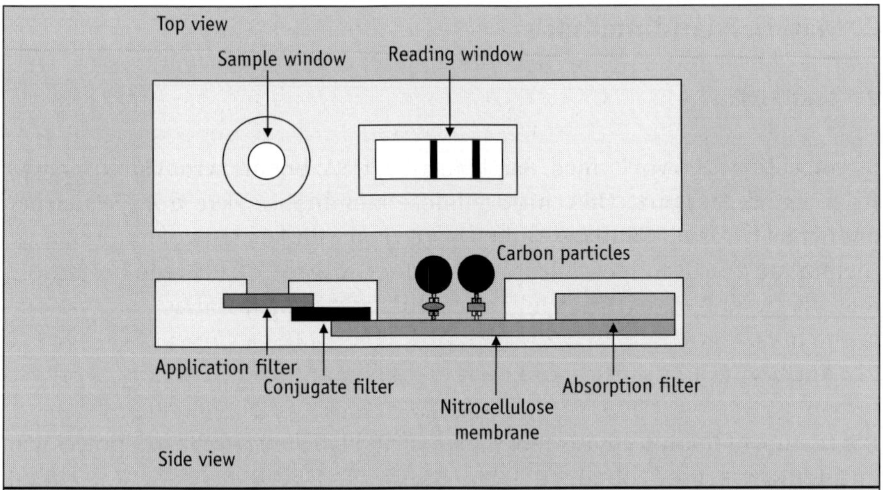

Figure 1. Top and side view of a one-step lateral flow immunoassay (LFIA). The final situation, after running the sample, is depicted. Two black lines are visible in the reading window, based on sandwich formation with specific antibodies and colloidal carbon particles.

suited for low-facility laboratory and field settings and do not require skilled personnel. To further broaden application possibilities, the LFIA procedure has been adapted such that the presence or absence of specific amplified genetic material can be determined. Following PCR, NASBA, or other amplification protocols a small volume of the final solution is added to the one-step device and the appearance of a coloured line is indicative of the presence of the specific amplicon.

In addition to this Nucleic Acid LFIA (NALFIA), the colloidal particles have been applied in a new Mini-array procedure on nitrocellulose coated slides arrayed with proteins. Flatbed scanning and image analysis were performed for quantification of the results.

2. Materials and methods

2.1 Materials

Streptavidin was from Zymed, San Francisco, USA; bovine serum albumin was from Sigma, St. Louis, USA; nitrocellulose membranes were from Schleicher and Schuell, Dassel, Germany; slides were coated with 16 pieces of nitrocellulose membrane of 6x6 mm each; running buffer consisted of 100 mM borate, pH 8.8, 1% (w/v) BSA, 0.05% (v/v) Tween-20, 0.02% (w/v) NaN_3.

2.2 Antibodies

For capture of fluorescein isothiocyanate (FITC) labelled amplicon a polyclonal antibody was used obtained from Biomeda, Hayward, USA. Polyclonal antibodies to detect digoxigenin-labelled amplicons were from Roche, Mannheim, Germany.

2.3 Amplification of genetic material

A specific part of the gene encoding *pha*C1-synthase from *Pseudomonas putida* (*P. putida*) was amplified by PCR using specific primers (Wolbert *et al.*, 2002), one labelled with biotin and the other with FITC. A set of labelled primers specific for *Salmonella* was used to amplify genetic material (429 base pairs) from this food pathogen (Aabo *et al.*, 1993). Primers for the multi-analyte NALFIA specific for *Escherichia coli* (*E. coli*) lamB gene (345 bp) and *Bacillus cereus* (*B. cereus*) gyrase subunit gene (364 bp) were based on publications by Purohit and Kapley (2002) and Chen and Tsen (2002), respectively. Isolated genomic DNA from the individual microorganisms was used as a template for amplification. In the case of *P. putida* the specific gene was inserted in a plasmid and this construct was used for PCR. Amplification reactions were performed with a GeneAmp PCR System 9700 from Applied Biosystems, Foster City, CA, USA.

2.4 Colloidal carbon particles - protein conjugates

Protein (350 µg/mL) in a low-ionic strength buffer (5 mM borate, pH 8.8) was incubated overnight with a 0.2% colloidal carbon particles solution on an end-over-end rotor at 4 °C. Particle - protein conjugates were washed three times in a 5 mM borate buffer, pH 8.8, 1% (w/v) bovine serum albumin (BSA), 0.02%

(w/v) NaN_3. Conjugates were stored in a 100 mM borate buffer, pH 8.8, 1% (w/v) BSA, 0.02% (w/v) NaN_3.

2.5 Immobilisation of proteins on nitrocellulose membranes

Proteins were sprayed onto plastic backed nitrocellulose membranes (AE99, Schleicher and Schuell, Dassel, Germany) by a CAMAG TLC spotter in concentrations between 125 to 1000 ng per µL and per strip. Membranes were dried and cut to strips of 5 mm width.

2.6 Immobilisation of proteins on FAST slides

Polyclonal anti-FITC antibody was spotted in a serial dilution (1000, 500, 250, 125 µg/mL) on FAST slide nitrocellulose membranes with the MicroCASTer (Schleicher and Schuell, Dassel, Germany) according to a protocol from the manufacturer. Spot diameter was approximately 0.5 mm and spotted volumes ranged from 30 to 70 nL. Membranes were blocked in a 100 mM borate buffer, pH 8.8, containing 1% (w/v) BSA.

2.7 LFIA specific for *E. coli* O157 and verotoxins

A LFIA for the detection of *E. coli* O157 and verotoxins (VT) 1 and 2 was developed (VTEC-LFIA). The procedure was essentially as published by Aldus *et al.* (2003) and Capps *et al.* (2004), and is briefly described here. Bacteria: Strains of the five WHO VTEC serotypes to be used for development of the dipstick and to determine performance characteristics were selected to be of differing VT-producing capabilities. Verotoxins: Lyophilised purified toxins were kindly donated by D. Acheson (Boston, USA), or purchased from Toxin Technology (Sarasota, USA). Crude preparations were also produced in-house by culture of VT-producing strains in TSB at 41.5 °C overnight. Food samples: Aliquots (25g or 25ml) of foods relevant to human infection by VTEC (unpasteurised whole milk, raw frozen minced beef, unpasteurised fresh pressed apple juice, and salami) were weighed into separate containers and stored at -20 °C. A number of aliquots for each food type were enriched and tested by immuno-magnetic separation (IMS) for relevant *E. coli* serogroups and for all VT-producing organisms. The test materials were verified as being free from *E. coli* of these serogroups and VT-producing organisms. Food trial: In the participating laboratories, test materials (25g) were dispersed in 225 ml modified tryptone soya broth (MTSB; Oxoid) containing novobiocin (20 mg/L), with blending as necessary. The food/broths were incubated at 41.5 °C for 21 +/-1 h. Aliquots of the enriched cultures were treated post-enrichment with polymyxin B sulphate (Sigma P1004; 5000 USP units per millilitre) for 30 min at 37 °C to enhance release of toxin (Ashkenazi and Cleary, 1990). Extracted sub-sample (150 μl) was mixed with running buffer (150 μl) and an aliquot of the modified sample (100 μl) was pipetted into the sample window of the final test devices housed in plastic cassettes (Figure 1). The strips were incubated on a flat surface for 20-40 min at room temperature. Each dipstick was examined for the development of grey/black lines in the reading window indicating the presence of cells and/or VT; for the test to be valid, the control line had to be present as well.

2.8 Nucleic Acid Lateral Flow ImmunoAssay

One-step lateral flow immunoassays (Figure 1) were adapted to detect amplified genetic material in a so-called nucleic acid lateral flow immunoassay (NALFIA). Of each set of primers used for specific amplification in PCR one was 5′-labeled with biotin and the other 5′-labeled with FITC or digoxigenin. Hereto, streptavidin was adsorbed onto colloidal carbon particles and antibodies against the tag at the other end of the amplicon, anti-FITC and/or anti-digoxigenin polyclonal antibodies, were immobilised on nitrocellulose membranes.

2.9 NALFIAs specific for *P. putida pha*C1-synthase, *Salmonella*, and the multi-analyte for *E. coli* and *B. cereus*

Following PCR, 0.015 to 3 μL of the reaction mixture containing the specific amplicons was mixed with running buffer (100 mM borate buffer, pH 8.8, 1% (w/v) BSA, 0.05% (v/v) Tween 20, 0.02% (w/v) NaN_3) to a total volume of 100 μL and added into the sample window of the NALFIA device. Results were scored after 15 minutes. Positive samples, containing double-labelled amplicons, were indicated by the appearance of a black line in the reading window. In the case of the multi-analyte NALFIA for *E. coli* and *B. cereus*, PCR reactions for the amplification of specific gene parts were carried out individually. Genomic DNA was isolated from cultured bacteria and was used as a template for PCR. The resulting DNA product for *E. coli* (345 bp) was FITC-biotin labelled and the *B. cereus* product digoxigenin-biotin labelled. The PCR products (0.5 μL of each) were mixed with streptavidin coated carbon particles and applied to a NALFIA device, separate and combined.

2.10 Mini-array for detection of PCR from *P. putida pha*C1-synthase gene

Serial dilutions of anti-FITC antibodies (250-1000 µg/mL; approximately 50 nL per spot) were applied on 6x6 mm NC-membranes (FAST slides, Schleicher and Schuell) and mounted with an incubation chamber surrounding the nitrocellulose membrane and having a volume of 80 µL. After blocking (100 mM borate, pH 8.8, 1% (w/v) BSA, 0.02% (w/v) NaN_3) for 3 hours at room temperature and washing, the membrane was incubated for 2 hours with PCR amplified material from the gene encoding the *pha*C1-synthase enzyme of *P. putida* and with colloidal carbon particles coated with streptavidin. In the PCR with the end-labelled primers the resulting amplicons contained specific tags, biotin and FITC, respectively. Subsequently, the membranes were washed with running buffer and scanned with an EPSON-Perfection-3200-Photo flatbed scanner. The scans were analysed for pixel grey volume with TotalLab image analysis software (Nonlinear Dynamics Ltd., Newcastle upon Tyne, UK).

3. Results

3.1 LFIA specific for *E. coli* O157 and verotoxins (VTEC-LFIA)

The LFIA specific for the detection of *E. coli* O157 and VT1/2 was validated by a 14-laboratory collaborative trial in food products (apple juice, beef and milk) after overnight culture of food samples (Capps *et al.*, 2004). Examples of LFIAs used in the 3-laboratory and the 14-laboratory trials are shown in Figure 2. In the 3-laboratory trial a multi-analyte LFIA was used with four lines; one control line, a line specific for an *E. coli* O157 cell membrane antigen, a line specific for VT2, and a line specific for both VT1 and VT2. A double-lane LFIA was used in the 14-laboratory trial. Part of the statistical results of this trial, based on the analysis of 1,200 samples, is presented in Table 1. For the *E. coli* O157 and the VT detection the specificity was 92 and 94%, respectively. The sensitivity of the *E. coli* O157 detection was 94%, whereas the sensitivity of the VT detection was 55%.

A comparison of LFIA collaborative trial results for detection of *E. coli* O157 with results for the collaborative trial of the ISO reference method for detection of *E. coli* O157 (Scotter *et al.*, 2000; ISO, 2001) indicated that 'like for like' the LFIA test performed equally as well, or much better, than the reference method (Clare Aldus, Institute of Food Research, Norwich, UK; personal

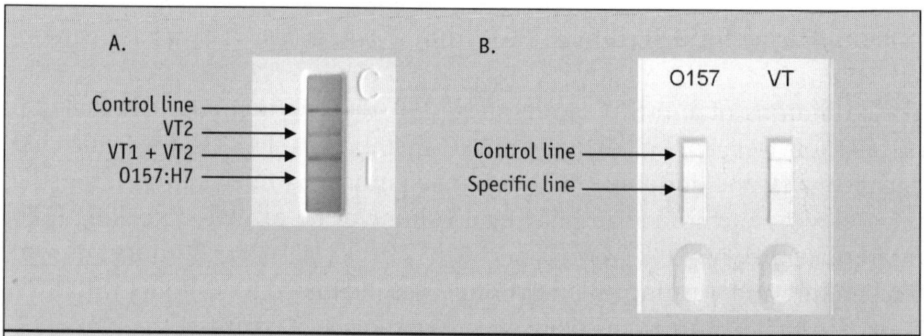

Figure 2. Examples of VTEC-LFIAs used in the 3-laboratory (A.) and the 14-laboratory (B.) food trial. The multi-analyte LFIA in A shows a line specific for *E. coli* O157, a line specific for VT1 and VT2, a line specific for VT2 and a test control line. The double-lane LFIA in B shows a specific line indicating the presence of *E. coli* O157 (left lane) and a specific line for VT detection (right lane). Both lanes show a test control line.

Table 1. *E. coli* O157 collaborative trial results showing statistical data for the *E. coli* O157 and VT detection lateral flow immunoassays.

	E. coli LFIA	
	O157 detection	VT detection
Specificity[a]	92 %	94 %
Sensitivity[b]	94 %	55 %

[a]Percentage of samples correctly identified as negative
[b]Percentage of samples correctly identified as positive

communication). In Table 2 sensitivity percentages are given for the detection of *E. coli* O157 in apple juice, beef and milk. In apple juice and beef the LFIA scored much better than the ISO method (97.5% compared to 83% and 100% compared to 78%, respectively), and in milk the data were quite similar, i.e., 96.2% compared to 98.1%.

Table 2. *E. coli* O157 method collaborative trial results; comparison between the official International Standard reference method BS EN ISO16654:2001 (2001) and the lateral flow immunoassay for *E. coli* O157 detection[a].

	Apple juice	Beef	Milk
Sensitivity[b] ISO method	83.0 %	97.5 %	78.0 %
Sensitivity[b] LFIA	100.0 %	98.1 %	96.2 %

[a]Data analysed by Clare Aldus, Institute of Food Research, Norwich, UK
[b]Percentage of samples correctly identified as positive

3.2 NALFIA for *P. putida pha*C1-synthase

The principle, performance and sensitivity of the NALFIA were assessed in preliminary experiments in which the amounts of the carbon-streptavidin conjugate and/or the volume of the *P. putida pha*C1-synthase PCR product were varied. In Figure 3 (Panel A) an example is shown in which various amounts

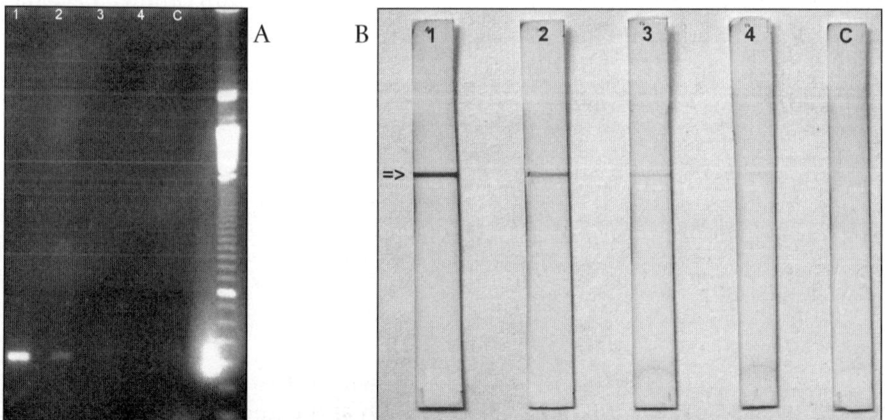

Figure 3. Sensitivity of NALFIA detection of the amplified genetic material of the *pha*C1-synthase gene from *P. putida*. Panel A shows an agarose gel with 1, 0.25, 0.062 and 0.015 µL PCR product (lanes 1 to 4), a negative control (C) and a lane with a marker Gene ruler. Panel B shows the same volumes of PCR product run in NALFIAs. The arrow indicates the position of the specific line.

of the PCR product, ranging from 1 down to 0.015 µL, were run in an agarose gel and stained with ethidium bromide. Specific bands can be observed in lane 1 and 2. The same amounts of PCR product (Figure 3, Panel B) and 1 µL of carbon-streptavidin conjugate diluted to 100 µl with running buffer were run through nitrocellulose strips. In NALFIA also on the fourth membrane a faint line is visible (see the line at the level of the arrow in panel B). This line results from the addition of 0.015 µL of PCR product.

3.3 NALFIA specific for *Salmonella*

Figure 4 shows the PCR result of a *Salmonella* specific amplicon consisting of 429 base pairs both on agarose gel stained by ethidium bromide (left panel) and in NALFIA (right panel). For the NALFIA 1 µL of the PCR product was mixed with 1 µL of colloidal carbon-streptavidin conjugate in a total volume of 100 µL running buffer and the solution was pipetted into the sample window of the one-step LFIA device. The positive line in NALFIA was visually scored after 15 minutes.

3.4 Multi-analyte NALFIA specific for *E. coli* and *B. cereus*

In Figure 5 a prototype NALFIA is presented for the simultaneous detection of *E. coli* and *B. cereus* amplicons. Genomic DNA was isolated from cultured bacteria and was used as a template for PCR. The FITC-biotin labelled *E. coli* and the

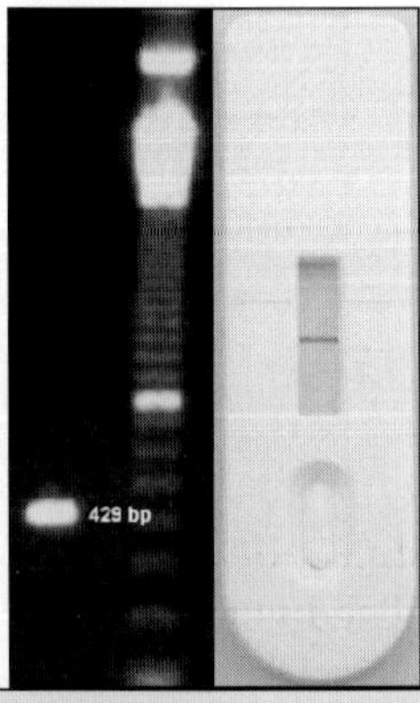

Figure 4. Detection of a *Salmonella* specific 429 bp amplicon by agarose gel (left panel, lane 1) with a Gene ruler in lane 2 and by NALFIA (right panel) showing a carbon line that indicates the presence of the amplicon in the PCR product.

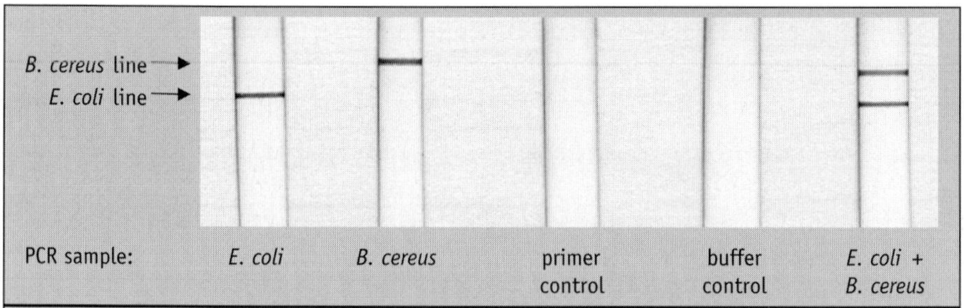

Figure 5. Prototype multi-analyte NALFIA for the simultaneous detection of PCR amplified material from *E. coli* and *B. cereus*. The specificities of the lines are indicated by the arrows. From left to right the NALFIAs were run with 0.5 µL of *E. coli* amplified material, *B. cereus* amplified material, a primer control, a buffer control and combined amplified material from *E. coli* and *B. cereus*.

digoxigenin-biotin labelled *B. cereus* products were run in LFIA with streptavidin coated carbon particles, and specific polyclonal antibodies were used as capture ligands. Both products resulted in the appearance of a black line at the correct and indicated specificities (left two membranes). The primer control (PCR without template DNA) and the buffer control (running buffer without PCR material) were negative (third and fourth membrane from the left). Running the products from the *E. coli* and *B. cereus* PCRs simultaneously resulted in the appearance of two lines, indicating the presence of both amplicons, i.e., microorganisms.

3.5 Mini-array for detection of PCR product from the *P. putida phaC1*-synthase gene

Part of the gene encoding the *pha*C1-synthase enzyme from *P. putida* was amplified by PCR and double-labeled with biotin and FITC. Serial dilutions of anti-FITC antibodies (250-1000 µg/mL) were spotted on 6x6 mm NC-membranes and after blocking incubated for 2 hours with colloidal carbon particles coated with streptavidin. Spots were analysed by flatbed scanning and pixel grey volumes were calculated with TotalLab analysis software. In Figure 6 the results of this preliminary experiment are depicted. In panel A the slide with 16 nitrocellulose membranes is shown and an enlarged part of the scanned image of one of the membranes after incubation with PCR product and streptavidin labelled colloidal carbon particles. In panel B, the means and

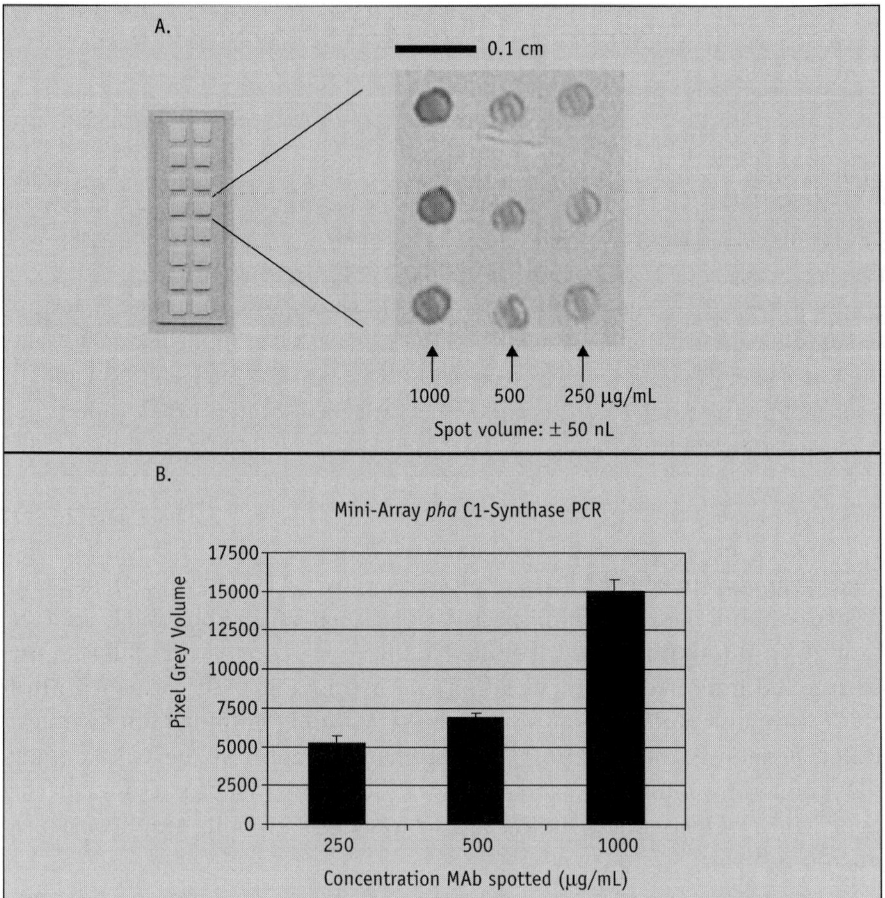

Figure 6. Mini-array for the detection of PCR product from the *P. putida pha*C1-synthase gene. Panel A shows a slide with nitrocellulose membrane coating and an enlarged part of a final result; panel B shows the calculated pixel grey volumes after flatbed scanning and image analysis.

standard deviations of 8 spots of each amount of FITC antibody are shown after image analysis with TotalLab software. A clear relationship with the amount of spotted antibody can be observed. Error bars indicate the standard deviation per concentration.

3.6 Appraisal of the contributions to speed and ease of performance of the colloidal particles based assays

In Table 3 an appraisal is given of the contributions to speed and/or ease of performance of the various colloidal particles based assays with respect to the various phases in bacterial screening. For each phase in conventional bacterial screening methods, i.e., sampling, pre-enrichment, selective enrichment, detection and identification, a score is given indicating whether the particular method is not or positively contributing to speed and/or ease of performance.

For LFIA especially the last three phases can be speeded up and simplified with respect to handling. In NALFIA the improved sensitivity of the PCR procedure leads to a shorter incubation time for the pre-enrichment phase as well. In particular cases the selective enrichment phase may even be skipped. As compared to the Mini-array the detection in NALFIA is much faster, i.e., 10 to 15 minutes, whereas in Mini-array the incubation of the membranes with the specifically labelled colloidal particles was around 2 hours in this non-optimised experiment, followed by flatbed scanning and image analysis. However, NALFIA is limited to the detection of 4 or 8 different amplicons (in single- and double-lane devices, respectively). With the Mini-array method it is possible to screen up to 30 spots per 6x6 mm nitrocellulose membrane. In addition, a one-slide incubation chamber accommodates 16 membranes that can be used simultaneously, increasing the number of spots up to 480. It will

Table 3. Appraisal[1] of contributions to speed and ease of performance of the colloidal particles based assays with respect to the various phases in bacterial detection.

Method	LFIA	NALFIA	Mini-array
Screening step			
Sampling	0	0	0
Pre-enrichment	0	+	+
Selective enrichment	+	+	+
Detection	+	+	+
Identification	+	+	+

[1]0 = no contribution to speed and/or ease of performance; + = positive contribution to speed and/or ease of performance.

depend on the particular application whether the NALFIA or the Mini-array method will be more appropriate.

4. Discussion

The ideal method for detection of microorganisms offers real-time detection of low numbers of living cells. Although such a method is far from reality a lot of present research is directed at shortening total assay times, simplifying performance and/or automising data recording, and processing. Nevertheless, current detection is still often based on culture plate methods. These methods involve various phases, such as sampling, resuscitation of injured cells, pre-enrichment, selective enrichment, detection and identification. We have developed a number of colloidal carbon particles based methods to improve total assay duration and/or ease of performance.

An example of a phenotypic detection method is the VTEC-LFIA specific for the detection of *E. coli* and the verotoxins 1 and 2 (Aldus *et al.*, 2003; Capps *et al.*, 2004). The verotoxin-producing *Escherichia coli* O157 is an important foodborne pathogen (Neill, 1997). Since infection may lead to severe clinical symptoms, surveillance studies are used to combat this microorganism. These studies are also important for setting up hazard analysis critical control point (HACCP) plans and other control measures (Capps *et al.*, 2004). Rapid and user-friendly methods to detect *E. coli* O157 and the verotoxins would be beneficial with respect to assay duration and cost and with regard to the minimal skills needed to execute these assays. A more-laborious and time-consuming confirmatory test would only be necessary for the samples tested positive with the rapid assay. The VTEC-LFIA only requires the addition of a small volume of culture supernatant into the sample window of the device, followed by the visual examination of the appearance of a grey/black line. This indicates the absence or presence, with respect to particular cut-off levels, of the *E. coli* specific antigen and/or the verotoxins produced by this organism. No expensive equipment is necessary and the total duration of the screening protocol is reduced to 24 hour.

As compared to the 3-laboratory study a double-lane device had to be used for the 14-laboratory trial. It appeared that delivery of one of the commercially acquired antibodies was not possible anymore. Alternative antibodies that were tested in the relatively short time available for the assemblage of the devices for the 14-laboratory trial all resulted in cross-reactive signals. Therefore, the *E. coli* O157 and the VT detection were separated by use of a double-lane device and of each culture supernatant two samples had to be pipetted into the two sample windows of the device. The statistical results of the *E. coli* O157 detection were good with 92% and 94% for specificity and sensitivity, respectively (Capps *et al.*, 2004). Surprisingly, the sensitivity results were equally well or much better than the official ISO reference method for detection of this micro-organism (Clare Aldus, IFR, Norwich, UK, personal communication; Scotter *et al.*, 2000; ISO, 2001).

The low sensitivity of the VT detection appeared to be mainly due to the cut-off level of the VT specific LFIA that could be obtained with the antibodies applied. In the same food trial VT levels were also determined with an ELISA and a high percentage of ELISA-positive samples with a low optical density value (i.e., < 0.2) were scored negative with the VT-LFIA. In general an ELISA is 2 to 10 times more sensitive than an LFIA and for the verotoxin detection this appeared to influence the outcome for the samples with low VT concentrations. Further improvement of VT detection may be expected by applying more sensitive antibodies.

Alternatively, an increase in the duration of the enrichment step will consequently lead to higher VT concentrations in the culture supernatant and to an increased scoring of true-positives.
Overall the application of the VTEC-LFIA could be attractive in terms of a much more extensive and cost-effective surveillance of food products for the presence of these microorganisms. Since results are available in just over half an hour after enrichment, risks presented by VTEC can be more thoroughly assessed (Aldus *et al.*, 2003; Capps *et al.*, 2004).

In addition to the phenotypic assays a number of genotypic detection methods have been presented in this study. Also these methods are based on signal generation by coloured colloidal particles. Although genotypic detection is not yet accepted as part of reference methods (Johnson, 2000; Ellis and Goodacre, 2001), the acceptance of this technology will be a matter of five to ten years. It is evident that there is an ever increasing force to rapidly report on possible contamination of food products with pathogens (Fung, 2002; Sachse, 2004). The *pha*C1-synthase specific NALFIA shows the feasibility of rapid and simple detection of amplified gene products. In a preliminary optimisation this NALFIA was compared to conventional detection by agarose gel and ethidium bromide staining. From the results it can be calculated that the NALFIA procedure is substantially more sensitive than the detection by agarose gel staining, i.e., approximately a factor 15. However, the more recently used fluorescent staining dyes have a considerably increased sensitivity and NALFIA's sensitivity will most probably be similar. Moreover, it may be an advantage that in NALFIA no potential toxic chemicals are used.

The prototype NALFIAs developed for the detection of *Salmonella* and the multi-analyte approach for the simultaneous detection of *E. coli* and *B. cereus* show the applicability of the method in the field of food pathogen monitoring. Especially in cases where only low-level laboratory conditions are available and numbers of food samples to be tested are low or intermediate, PCR and rapid NALFIA detection would be a preferred combination.

Other applications for the NALFIA procedure can be foreseen:

- detection of micro-organisms and their toxins (food spoilage and poisoning);
- detection of GMOs (food safety and identification; see Koets *et al.*, 2003, for a prototype NALFIA to detect genetically modified Roundup Ready soybeans);
- detection of allergenic products / contaminations (complementary to immunoassays);
- assessment of adulteration / fraud (determination of foreign DNA (milk, fish));
- assessment of active genes (mRNA) in crops and fruits ("Genomics" results; user-friendly tests to screen for the major key-genes with respect to quality and/or safety);
- medical applications (bacterial and viral infections);
- detection of antibiotics resistance genes.

Especially in relation to the assessment of active genes (detection of mRNA) it is of interest that the NALFIA can be adapted to also detect amplified genetic material via the NASBA procedure (Compton, 1991). Moreover, since NASBA is an isothermal amplification technology, this approach would further simplify the total procedure of amplification and detection (NASBA-NALFIA).

The applications as mentioned above would also be possible with the third colloidal particles based format described, i.e., the Mini-array method. The cheap procedure and simple performance of the Mini-array approach with a handheld arrayer on slides with nitrocellulose membrane layers, followed by incubation with specifically labelled colloidal carbon particles opens various interesting diagnostic possibilities. With the handheld arrayer over 700 spots can be positioned per slide, or approximately 30 spots on each of 16 membranes with a size of 6x6 mm. The method is ideally suited for semi-quantitative purposes, such as detection of a mixture of food pathogens, the analysis of microorganisms causing food spoilage, the assessment of food adulteration, or the presence of allergenic material. As compared to the flat DNA-micro-array chip surfaces, the 3-dimensional structure of the nitrocellulose coating (thickness approximately 15 μm) provides a large surface area for protein binding. In addition, covalent binding is possible as well. Especially the format with 16 membranes (6x6 mm) per slide is perfectly suited to screen a maximum of 16 different samples for up to 30 different specificities (spots). In addition, based on the signal generation with colloidal particles, it is possible to combine detection of various target compounds on a single membrane; protein detection can be done simultaneously with DNA detection, and/or with detection of haptens like antibiotics and mycotoxins by incubating with a mixture of carbon-ligand conjugates with the required specificities. Furthermore, data recording and processing is simple and cheap. Digitising by conventional flatbed scanning and image analysis by standard analysis software is an inexpensive form of automation. Especially for intermediate numbers of samples this procedure is preferred over the sophisticated micro-array technology, for which expensive apparatus are needed to process the data.

In Table 3 an appraisal is given of the contributions of the various rapid colloidal particles based methods to the speed and ease of performance in bacterial detection. For detection of bacterial cells with LFIA between $5x10^4$ to 10^6 cells per mL are necessary to give a positive signal. This means that in most cases a pre-enrichment step is needed, although a rapid and optimised protocol may lead to an incubation time between 18 and 22 hours. Followed by the LFIA detection step (15 minutes) the result is a substantial increase in overall assay time. In principle, the methods based on detection of specific DNA or RNA are more sensitive than the immunochemical methods (Johnson, 2000). International standards derived from traditional detection methods require a detection threshold of one cell per 25 gram of food sample. Malorny *et al.* (2003) have calculated what the consequences would be for a molecular test: "The theoretical detection limit of one microbial cell per PCR reaction can usually be translated in practice to 10^3 to 10^4 cells per mL of pre-enriched sample, provided a small volume of initial matrix is used in the PCR reaction. Therefore, a PCR assay, usually preceded by an enrichment step for a primary multiplication of bacterial cells, should itself detect at least 10 to 100 copies of the bacterial target DNA in the reaction". The number of cells sufficient for detection can be translated into an accelerated pre-enrichment step of 4 to 8 hours, followed by the DNA / RNA amplification procedure and NALFIA or Mini-array detection. This results in an overall assay duration of 6 to 10 hours, which is possible in a single working day. The Mini-array as presented in this study was not optimised yet. Therefore, it is unknown whether the incubation time of 2 hours is necessary; perhaps shorter incubation times will be possible for particular applications. Moreover, since NALFIA is limited to the detection of 4 or 8 different amplicons (in single- and double-lane devices, respectively) up to 30 spots per 6x6 mm nitrocellulose membrane may be screened in the Mini-array method. In addition, a one-slide incubation chamber accommodates 16 membranes that can be used simultaneously, increasing the number of spots up to 480. Moreover, data processing can be automised for the Mini-array method. Therefore, it will depend on the particular application whether the NALFIA or the Mini-array method will be more appropriate.

'Rapid' is a subjective term used loosely to describe a vast array of methods, including miniaturised biochemical kits, antibody- and DNA-based kits and assays that are modifications of conventional tests to speed up analysis (Zhang, 2003). In bacterial diagnostics 'rapid' can be translated into any time frame between real-time and one day. Speed as a leading parameter in assay development is counterbalanced by parameters such as required test sensitivity, especially at critical points in the procedure. Furthermore, requirements as laid down in legislation and regulation, including GMP and HACCP protocols limit a further decrease in total assay time. Moreover, rapid methods have not been generally accepted as standard methods. In the Bacteriological Analytical Manual the U.S. Food and Drug Administration (FDA) states that a positive result by a rapid method is only regarded as presumptive and must be confirmed by standards methods. A screening test indicates if a sample is potentially positive for a pathogen. In food analysis, rapid methods still lack sufficient sensitivity and specificity for direct testing, so foods still need to be culture-enriched before analysis (U.S. Food and Drug Administration, 2001). However, the FDA indicates that enrichment is a limitation in terms of assay speed, but also provides essential benefits such as diluting the effects of inhibitors, allowing the differentiation of viable from non-viable cells and allowing for repair of cell stress or injury (resulting from food processing). Evaluation of rapid methods show that some perform better in some foods than others (U.S. Food and Drug Administration, 2001). This can be attributed mostly to interference by food components. The large number of rapid methods for a particular pathogen can be confusing and overwhelming to the user and has limited the effective evaluation of these methods. As a result, only few methods have been officially validated for use in food testing. Furthermore, sample preparation and (pre-)enrichment are still important and crucial steps in microbiological analysis. A major point concerns the statistically relevant way of sampling. Does it provide sufficient confidence that a negative test result really indicates the safety of the tested food product? In a number of traditional and recent (rapid) methods the procedure from sampling to (pre-)enrichment is still limiting a real break-through. In addition to developments with respect to a decrease of total assay time, simpler assay protocols that require low or no skills are being introduced as well. This approach may be directed to the total procedure, or to (a number of) individual steps (see the rapid methods as presented in this study; Table 3). Another way of further simplifying methods is maximising automation, i.e., irrespective format and technology of the method, reducing hands-on time and, in addition, labour costs. A third possibility is a shift of the test procedure to earlier stages in the production process, based on GMP, HACCP and/or tracking & tracing protocols, and often resulting in a reduced number of samples to be analysed.

New research areas such as nano(bio)technology, genomics, proteomics and metabolomics will result in miniaturised methods, lab-on-a-chip solutions, that enables multiple and simultaneous pathogen detection in almost real-time (Rider *et al.*, 2003; Wu *et al.*, 2003; Terletski *et al.*, 2003; Leonard *et al.*, 2003). In this respect it is extremely important to manage the overwhelming amount of data that may be produced in these assays, especially relevant for technologies that are based on a fingerprinting approach. The software should not only give the correct answers, but also identify those variables that optimally describe the microbiological status (Ellis and Goodacre, 2001). It should give understandable and practical answers, instructions, or suggestions that result in the right measures to be taken. Simplification for the user is anticipated, going hand in hand with a further decrease in total assay duration.

5. Conclusions

The lateral flow principle was used to develop rapid and simple methods to detect bacterial cells (LFIA) and genetic material (NALFIA). In both cases sampling and (pre-)enrichment is necessary, although the total procedures can be accelerated to 18 to 22 hours for the protein-directed LFIA and 6 to 10 h for the DNA/RNA-directed NALFIA. Overall handling is simplified and the signal can be read by the appearance of a black line in the assay device.

The Mini-array procedure on nitrocellulose coated slides with colloidal carbon particles is cheap and easy to perform. The method is ideally suited for semi-quantitative purposes such as detection of genetic material of a mixture of food pathogens, if desired combined with detection of bacterial proteins or other metabolites as well. The possibility to digitise the spots on the membrane by flatbed scanning, followed by analysis of the image, enables future automation in data processing.

Acknowledgements

The authors acknowledge the collaboration with Clare Aldus, Mike Peck and Gary Wyatt from the Institute of Food Research, Norwich, UK, and the financial support from the Food Standards Agency, UK (*E. coli* O157 LFIA); Henk Aarts and Esther Kok from RIKILT-Institute of Food Safety, Wageningen, the Netherlands (*Salmonella* NALFIA); Renata Ariëns and Jan Wichers (*E. coli* O157 LFIA), Luc Hornstra (*B. cereus* / *E. coli* NALFIA), Hans Mooibroek and Emil Wolbert (*Pseudomonas* NALFIA and Mini-array) from Agrotechnology & Food Innovations, Wageningen, the Netherlands.

References

Aabo, S., O.F. Rasmussen, L. Rossen, P.D. Sorensen and J.E. Olsen, 1993. Salmonella identification by the polymerase chain reaction. Molecular and Cellular Probes 7, p. 171-178.

Aldus, C.F., A. van Amerongen, R.M.C. Ariëns, M.W. Peck, J.H. Wichers and G.M. Wyatt, 2003. Principles of some novel rapid dipstick methods for detection and characterisation of verotoxigenic *Escherichia coli* (VTEC). Journal of Applied Microbiology 94, p. 1-10.

Ashkenazi, S. and T.G. Cleary, 1990. A method for detecting Shiga toxin and Shiga-like toxin-1 in pure and mixed culture. Journal of Medical Microbiology 32, p. 255-261.

Birch, L., C.E. Dawson, J.H. Cornett and J.T. Keer, 2001. A comparison of nucleic acid amplification techniques for the assessment of bacterial viability. Letters in Applied Microbiology 33, p. 296-301.

Capps, K.L., E.M. McLaughlin, A.W.A. Murray, C.F. Aldus, G.M. Wyatt, M.W. Peck, A. van Amerongen, R.M.C. Ariëns, J.H. Wichers, C.L. Baylis, D.R.A. Wareing and F.J. Bolton, 2004. Validation of three rapid screening methods for detection of verotoxin-producing *Escherichia coli* in foods: interlaboratory study. Journal of AOAC International 87, p. 68-77.

Chen, M.L. and H.Y. Tsen, 2002. Discrimination of *Bacillus cereus* and *Bacillus thuringiensis* with 16S rRNA and *gyrB* gene based PCR primers and sequencing of their annealing sites. Journal of Applied Microbiology 92, p. 912-919.

Compton, J., 1991. Nucleic acid sequence-based amplification. Nature 350, p. 91-92.

Ellis, D.I. and R. Goodacre, 2001. Rapid and quantitative detection of the microbial spoilage of muscle foods: current status and future trends. Trends in Food Science & Technology 12, p. 414-424.

Fung, D.Y.C., 2002. Rapid methods and automation in microbiology. Comprehensive Reviews in Food Science and Food Safety 1, p. 1-22.

ISO, 2001. Microbiology of food and animal feeding stuffs - Horizontal method for the detection of *E. coli* O157. BS EN ISO 16654: 2001.

Johnson, J.R., 2000. Development of polymerase chain reaction-based assays for bacterial gene detection. Journal of Microbiological Methods 41, p. 201-209.

Koets, M., N. Barbier, E.J.H. Wolbert, A. Mooibroek and A. van Amerongen, 2003. Rapid and simple one-step and Mini-array methods to detect and quantify amplified DNA and proteins using ligand-labelled colloidal particles. In: Proceedings EURO FOOD CHEM XII - Strategies for Safe Food, 24-26 September 2003, Brugge, Belgium, p. 121-124.

Leonard, P., S. Hearty, J. Brennan, L. Dunne, J. Quinn, T. Chakraborty and R. O'Kennedy, 2003. Advances in biosensors for detection of pathogens in food and water. Enzyme and Microbial Technology 32, p. 3-13.

Malorny, B., P.T. Tassios, P. Rådström, N. Cook, M. Wagner and J. Hoorfar, 2003. Standardisation of diagnostic PCR for the detection of foodborne pathogens. International Journal of Food Microbiology 83 39-48 (EU project FOOD-PCR, www.PCR.dk).

Neill, M.A., 1997. Overview of verotoxigenic *Escherichia coli*. Journal of Food Protection 60, p. 1444-1446.

Niessen, M.J.F., J.H. Wichers, H.A. Lee, M. Alcocer, M.R.A. Morgan and A. van Amerongen, 1999. Rapid sol particle immunoassay for the detection of aflatoxin in food products. In: V. Gaukel and W.E.L. Spiess (Eds.), Proceedings Part 3, 3rd Karlsruhe Nutrition Symposium, Eur. Res. towards Safer and Better Foods, 18/20 October 1998, Karlsruhe, Germany, p. 23-29.

Purohit, H.J. and A. Kapley, 2002. PCR as an emerging option in the microbial quality control of drinking water, Trends in Biotechnology 20 325-326.

Rider, T.H., M.S. Petrovick, F.E. Nargi, J.D. Harper, E.D. Schwoebel, R.H. Mathews, D.J. Blanchard, L.T. Bortolin, A.M. Young, J. Chen and M.A. Hollis, 2003. A B cell-based sensor for rapid identification of pathogens. Science 301, p. 213-215.

Sachse, K., 2004. Specificity and performance of PCR detection assays for microbial pathogens. Molecular Biotechnology 26, p. 61-79.

Scotter, S., M. Aldridge and K. Capps, 2000. Validation of a method for the detection of *E. coli* O157 in foods. Food Control 11, p. 85-95.

Terletski, V., S. Schwarz, J. Carnwath and H. Niemann, 2003. Subtracted restriction fingerprinting - A tool for bacterial genome typing. BioTechniques 34, p. 304-313.

U.S. Food and Drug Administration, Center for Food Safety and Applied Nutrition, 2001. Bacteriological Analytical Manual Online, January 2001, Appendix 1, Rapid Methods for detecting foodborne pathogens; www.cfsan.fda.gov/~ebam/bam-toc.html.

Van Amerongen, A., J.H. Wichers, L.B.J.M. Berendsen and A.W.J. Van Doorn, 1994. Immunoassays using a carbon sol label. Patent application and patented in South Africa (95/3880, 1996), New Zealand (272119, 1997), Australia (693523, 1998), Israel (113,682, 2001), Taiwan (136446, 2001), China (95105463.5, 2003).

Van Dam, G.J., J.H. Wichers, T.M. Falcao Ferreira, D. Ghati, A. van Amerongen, A.M. Deelder, 2004. Diagnosis of schistosomiasis by reagent strip test for detection of CCA. Journal of Clinical Microbiology, in press.

Wichers, J.H., R.M. van Es, G.D. Keizer, A.W.J. van Doorn and W.M.J. van Gelder, 1987. Stained sols of non-metallic elements or compounds, their preparation and use. Patent filed 1987, European Patent 0370561, June 24 1992.

Wolbert, E.J.H., J.P. Gorissen, M.C.M. Dielissen, R. Weusthuis, G. Eggink and H. Mooibroek, 2002. Isolation and initial characterisation of microbial species with high mcl-PHA production. Abstract Int. Symp. Biological Polyesters (ISBP2002), 22 - 26 Sept. 2002 in Münster (Germany). Book of abstracts, p. 107.

Wu, C.-F., J.J. Valdes, W.E. Bentley and J.W. Sekowski, 2003. DNA microarray for discrimination between pathogenic O157:H7 EDL933 and non-pathogenic *Escherichia coli* strains. Biosensors and Bioelectronics 19, p. 1-8.

Zhang, J., 2003. Food safety and beyond. http://www.glue.umd.edu/~nsw/enpm808b/zhang-j.ppt.

A rapid nucleic acid detection method for specific bacteria: The case of *Listeria monocytogenes*

T. Wikman, K. Antonen, T. Korpimäki, T. Lövgren and J. Nurmi
University of Turku, Department of Biotechnology, Tykistökatu 6A, 20520 Turku, Finland

Abstract

Commonly used methodologies for the detection of specific organisms are severely compromised by the laborious and time consuming steps of nucleic acid extraction, amplification and amplification product identification. This is why we have developed a rapid nucleic acid detection method amenable to automation. The monitoring of our model analyte, *Listeria monocytogenes*, is accomplished by using a novel, rapid sample pretreatment method, fast thermal cycling and homogeneous real-time PCR based on environment sensitive lanthanide chelates and time-resolved fluorescence. *Listeria* cells were monitored from spiked milk and cheese samples. The method can detect as little as 2-20 CFUs from a crude sample (confirmed with platings). After enrichment the whole process of sample preparation, target sequence amplification and identification can be completed in less than 60 minutes.

Keywords: *L. monocytogenes*, real-time PCR, lanthanide chelate, time-resolved fluorescence

1. Introduction

Listeriae are gram-positive bacteria that are widely distributed in nature. There are six *Listeria* species and the only harmful member of the genus is *L. monocytogenes*. It is a major human food pathogen capable of causing a condition called listeriosis. The fatality rate of the disease is 20-30%. Listeriosis is found both in outbreaks and as single cases. There are also two forms of listeriosis. Non-invasive listeriosis has self-limiting gastrointestinal symptoms. Invasive listeriosis almost entirely affects newborns, pregnant women, immunocompromised individuals and the elderly. The symptoms of listeriosis appear after several weeks of exposure and are meningitis, septicemia and abortion.

L. monocytogenes has exceptional growth conditions: it is able to grow over a temperature range from 1 to 45 °C and at a pH range of 4.4 to 9.6. This is why it can be found as a contaminant in various foods. It is most often encountered in fresh foods of both plant and animal origin like raw milk, soft cheeses, fresh and frozen meat, poultry, seafood, fruits and vegetable products. Due to its prevalence several listeriosis outbreaks are reported annually (Rocourt *et al.*, 2000).

Listeria is an emerging food safety concern because of its ability to grow at refrigeration temperatures. *Listeriae* are also quite resistant to preservation methods such as high salt concentrations or acidic conditions. Current food processing methods aim to reduce the number of spoilage bacteria on meats and cheeses in order to prolong refrigerated storage time. When the number of spoilage bacteria are reduced, any *Listeria* which survive have a competitive advantage and could proliferate to an infective dose when stored for more than three weeks. *Listeria* is killed by pasteurisation and cooking, however, in certain ready-to-eat foods such as hot dogs and deli meats, contamination may occur after cooking but before packaging. Processed meats which are consumed without cooking have become a high risk food. Because of these facts there is a requirement for fast and reliable *Listeria monocytogenes* detection methods.

The traditional culture-based detection methods for *Listeria* in food samples involve selective enrichments with subsequent culturing on selective media, followed by serological and/or biochemical species identification (Cox *et al.*, 1998). These methods are laborious and time-consuming: it takes from 3 to 7 days to conclude with certainty whether a food sample has been contaminated with *Listeria*. Due to possible interpretation difficulties of these methods they require skilled analysts.

The API-Listeria test (bioMerieux, Marcy-l´Etoile, France) is specifically designed for the genus *Listeria* and includes 10 biochemical differentiation tests in a microtube format. The CAMP (Christie, Atkins, Munch-Petersen) test is used to differentiate between *L. monocytogenes* and *L. innocua* which is the most frequently encountered non-pathogenic *Listeria* species. The test is based on the haemolytic property of *L. monocytogenes* which is sometimes difficult to read on a blood agar. In addition there are culture tests that are based on chromogenic media like the Rapid´L.MONO® (Bio-Rad, Marnes la Coquette, France) which is a selective agar medium, that only *Listeria* species can grow on. This test gives results within 24-48 h.

Also immunoassays have been employed to the monitoring of *L. monocytogenes* from different food samples. *L. monocytogenes* immunoassays use genus-specific monoclonal antibodies to recognize the antigens on bacterial cells. Immunoassays can be automated and they are not as labour-intensive as culturing methods. False positive results may be caused by antibodies which may not be 100% specific. Commercially available *L. monocytogenes* tests include the VIDAS LMO™ -(bioMérieux, Marcy l'Etoile, France) assay, which is an enzyme-linked fluorescent immunoassay (ELFA) for use on the VIDAS system for the qualitative detection of *L. monocytogenes*. The procedure involves two enrichments and an immunoassay and takes 48 hours to complete. ListerTest™ (Vicam, MA,USA) is based on immunomagnetic separation of *L. monocytogenes* cells from a food sample. In the assay microscopic magnetic beads that have been coated with *L. monocytogenes* specific antibodies capture *L. monocytogenes* cells. The complexes are separated from the sample and plated on a selective medium. *L. monocytogenes* -colonies are observed after 22 hours and the whole procedure lasts approximately 24 hours. The sensitivity of the ListerTest™ reported by the manufacturer is the same as with culturing methods, which might not entirely be the case because of the lack of the pre-enrichment step.

The invention of PCR has enabled the detection of any specific nucleic acid sequence. The first nucleic acid monitoring methods were severely limited by the tedious and time-consuming post amplification steps, e.g. agarose gel electrophoresis and/or different kinds of hybridisations. Nowadays several alternatives exist for the monitoring of amplification product accumulation in real-time using fluorescent dyes, probes or primers.

When genuinely accurate and robust detection technologies are needed nucleic acid assays are the methods of choice. A problem that is frequently encountered with culture methods and immunoassays is that they cannot always make a difference between *L. monocytogenes* and its close relative *L. innocua*. This problem is alleviated with PCR-based detection methodologies. *L. monocytogenes* can be distinguished from other *Listeria* species with for example the virulence factor Listeriolysin O (LLO). This is a hemolysin that all virulent *L. monocytogenes* strains contain. That is why the *hlyA* gene is the target of many molecular screening methods of *L. monocytogenes*. Designing nucleic acid-based assays for *L. monocytogenes* may pose a problem because several *Listeria* genome sequences are patented.

Confirmation of *L. monocytogenes* from colonies on primary isolation plates can be done with chemiluminescent DNA probes. AccuProbe® from Gen-Tech (CA, USA) hybridises to a specific ribosomal RNA sequence that is unique to *L. monocytogenes*. The total assay time varies between 18-48 hours of which the duration of the hybridisation assay reported by the manufacturer is 50 minutes. A qualitative PCR-based detection method for *L. monocytogenes* in food samples is the Bax® System (Qualicon, DE, USA). It comprises of two enrichments followed by one hour sample preparation and an automated PCR procedure which delivers results four hours later. Also real-time PCR has been applied to the detection of *L. monocytogenes* from food samples. For example Light Cycler® *Listeria monocytogenes* Detection Kit (Roche Applied Science, USA) uses hybridisation probes to produce qualitative results in 65 minutes from purified DNA.

The real-time PCR detection chemistry used in this study has been published by Nurmi *et al.* (2002). It is based on environment sensitive lanthanide chelates and time-resolved fluorometry. The fluorescence signal in the method is generated via a 5´-3´-exonucleolytic DNA polymerase that in the extension phase of the PCR digests the lanthanide chelate labeled probe hybridized to its specific target sequence. The chelate is released into solution and fluorescence increased because the chelate has a higher fluorescence intensity when it is free in solution than when it is attached to single-stranded DNA. The unspecific background signal is even further decreased by a short quencher probe that is complementary to the detection probe.

2. Materials and methods

The food samples were artificially contaminated by mixing 1 mL of milk (skimmed, semi-skimmed and whole) or 1 g of cheese (13% and 44% fat soft cheese and 29% fat blue cheese) with *L. monocytogenes* cells grown overnight. The spiked samples were pre-enriched as duplicates in a total volume of 10 millilitres in half-Fraser for 16-20 h at 30 °C. The artificially contaminated samples were treated for PCR with a proprietary sample preparation method and analysed in real-time PCR.

3. Results

The rapid *L. monocytogenes* detection technology, AmpliFast, was tested with six dairy products: three different milks and three different cheeses. The standard curves for the different sample materials were generated by plotting threshold values gained from the amplification curves against the logarithm of colony forming units (CFU) determined with platings.

The artificially contaminated milk samples generated linear standard curves with a large dynamic range (figure 1). The threshold cycles were lower for the whole milk samples than for the semi-skimmed or skimmed milk samples probably because the enrichment is more efficient for whole milk. The detection limits of these assays were 3 CFUs per 1 milliliter of sample for semi-skimmed and whole milk and 20 CFUs for skimmed milk.

The cheese samples tested produced results with greater diversity than milk samples. This can be due to different textures and fat contents of different cheeses. Nevertheless the spiked cheese samples were also found to produce linear standard curves (figure 2). Even though blue cheese has bigger fat content than

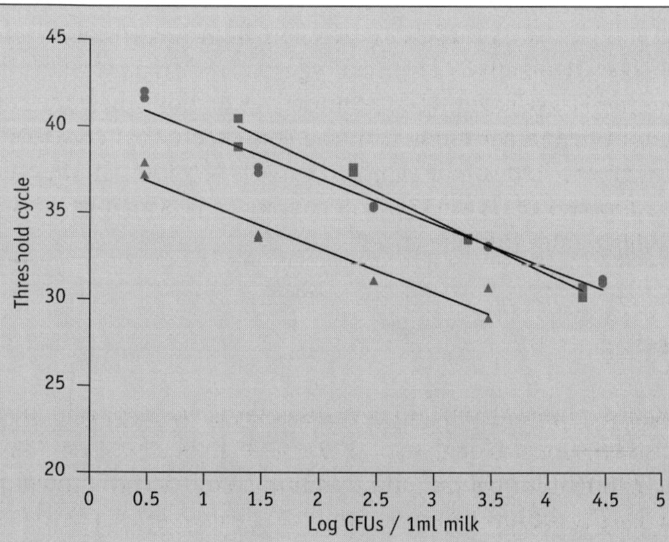

Figure 1. Standard curves generated from skimmed (squares), semi-skimmed (dots) and whole milk (triangles) samples.

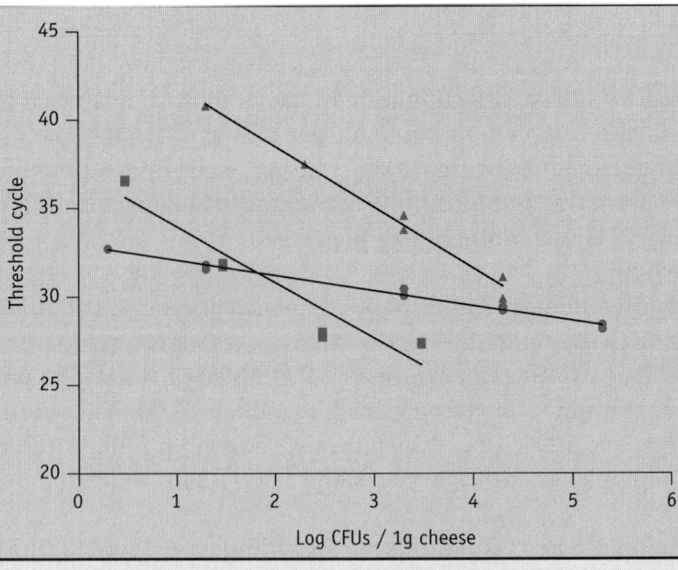

Figure 2. Standard curves generated from the 13% soft cheese (squares), 44% fat soft cheese (dots) and blue cheese (triangles) samples.

the 13% fat soft cheese its threshold cycles are higher. A reason for this might be that blue cheese contains lots of other organisms which compete for nutrients with *L. monocytogenes* cells during enrichment. A situation where the enrichment is perhaps not linear anymore is with the soft cheese that contains 44% fat. This may be why the slope of the standard curve is very low. The detection limits in these assays were 3 CFUs for 13% soft cheese, 2 CFUs for 44% fat soft cheese and 20 CFUs per gram for blue cheese.

4. Discussion

PCR is the only test utilized for rapid detection of *L. monocytogenes* in food samples and in clinical specimens (Allerberger, 2003). The phase of *Listeria* detection that most severely limits throughput and duration of different methods is sample enrichment. Due to the low detection limits of *L. monocytogenes* in foods in many countries a sample pre-enrichment step is needed to reach the required sensitivity and to ascertain negative samples. Most detection protocols for *Listeria* include pre-enrichment, selective enrichment and isolation on selective media

(Beumer and Hazeleger, 2003). In the method developed the detection of *L. monocytogenes* can be carried out in 17-20 hours from which sample treatment, target sequence amplification and detection take less than one hour (figure 3). If a more abundant analyte which does not need enrichment is monitored the whole procedure of sample treatment and analysis can be completed in less than one hour depending on the PCR cycling parameters.

From the linearity of the standard curves it can be deduced that the quantitativity between initial cell amounts is well maintained throughout the whole procedure of enrichment, sample treatment and amplification. The detection limit of the method developed is very low. It varies between 2 and 20 *L. monocytogenes* cells per one gram or millilitre of sample. The linear range of the assay is at least five orders of magnitude which assures quantitative detection of both small and large amounts of bacterial cells.

The real-time PCR detection chemistry used benefits from low background fluorescence and high signal-to-noise ratios. Typical signal-to-noise ratios in the method are approximately 10. In conclusion it can be stated that the different areas of the technology developed, sample treatment, fast cycling or real-time PCR detection chemistry, can be used together or separately for the detection of any nucleic acid analyte.

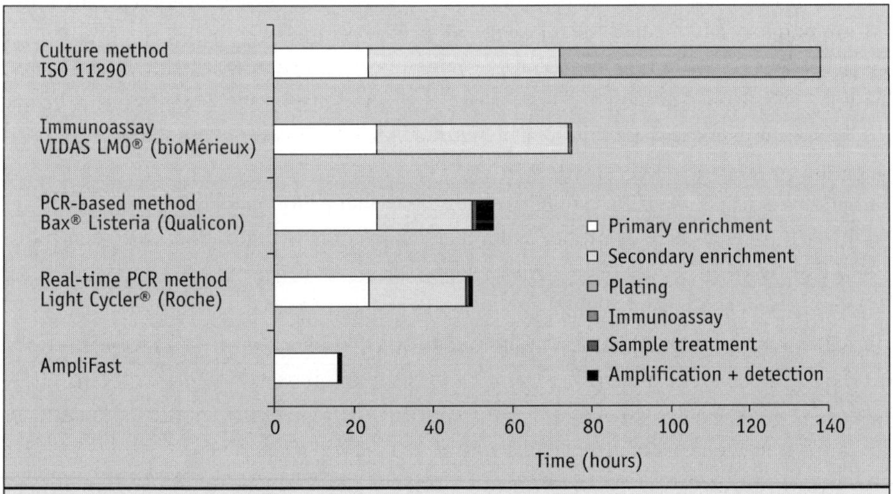

Figure 3. A comparison of the duration of different *L. monocytogenes* detection methods.

5. Conclusions

The polymerase chain reaction is a valuable tool for the monitoring of food-borne pathogens. Using PCR-based assays it is possible to shorten the time needed for the detection and confirmation of micro-organisms. The quantitative detection of *L. monocytogenes* can be completed in approximately 20 hours from which the sample treatment, target sequence amplification and detection only take approximately one hour. The method is very sensitive (2-20 CFUs / 1 gram or millilitre of sample), specific and has a large linear range over at least 4 logarithms.

Acknowledgements

We would like to thank the National Technology Agency of Finland (Tekes) for financial support.

References

Allerberger, F., 2003. Listeria: growth, phenotypic differentiation and molecular microbiology. FEMS Immunology and medical Microbiology 35, p. 183-189.

Beumer, R. and Hazeleger, W., 2003. *Listeria monocytogenes*: diagnostic problems. FEMS Immunology and medical Microbiology 35, p. 191-197.

Cox, T., Frazier, C., Tuttle, J., Flood, S., Yagi, L., Yamashiro, C.T., Behari, R., Paszko, C., and Cano, R.J., 1998. Rapid detection of *Listeria monocytogenes* in dairy samples utilizing a PCR-based fluorogenic 5´nuclease assay. Journal of Industrial Microbiology & Biotechnology 21, p. 167-174.

Nurmi, J., Wikman, T., Karp, M., and Lövgren, T., 2002. High-performance real-time quantitative RT-PCR using lanthanide probes and a dual-temperature hybridization assay. Analytical Chemistry 74, p. 3525-2532.

Rocourt, J., Jacquet, Ch., and Reilly, A., 2000. Epidemiology of human listeriosis and seafoods. International Journal of Food Microbiology 62, p. 197-209.

Norovirus in food and water: Isolation and quantification

H. Schnerr[*1], A. Knight[1], P. Patel[1], M. Carter[2], M. Adams[2] and A.P. Wyn-Jones[3]*
[1]Leatherhead Food International Ltd., Randalls Road, Leatherhead, Surrey, KT22 7RY, United Kingdom, [2]School of Biological Sciences, University of Surrey, Guildford, United Kingdom, [3]School of Science, University of Sunderland, Sunderland, United Kingdom

Abstract

Noroviruses (NVs) are the most common cause of non-bacterial gastro-enteritis in adults worldwide. However, NVs cannot be grown *in vitro*, making their detection difficult. A real-time reverse transcription-polymerase chain reaction assay was coupled with SYBR®Green intercalation for the detection and quantification of Norovirus. A new primer pair was designed for genogroup II based on the highly conserved region of the RNA polymerase gene. The NV real-time RT-PCR assay developed had a detection limit of 10^1 RNA copies per reaction and showed a dynamic linear detection range between 10^2 and 10^8 RNA copies. The coefficients of variation of threshold cycle values in five independent runs were less than 4.0 % for different concentrations of plasmid DNA. Real-time RT-PCR was used to examine the NV RNA levels in human faeces. Negative controls showed no specific amplification. The approach provides a simple and useful tool for the development of NV concentration procedures for the analysis of food and water samples.

Keywords: Norovirus, quantitative RT-PCR, SYBR®Green I

1. Introduction

The non-cultivable Noroviruses are members of the family *Caliciviridae* (Jiang *et al.*, 1993) and have emerged as the single most common cause of outbreaks of acute non-bacterial gastro-enteritis throughout the world. Infection may be person-to-person or foodborne and infection is frequently associated with consumption of shellfish (Richards, 1987; Rippey, 1994; Hale *et al.*, 2000). Food produce could be contaminated either at source or at any stage during processing, packaging and distribution and it is recognised that the risk of foodborne disease has been increased in recent years by the increase in consumption of pre-prepared foods, particularly those consumed without further cooking.

NVs cannot be grown *in vitro*, making their detection difficult. Current methods are based on immunoassay or sensitive generic reverse transcription polymerase chain reaction (RT-PCR) amplification with short segments of the viral genome as target (Lopman *et al.*, 2002). These methods have largely replaced diagnosis by electron microscopy.

Noroviruses are diverse and have been genetically classified into three major genetic groups consisting of 17 genetic subgroups (Schuffenecker *et al.*, 2001). Between July 1992 and June 1998, Bristol/93/UK (Grimsby) virus, a genogroup II representative, was the predominant strain detected in UK outbreaks and overall was associated with 61 % of NV outbreaks (Hale *et al.*, 2000).

Fortin *et al.* (2001) showed that the length of the amplicon has a dramatic effect on the real-time PCR assay. Smaller amplicons give more consistent results because PCR was more efficient. Furthermore, reducing the amplicon size reduced the C_T value and increased the fluorescence signal. We have designed a new pair of primers for genogroup II based on the RNA polymerase gene, the most conserved region of the genome.

Fluorescence based real-time RT-PCR is widely used for the quantification of mRNA levels and is a critical tool for basic research, molecular medicine and biotechnology (Bustin, 2002; Mackay *et al.*, 2002).

We have developed a rapid and highly sensitive real-time quantitative PCR method using the iCycler iQ™ detection system (Bio-Rad) coupled with a fluorescence technique using the fluorescence dye SYBR®Green I. SYBR®Green I possesses the selective affinity to bind to double-stranded DNA (*ds*DNA). But does not react with single-stranded DNA or with RNA present in a sample. The dye bound to *ds*DNA has a greatly enhanced fluorescence emission at 521 nm. The intensity of fluorescence is proportional to the concentration of *ds*DNA and thus provides a measure for the quantity of newly synthesised product during PCR. The method does not require any post-PCR analyses. The specificity of the product is monitored by analysing the melting curve of the product DNA (Ririe *et al.*, 1997).

Assays to detect virus in food need to be adapted for each food substance i.e. shellfish (Formiga-Cruz *et al.*, 2002) or fruits and vegetables (Dubois *et al.*, 2002). Water can be tested for NVs by using RT-PCR to detect virus when large volumes of water are processed through specially designed filters (Lamothe *et al.*, 2003). Efficient concentration methods represent a prerequisite for sensitive detection methods. Although a number of concentration methods exist for NVs, data of the efficiency of the concentration steps is difficult to achieve. Although ELISA and viral plaque assays can be used, they may not represent the true viral load or be suitable models for NVs. The objectives of the present study were to develop a real-time SYBR®Green I method for the detection of Noroviruses genogroup II, to determine the sensitivity of the method, and to investigate its use as a rapid method for investigating the efficiency of different concentration steps in virus isolation.

2. Materials and methods

2.1. Collection of samples

Stool specimens were collected from different gastro-enteritis outbreaks in UK. They were used directly in the RNA extraction procedure.

2.2 RNA extraction

Viral RNA was extracted from 250 mg of faecal material with the RNeasy™ Midi extraction kit (Qiagen) or from 100 mL of concentrates with the RNeasy™ Mini extraction kit (Qiagen), according to manufacturer's instructions. RNA was eluted with 250 µL or 50 µL RNase-free water, respectively, and stored at -80 °C until used in RT-PCR assays.

2.3 Oligonucleotide primers

The primers G2for and G2ref (see Table 1) were designed using the Beacon Design 2.0 software (Biosoft international) with the following primer parameters. The annealing temperature was set at 50.0 ± 5 °C, the primer length range at 18 to 22 bp and the specific amplicon length at 140 to 150 bp. The primers were synthesised by PE Applied Biosystems.

Table 1. Primer sequences and locations.

Primer	Orientation	Genomic location[a]	DNA sequence
G2for	sense	4406-4426	5'-CTTCTATCTCCTAGTGTGATG-3'
G2ref	anti-sense	4534-4553	5'-ACAGGTTTGTAACTTCAGAG-3'

[a] Genomic locations refer to Lordsdale virus (Liu *et al.*, 1995; Dingle *et al.*, 1995)

2.4 NV Plasmid standard for the quantification

The NV RNA standard was synthesised from plasmid DNA containing an inserted RNA polymerase sequence derived from the Middlesbrough outbreak (Willcocks *et al.*, 1993). We used the MEGAshortscript™ High Yield Transcription Kit (Ambion Inc.) to produce NV mRNA. The concentration of the RNA was calculated by densitometric quantification in ethidium bromide stained agarose gels. Using as external standard, the RNA was diluted serially from 10^{10} to 10^3 copies. Four replicates of each copy number were amplified to generate a standard curve.

2.5 Detection of NVs by RT-PCR

Viral nucleic acids recovered from the purification procedures above were used for the synthesis of complementary DNA. In each experiment, tubes to which no viral nucleic acids were added were used as negative controls. The reverse transcription and the amplification were performed using a single-tube RT-PCR method with the Qiagen one-step RT-PCR kit. The PCR amplifications were performed in an iCycler™ system (Bio-Rad) using the specific primers G2for and G2ref. The reactions were carried out in a 96-well plate in a 25 µL reaction volume containing 12.5 µL QuantiTect™ SYBR®Green RT-PCR Master Mix (Qiagen Ltd., U.K.), 0.25 µL QuantiTect™ RT Mix, 0.5 µM each primer and 1 µL template solution.

The thermal profile commenced with a 30 minute 50 °C step followed by heat activation of the enzyme at 95 °C for 15 min. The cDNA was amplified by the following cycle conditions: 50 cycles of template denaturation at 95 °C for 15 s, primer annealing at 50 °C for 30 s, primer extension at 72 °C for 5 s.

Fluorescence measurements were recorded at 72 °C for 5 s during each extension step. At the end of each PCR run, data were automatically analysed by the iCycler iQ™ software and amplification plots were generated.

A dilution series of the plasmid standard was analysed in each 96-well plate alongside the unknown samples. Each sample was tested in triplicate. Each triplicate reaction was repeated independently at least three times to assess the reproducibility of the results.

Following the final amplification cycle, a melting curve was acquired by one cycle of heating to 95 °C for 3 minutes, cooling to 55 °C and slow heating to 95 °C at 0.5 °C steps for 10 s with continuous measurement of fluorescence at 521 nm.

2.6 Data analysis

Data acquisition and subsequent data analyses were performed using the iCycler IQ™ real-time detection system (Version 3.0). A standard curve was generated using input target copy numbers versus the corresponding threshold cycle (C_T) values using the least squares fit method. The C_T is defined as the cycle at which the fluorescence acquired during amplification is significantly higher than the baseline level. We specified the threshold of not less than five standard deviations from the background level for determining a positive signal. The baseline level was calculated automatically between cycle 2 and 10. The C_T is inversely proportional to the log of the initial quantity of target molecule. The greater the amount of target copies in the reaction, the lower the C_T value (Higuchi *et al.*, 1993). The copy numbers for RNA contained in experimental samples can then be calculated after real-time amplification from the linear regression of the standard curve.

The quality of the product was monitored by a melting curve. The first negative derivative of fluorescence over temperature (-dF/dT) was plotted against temperature to result in a characteristic melting peak enabling differentiation of secondary products and primer dimers since these show broader peaks with lower melting temperatures.

2.7 Evaluation of the filtration procedures

Filtration efficiency was determined by RNA quantitation of two dilutions (1.5 x 10^6 and 1.5 x 10^3) from a NV genogroup II virus stock. For this, each viral dilution was either extracted directly or spiked into 1 litre of sterile water and filtered before extraction. The relative quantitation (filtered versus non-filtered virus samples) allowed us to calculate the fraction of virus recovered from the inoculum and to deduce the filtration efficiency. All assays were performed in triplicate.

The water samples were vigorously shaken and filtered through a positively charged membrane (Zetapor filter membrane, Cuno Inc.) according to previously described methods (Gilgen *et al.*, 1997; Beuret, 2003). Prefilters were omitted. Furthermore we tested different centrifugal concentrators (Vivaspin 2-100000, Sartorius; Centricon-100 microconcentrator, Millipore) further to reduce the volume of filtrate.

The loss of viral RNA during the isolation procedure was calculated from the difference between copy number mean values of filtered samples and of non-filtered ones. The non-filtered samples were set to 100 %. From this, the percent viral recovery (R) from the filtration step is deduced as:

$$R = \frac{\text{copynumber}_{\text{filtered}} \cdot 100\%}{\text{copynumber}_{\text{nonfiltered}}}$$

3. Results

3.1 Sensitivity of the SYBR®Green PCR

The sensitivity of SYBR®Green PCR step was determined by using a serial dilution of NV plasmid DNA as template for amplification. A linear relationship was obtained between the input copy number of the plasmid DNA and the C_T values with a regression coefficient (r^2) greater than 0.99 over 8 $\log^{10}$ dilutions, ranging from 1.5x10^8 copies to 1.5x10^1 copies (Figure 1). The sensitivity of SYBR®Green RT-PCR using a dilution series of NV RNA was 3.5x10^1 RNA copies. The linear relationship ranged from 3.5x10^{10} copies to 3.5x10^2 copies with r^2, again greater than 0.99 (data not shown).

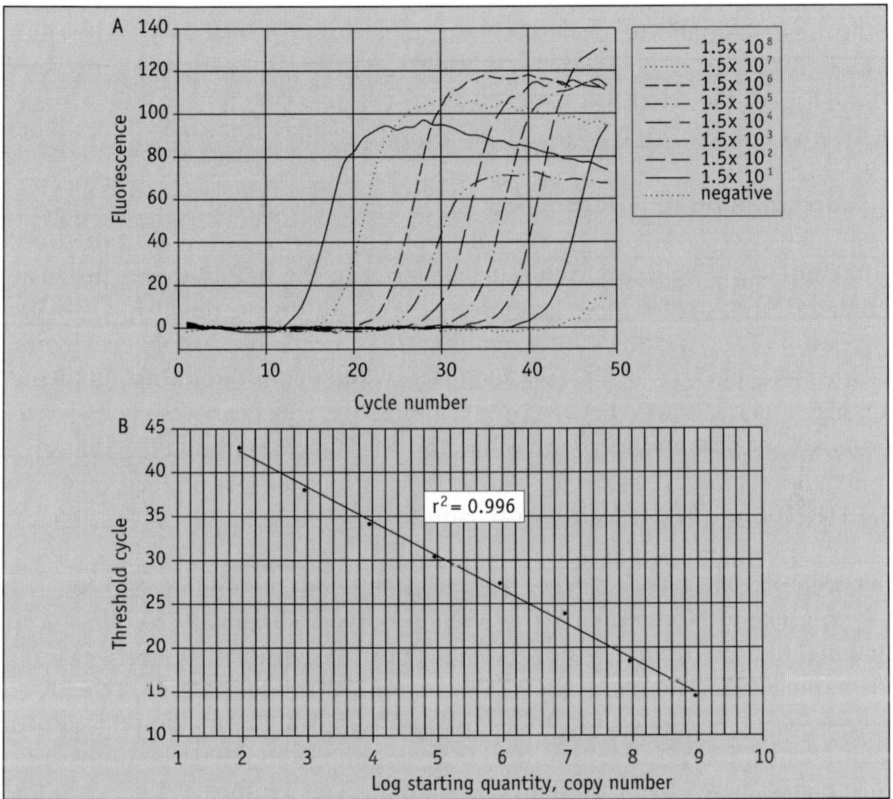

Figure 1. Analysis of the real-time RT-PCR sensitivity and linearity with NV plasmid DNA. (A) Amplification plot of 10-fold serial dilutions of the plasmid DNA. (B) The standard curve of 10-fold serial dilutions with a correlation coefficient (r^2) of 0.996.

3.2 Reproducibility

In order to determine the reproducibility of the real-time PCR amplifications were carried out in four parallel experiments and were examined on five consecutive days. These were performed using 1.5×10^8, 10^5 and 10^1 copies of NV plasmid DNA target per reaction.

The coefficient of variation (CV) was determined in the log-linear phase of PCR used for calculation of quantitative data. This was found to be 4.0, 3.4 and 2.4

% at $1.5x10^8$, 10^5 and 10^1 copies/reaction, respectively. The threshold cycle values were as follows: for $1.5x10^8$, at 16.78 cycles (±0.67); for $1.5x10^5$, at 29.53 cycles (±1.01); and for $1.5x10^1$, at 45.90 cycles (±1.08). This is equivalent to a standard error of 0.15, 0.23 and 0.24, respectively.

3.3 Identification of PCR products

A melting curve was determined to characterise the amplification products produced from each sample and external standard during the PCR. Gaussian regression of the melting curves resulting from the various external standards showed a T_m peak at 88.5 °C (see Figure 2). A second peak was visible showing a range of temperatures between 75 °C and 83 °C. This peak was attributed to formation of primer dimers and other non-specific amplification products.

3.4 Detection of NV RNA in faecal samples

In order to investigate the practical application of our method, we investigated the presence of Norovirus RNA in virus positive and negative faecal sample material in a preliminary study. Twenty positive samples were tested and all displayed a signal in the iCycler™ analysis. No signal was detected in the stool samples used as negative control or in mock samples. The method developed allowed quantification of NV content in these samples and the determined virus load ranged from 10^4 to 10^9 RNA copies. Table 2 shows the level of NV in the samples. Specificity of the amplicons was verified with the help of melting curve analyses (T_m=88.5 °C). These showed exactly the behaviours predicted for the duplex of known size and sequence.

3.5 Application in determining concentration efficiencies

Using this approach we have been able to investigate the retention and release of NVs from different filtration matrices. Current methods for virus concentration commonly include the use of Zetapor filter membranes (CUNO Inc.) (Beuret, 2003) and centrifugation concentrators e.g. Vivaspin 2-100000 (Sartorius) or other suppliers like Millipore (Gilgen *et al.*, 1997; Lamothe *et al.*, 2003).

Following the Beuret protocol (2003) to isolate the virus RNA directly from the membrane, we found a recovery of 23.2 % for 10^6 and 22.7 % for 10^3 virus particles. Eluting the virus from membrane first and concentrating the eluate using centrifugal concentrators we found recovery rates between 5.1 and 16.2 % depending on the manufacture.

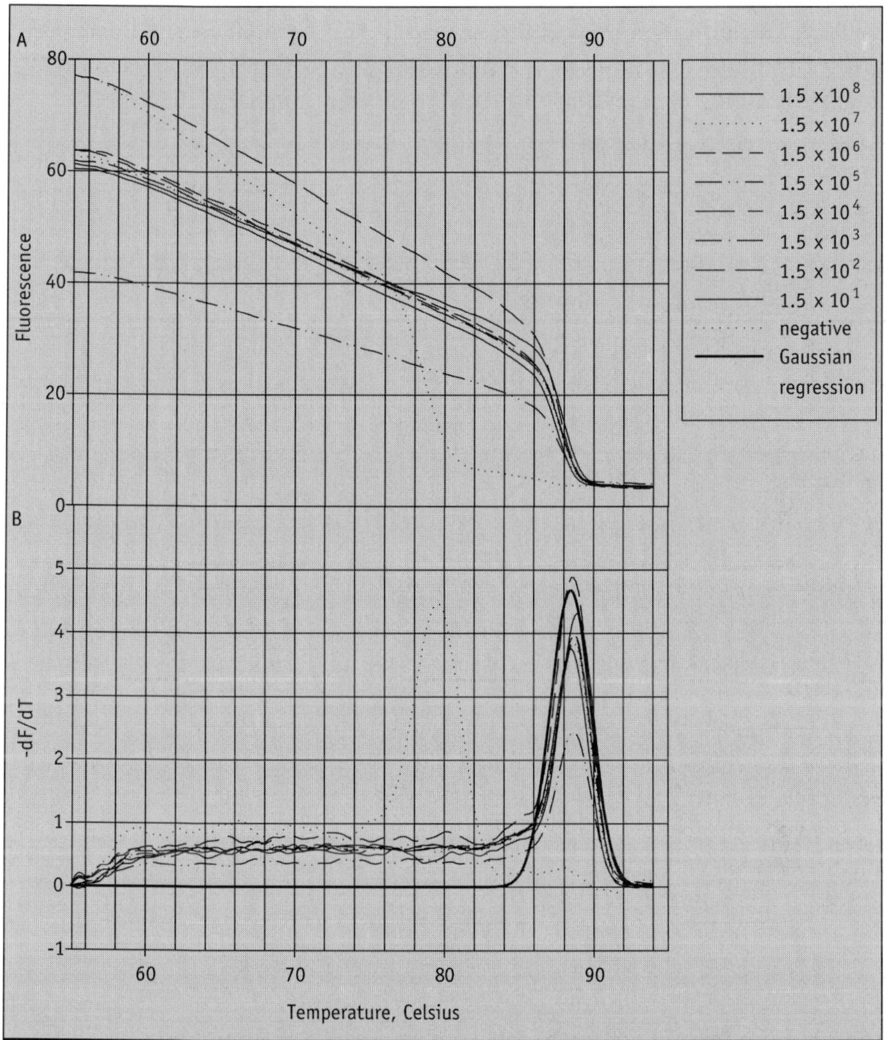

Figure 2. (A) Melting curves of specific product from NV Plasmid DNA and primer dimer are shown. (B) Through the first negative derivative of fluorescence (-*d*F/*d*T) the differences between specific and non-specific product become more obvious. The specific melting temperature was determined over a Gaussian regression at 88.5 °C.

Table 2. Quantification of Noroviruses load in faecal sample material using the SYBR®Green real-time RT-PCR. The table shows the coefficients of variation (CV) for the threshold cycle and the calculated input copy number in the RT-PCR.

Samples	C_T	Mean	CV (%)	Mean copies	Standard deviation	CV (%)
02-01152	34.38	34.35	0.16	2.74×10^4	5.50×10^2	2.01
	34.39					
	34.29					
02-03973	16.74	16.98	1.23	9.70×10^8	1.33×10^8	13.69
	17.12					
	17.08					
02-04081	31.41	31.29	2.45	1.82×10^5	0.77×10^5	42.40
	31.99					
	30.47					
02-06521	29.93	29.35	1.76	5.71×10^5	1.66×10^5	29.00
	28.93					
	29.20					
02-12309	16.70	16.85	0.79	1.04×10^9	9.91×10^7	9.51
	16.90					
	16.95					
02-29681	26.63	27.21	1.95	2.08×10^6	0.61×10^6	29.36
	27.33					
	27.67					
ML1	33.05	33.46	1.13	1.28×10^5	0.31×10^5	23.89
	33.80					
	33.52					
ML2	34.93	34.55	1.16	6.64×10^4	1.62×10^4	24.41
	34.13					
	34.58					
ML3	N/A					
ML4	N/A					
SRS1	18.29	18.26	0.23	1.30×10^8	3.54×10^6	2.73
	N/A					
	18.23					

Table 2. Continued.						
SRS2	24.43	24.53	0.45	2.86×10^7	0.20×10^7	6.90
	24.51					
	24.65					
MG 2736	31.00	30.84	0.81	6.24×10^5	0.96×10^5	15.45
	30.96					
	30.55					
MG 2748	30.04	30.24	1.78	9.18×10^5	2.80×10^5	29.75
	29.83					
	30.85					
MG 2759	26.53	26.37	0.52	9.37×10^6	0.75×10^6	8.05
	26.32					
	26.27					
MG 2767	31.45	31.49	1.50	4.29×10^5	1.20×10^5	27.86
	31.98					
	31.04					
V 1431	31.11	30.69	1.19	6.88×10^5	1.43×10^5	20.81
	30.49					
	30.47					
V 1434	30.58	30.81	0.84	6.35×10^5	0.97×10^5	15.24
	30.76					
	31.09					
V 1446	30.40	30.26	0.45	8.84×10^5	0.71×10^5	8.05
	30.13					
	30.24					
NLV II tue	31.82	31.49	1.35	4.26×10^5	1.14×10^5	26.77
	31.63					
	31.01					

4. Discussion

Noroviruses are important economically as a cause of both sporadic gastro-enteritis in the community and large outbreaks in hospitals and other institutional settings (Hale *et al.*, 2000) Of all gastro-enteritis outbreaks reported to the surveillance unit for England and Wales, nearly 50 % are thought to be due to Norovirus (Dedman *et al.*, 1998). Historically, Noroviruses were diagnosed by electron microscopy (EM) (Atmar and Estes, 2001), but EM is rather insensitive and requires highly skilled operators and expensive equipment. Viruses can also be detected by RT-PCR. However, differences in the detection limit of the different assays may have direct consequences for the interpretation of the results. A person found to be positive by EM test must be shedding a large number of particles (usually assumed to be around 10^6 per mL). In contrast a person found positive by PCR might be shedding far fewer and thus pose less of a risk of transmission of infection. No clear guidelines are yet available on interpretation of these different test results (Koopmans and Duizer, 2002).

This problem can now be addressed by the development of quantitative methods such as that presented here.

The use of molecular techniques has dramatically improved the ability to identify human pathogens from clinical and environmental samples. In recent years, real-time RT-PCR based on different fluorescence formats has been developed for the detection of RNA viruses in various fields of virology (Mackay *et al.*, 2002). The SYBR®Green approach has provided a rapid and accurate analysis of many different targets.

We have described improvements to conventional and competitive RT-PCR amplification in the detection of NVs. Correct measurements can be achieved if the linear part of the sigmoidal curve is used for quantification. The combination of SYBR®Green chemistry with the iCycler™ avoids the limitations of conventional end point determination in quantitative PCR analysis since cycle-by-cycle monitoring of the amplification permits accurate identification of the onset of the linear amplification phase. This allows accurate determination of the template concentration from the amount of PCR product induced by reference to an external standard. The real-time RT-PCR described is highly sensitive; a linear relationship between input DNA template and C_T values over a range from approximately 10^8 copies down to 10^1 copies was detected. Detection of virus over such a wide dynamic range is very useful for measuring viral loads in samples with various levels of infection (Dhar *et al.*, 2001).

Owing to the sensitivity of this method, it is highly susceptible to PCR carry-over or other contamination. Therefore, laboratory hygiene practices should be followed very strictly to prevent any potential contamination that may give false positive results (Dhar *et al.*, 2002), e.g., strict local separation of isolation, preparation, and amplification. Furthermore, in respect of contamination, the one-tube format is preferable to nested reactions.

This method offers great advantage over conventional quantitative PCR since agarose gel electrophoresis or Southern blots are not required. Beyond this, the method provides a highly reproducible measurement. The coefficient of variation of threshold cycle values in independent inter-runs was less than 4.0 %. The real-time RT-PCR assay on the NV positive samples yielded reproducible data, with less than 2.0 % of the assay variation. However, a CV of threshold cycle less than 2.5 % can result approximately in a CV of copy number in the samples of 40.0 %. A reason for this is that a small change in the C_T value results in a large change in the input copy number value. Therefore, it is not possible to quantify the virus with absolute certainty. A disadvantage of this approach is the fact that internal standards cannot be used. Such standards could control for the efficiency of RNA isolation and RT steps as well as identifying any samples rendered falsely negative by amplification inhibitors (Cone *et al.*, 1992; Arnal *et al.*, 1999).

A significant advantage of this approach is its application to the measurement of concentration efficiency. Currently only recently developed semi-quantitative (and less sensitive) ELISA assays or model viral systems such as Feline calicivirus are available for this application. Our data demonstrate the utility of this approach for determining the concentration efficiency of NVs. Using the centrifuge concentrators we achieved recovery rates of virus particles up to 16 %. A maximum recovery of about 25 % was achieved from water by membrane filtration and RNA-isolation directly from the membrane. These results show that the development of isolation and concentration procedures might be a main point in method development, specifically in view of very low virus concentrations in food and water samples.

5. Conclusions

In summary, the SYBR®Green RT-PCR method described above is a rapid, reproducible and highly sensitive method for the detection of genogroup II NVs. Thus the method can be applied to screening large numbers of specimens and measuring viral loads for diagnostic and epidemiological studies. Furthermore, we have demonstrated that commonly used separation and concentration methods used for NVs are relatively inefficient. We are currently developing more efficient procedures for the concentration of NVs from food and water samples.

Acknowledgements

This work was supported under the Food LINK programme by the Department for Environment, Food and Rural Affairs (Defra). H.S. is a recipient of a Marie Curie Fellowship financed by the European Commission.

References

Arnal, C., Ferre-Aubineau, V., Mignotte, B., Imbert-Marcille, B.M. and Billaudel, S., 1999. Quantification of Hepatitis A Virus in shellfish by competitive reverse transcription- PCR with coextraction of standard RNA. Appl. Environ. Microbiol. 65, p. 322-326.

Atmar, R.L. and Estes, M.K., 2001. Diagnosis of noncultivatable gastro-enteritis viruses, the human caliciviruses. Clin. Microbiol. Rev. 14, p. 15-37.

Beuret, Ch., 2003. A simple method for isolation of enteric viruses (noroviruses and enteroviruses) in water. J. Virol. Methods 107, p. 1-8.

Bustin, S.A., 2002. Quantification of mRNA using real-time reverse transcription PCR (RT-PCR): trends and problems. J. Molec. Endocrin. 29, p. 23-39.

Cone, R.W., Hobson, A.C. and Huang, M.L., 1992. Coamplified positive control detects inhibition of polymerase chain reactions. J. Clinic. Microbiol. 30, p. 3185-3189.

Dedman, D., Laurichesse, H., Caul, E.O. and Wall, P.G., 1998. Surveillance of small round structured virus (SRSV) infection in England and Wales, 1990-5. Epidemiol. Infect. 121, p. 139-149.

Dhar, A.K., Roux, M.M. and Klimpel, K.R., 2001. Detection and quantification of infectious hypodermal and hematopoietic necrosis virus and white spot virus in shrimp using real-time quantitative PCR and SYBR Green chemistry. J. Clin. Microbiol. 39, p. 2835-2845.

Dhar A.K., Roux M.M. and Klimpel K.R., 2002. Quantitative assay for measuring the Taura syndrome virus and yellow head virus load in shrimp by real-time RT-PCR using SYBR Green chemistry. J. Virol. Methods 104, p. 69-82.

Dingle, K.E., Lambden, P.R., Caul, E.O. and Clarke, I.N., 1995. Human enteric Caliciviridae: the complete genome sequence and expression of virus-like particles from a genetic group II small round structured virus. Gen. Virol. 76, p. 2349-2355.

Dubois, E., Agier, C., Traoré, O., Hennechart, C., Merle, G., Crucière, C., Laveran, H., 2002. Modified Concentration Method for the Detection of Enteric Viruses on Fruits and Vegetables by Reverse Transcriptase-Polymerase Chain Reaction or Cell Culture. J. Food Protect. 65, p. 1962-1969.

Formiga-Cruz, M., Tofino-Quesada, G., Bofill-Mas, S., Lees, D.N., Henshilwood, K., Allard, A.K., Conden-Hansson, A.C., Hernroth, B.E., Vantarakis, A., Tsibouxi, A., Papapetropoulou, M., Furones, M.D., Girones, R., 2002. Distribution of human virus contamination in shellfish from different growing areas in Greece, Spain, Sweden, and the United Kingdom. Appl. Environ. Microbiol. 68, p. 5990-5998.

Fortin, N.Y., Mulchandani, A. and Chen, W., 2001. Use of Real-Time Polymerase Chain Reaction and Molecular Beacons for the Detection of *Escherichia coli* O157:H7. Anal. Biochem. 289, p. 281-288.

Gilgen, M., Germann, D., Lüthy, J. and Hübner, Ph., 1997. Three-step isolation method for sensitive detection of enterovirus, rotavirus, hepatitis A virus, and small round structured viruses in water samples. Int. J. Food Microbiol. 37, p. 189-199.

Hale, A., Mattick, K., Lewis, D., Estes, M., Jiang, X., Green, J., Eglin, R. and Brown, D., 2000. Distinct epidemiological patterns of Norwalk-like virus infection. J. Med. Virol. 62, p. 99-103.

Higuchi R, Fockler, C, Dollinger G. and Watson, R., 1993. Kinetic PCR analysis: Real-time monitoring of DNA amplification reactions. Biotechnology 11, p. 1026-1030.

Jiang, X., Wang, M., Wang, K. and Estes, M.K., 1993. Sequence and genomic organization of Norwalk virus. Virology 195, p. 51-61.

Koopmans, M. and Duizer, E., 2002, Foodborn viruses: an emerging problem. ILSI Europe Report Series (ISBN 1-57881-130-9), 10.

Lamothe, G.T., Putallaz, T., Joosten, H., Marugg, J.D., 2003. Reverse Transcription-PCR Analysis of Bottled and Natural Mineral Waters for the Presence of Noroviruses. App. Environ. Microbiol. 69, p. 6541-6549.

Liu, B.L., Clarke, I.N., Caul, E.O. and Lambden, P.R., 1995. Human enteric caliciviruses have a unique genome structure and are distinct from the Norwalk-like viruses. Arch. Virol. 1140, p. 1345-1356.

Lopman, B.A., Brown, D.W. and Koopmans, M., 2002. Human caliciviruses in Europe. J. Clin. Virol. 24, p. 137-60.

Mackay, I.M., Arden, K.E. and Nitsche, A., 2002. Survey and summary: Real-time PCR in virology. Nuc. Acid. Res. 30, p. 1292-1305.

Richards, G.P., 1987, Shellfish associated enteric virus illness in the United States, 1934-1984. Estuaries 10, p. 84-85.

Rippey, S.R., 1994. Infectious diseases associated with molluscan shellfish consumption. Clin. Microbiol. Rev. 7, p. 419-425.

Ririe, K.M., Rasmussen, R.P. and Wittwer, C.T., 1997. Product differentiation by analysis of DNA melting curves during the polymerase chain reaction. Anal. Biochem. 270, p. 154-160.

Schuffenecker, I., Ando, T., Thouvenot, D., Lina, B., Aymard, M., 2001. Genetic classification of "Sapporo-like viruses". Arch. Virol. 146, p. 2115-2132.

Willcocks, M.M., Silcock, J.G. and Carter, M.J., 1993. Detection of Norwalk virus in the UK by the polymerase chain reaction. FEMS Microbiol. Lett. 112, p. 7-12.

Novel molecular and biochemical techniques for quality control and monitoring in the agrofood production chain

C.D. Schoen, M. Szemes, P.J.M. Bonants, A. Speksnijder, M.M. Klerks, P.H.J.F. van den Boogert, C. Waalwijk, J.M. van der Wolf and C. Zijlstra
Plant Research International, P.O. Box 16, 6700 AA Wageningen, the Netherlands

Abstract

An overview is given of rapid methods for the detection of plant-related organisms in plant material, soil, compost, water, etc. Protein-based detection assays such as isozyme analysis, ELISA, immunofluorescence colony-staining and future applications of immunological tests are described as well as nucleic acid-based tests. Examples of RNA amplification tests, such as RT-PCR, NASBA and AmpliDet RNA are given, and numerous DNA-based tests using PCR, either with or without the use of probes, are illustrated. The latest development in the field of multiplex detection using microarrays is presented as the pUMA technology. Most tests described are directed towards the detection of plant pathogens such as viruses, bacteria, fungi and nematodes. A test for the detection of mRNA of a mycotoxin producing fungus is also shown in this review. Similar tests have been developed for the detection of human pathogens such as *Escherichia coli* O157:H7 and *Salmonella enterica* that are able to contaminate plants via animal manure. This illustrates that the technology of many of the tests described can equally well be used for the development of assays to detect harmful organisms in food, feed, water, air or any other environment in the agrofood production chain.

Keywords: DNA, PCR, ELISA, AmpliDet RNA, diagnostics, detection, pathogens, microarrays, bacteria, viruses, nematodes, fungi, mycotoxins, multiplex, NASBA, DGGE, pUMA, TaqMan, Molecular Beacon

1. Introduction

Quality control and monitoring in the agrofood chain is necessary to ensure safe agriculture, horticulture and agrofood processing. Since growers, producers, processing industry and consumers increasingly demand higher quality for starting material and agrofood products, there is a rising interest for rapid, sensitive, specific and user-friendly diagnostic methods. Nucleic acid diagnostic technologies can be of great help nowadays as they have undergone huge innovations over the last two decades. While Enzyme-Linked Immunosorbent Assays (ELISA) were state of the art in the seventies and mid-eighties for the detection of human, animal and plant pathogens, nowadays molecular methods are able to offer better sensitivity, specificity and/or speed. Due to the elucidation of the genomic composition of organisms, insights in gene expression, genetics and host-pathogen interactions, molecular knowledge has explosively expanded. Consequently, DNA and RNA amplification methods can enable an accurate diagnosis, leading to an easier approach of detection, characterisation and identification of pathogens, genes and disease suppressing microorganisms or antagonists. These advancements also increasingly fulfil the needs required for microbiological quality control and monitoring in the agrofood chain.

This review describes some techniques that have been developed for detection of plant pathogens and other plant-related organisms, and that can be used in agrosystems. During culture and production of various agroproducts, such as potatoes, corn, tomatoes, sugar beet, but also flowers and flower bulbs, many plant pathogens cause enormous damage. Methods to detect these plant pathogens enable to check endproduct quality and to monitor pathogens during the whole production chain. Many of these techniques can easily be applied in detecting other organisms such as plant-related human pathogens causing human health problems, for use in agrofood processing systems.

2. Plant-related organisms in agrosystems

2.1. Plant pathogens

Because agrosystems are becoming less dependent on chemical control, there is an increasing demand for fast, reliable methods capable of detecting plant-related organisms. Such methods can be used for:

- Quality monitoring of substrates. The compost and potting soil industry requires guarantees that their product is free of plant parasitic organisms as well as meeting the manufacturer's claims that it contains plant health-promoting organisms.
- Quality control of products. 'Tracing and tracking' of plant-related organisms in products throughout the chain is becoming more and more important to assure quality for the consumer (Klerks and Leone, 2002). This concerns the monitoring of crop pests or diseases that may cause problems during the post-harvest phase (e.g. formation of mycotoxins). This topic requires special attention due to changes currently taking place in crop systems (less generic control of crop pests and diseases).
- Compliance with export regulations. The export of plant products is controlled by strict regulations. The presence of quarantine organisms is absolutely unacceptable. Tests currently in use to show the presence of quarantine organisms are generally very time consuming, keeping shipments that are tested under quarantine for weeks. Rapid tests could solve this problem.
- Prevention. Preventing contamination is important from the viewpoint of using healthy starting material (seeds, bulbs, cuttings). Diseased starting material in a crop production system is often the source of infection. Tests for detecting crop disease organisms have been used for years to select healthy starting material. Another application is the detection of crop pests and diseases in substrates. If the pests and diseases present in the soil of a specific plot of land are known beforehand, an appropriate cultivation plan can be drawn up, and the best crops and varieties can be selected. Detection can also play an important role in greenhouse horticulture by testing recirculating irrigation water. Such testing can prevent an entire crop from becoming infected due to recirculated disease organisms.

Many detection tests that have been developed, can be used for the above aims. They can be used on soil, compost, water, plant material, inocula, etc. In our laboratory tests have been developed that are directed towards different kinds of plant pathogens:

- Viruses and viroids (e.g. viruses/viroids of vegetables, fruits, ornamentals, potato).
- Bacteria (brown rot (*Ralstonia*), ring rot (*Clavibacter*), root knot (*Agrobacterium*), soft rot (*Erwinia*), *Xanthomonas* spp.).
- Phytoplasms (*Stolbur*).

- Fungi (*Phytophthora* spp., *Fusarium oxysporum, Rhizoctonia solani, Synchytrium, Olpidium, Phoma, Alternaria, Nectria, Guignardia*).
- Nematodes (*Meloidogyne chitwoodi, M. fallax, M. hapla, M. minor, M. naasi, M. incognita, M. javanica, M. arenaria, Globodera rostochiensis, G. pallida*).
- Insects (e.g. *Thrips* species).

In addition, tests are developed for the detection of mRNAs of certain organisms. This enables the detection of metabolically active pathogens. This is particularly interesting when it concerns toxin-producing plant pathogens, e.g. mycotxin-producing *Fusarium* strains.

2.2. Plant-related human pathogens

Some attention has been focused on the detection of human pathogens in agrosystems as well. Human pathogens can spread throughout the food production chain in various ways. Contamination with several pathogens can take place via animal manure, for example *Salmonella* spp. mainly associated with chicken and pig manure, and toxigenic *Escherichia coli* O157 with cow manure (Nicholson *et al.*, 2000). Most attention has been paid to contaminated animal products because a large proportion of animal products can carry particular pathogens. For example, 36% of poultry samples were contaminated with *S. enteritidis* in a Belgian retail market (Uyttendaele *et al.*, 1999), and 16% of dairy cattle produced faeces containing *E. coli* O157 in a British study (Nicholson *et al.*, 2000). Therefore, most research and regulation has been focused on the animal production and marketing chain. Recently, several outbreaks of intestinal infections have been associated with pathogens ingested with vegetables and fruits. Contrary to animal products, which are generally cooked, various fruits and vegetables are consumed raw. The best known examples are outbreaks of the deadly *E. coli* O157:H7 stemming from contaminated, unpasteurised apple juice, and salmonellosis associated with diced tomatoes (Phillips, 1999; Weissinger *et al.*, 2000). The original source has been traced back to animal manure used as fertiliser in the production of these fruits and vegetables.

Contamination with human pathogens not only occurs on the outside of plants or plant products, but intestinal human pathogens have also been found inside plant tissues (Ackers and Mahon, 1998; Guo *et al.*, 2001, 2002a, 2002b). When these plants or plant products are cut, the bacteria can spread throughout the food processing and packing operations, resulting in potentially high contamination levels in, for example, shredded lettuce, diced tomatoes, cut cabbage, etc. Thus, the risk of disease outbreaks may be increased by the expansion of ready-to-eat foods and fast-food chains. In our laboratory tests have been developed that are directed towards the plant related human pathogens *E. coli* O157:H7 and *Salmonella enterica*.

3. Method requirements for detection and identification

In the agrosector the need for reliable identification and sensitive detection of plant pathogens, beneficial microorganisms and, more recently, plant-related human pathogens is increasing. Important developments are nowadays:

- Quantitative detection to quantify inoculum densities of pathogenic variants to estimate risks for economic damage.
- Multiplex detection to detect multiple plant pathogens (targets) in the same sample in a single test.

Different aspects have to be considered and certain criteria have to be fulfilled for developing effective detection methods:

- Specificity. The test should be very specific and should only detect the target pathogen and not closely related species; no false positives or false negatives should be observed.
- Sensitivity. The test should be sensitive in detecting, for example, 1 zoospore; this goes especially for quarantine-organisms requiring a nil-tolerance.
- Costs. For routine testing the test should be cost effective.
- Expertise. The presence of sufficient technical and personal expertise to perform the test should be considered.
- Robustness. The test should be robust, including a high repeatability and reproducibility.
- High throughput. Sometimes the test should be suitable for high throughput screening.
- Speed. The results obtained should be available within a short time.

4. Methods for detection and identification

Previously, identification took place on the basis of physiological (Figure 1), biological and morphological (Figure 2) characteristics requiring specialised expertise. The procedure was usually very time-consuming and the results were often not unequivocal.

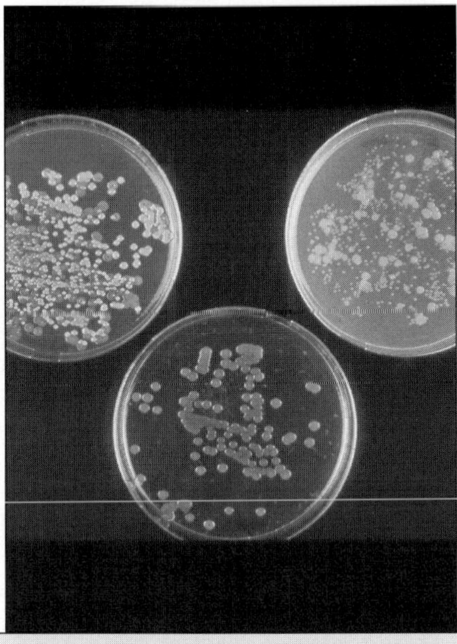

Figure 1. Growth of bacteria on selective media (Bonants *et al.*, 2000).

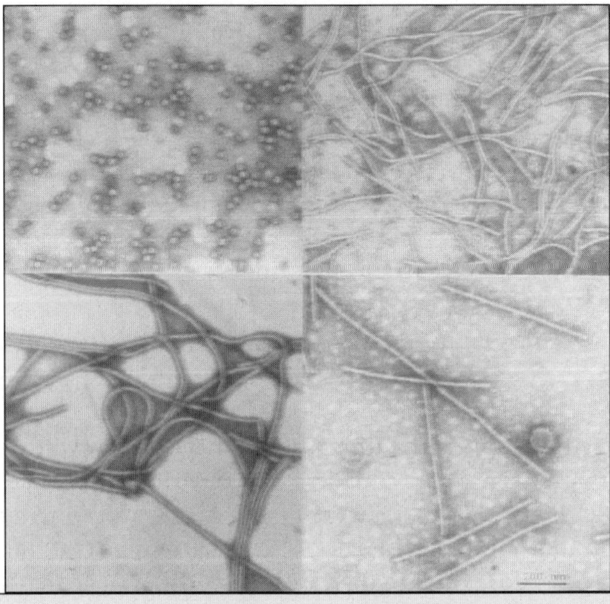

Figure 2. TEM (Transmission Electron Microscopy) picture of viruses.

4.1. Protein-based methods

Protein-based methods are much more accessible and can be carried out routinely. Isozyme analysis is a relatively fast way to identify species of the nematode genus *Meloidogyne* (Esbenshade and Triantaphyllou, 1985, 1990). The comparison of esterase and malate dehydrogenase patterns shows great consistency in the separation of the species. However, for clear and reliable results the isozyme analysis can only be done with females at a specific developmental stage.

The introduction of immunological techniques in the 1960s was a breakthrough. At present, the business community in the Netherlands and abroad conducts approx. 10 million ELISA tests each year to detect the presence of bacteria and viruses in agroproducts such as potatoes, ornamentals and flower bulbs.

For the detection of certain bacteria Immunofluorescence Colony-Staining (IFC) can be used (Figure 3). After incubation of the bacteria on agar fluorescently labelled antibodies are added. After washing unbound antibodies fluorescent colonies can be seen under a UV microscope and isolated for further studies, e.g. for confirmation by PCR (Van der Wolf *et al.*, 1998, 2000; Van der Wolf and Van Beckhoven, 2004). In the Netherlands most of the required antisera are supplied by Plant Research International. They still comprise largely polyclonal antibodies. The disadvantage of such antibodies is that they are unsuitable for the detection of economically important fungi, which have many dominant cell wall antigens in common resulting in aspecific reactions. During the 1990s a new method was developed to produce antibodies without immunisation: phage display (Griffiths *et al.*, 1994; Griffiths and Duncan, 1998). Antibodies (and recombinant antibodies) binding only to the correct property of the crop disease, are isolated from large gene banks by means of *in vitro* selection. There are several advantages of this method: the antibodies are monoclonal, so they can theoretically be produced in unlimited quantities, and they are of consistent quality. Furthermore, no animals are involved in the production. Specificity of these antibodies may allow us to solve cross-reaction problems or serotype diversity within targets. Phage display will probably allow us to meet the ever-increasing quality demands that clients are placing on diagnostic tools. Using this method monoclonal antibodies have already been selected for a number of viruses (Griep *et al.*, 1999, 2000; Toth *et al.*, 1999) and brown rot bacteria (Griep *et al.*, 1998).

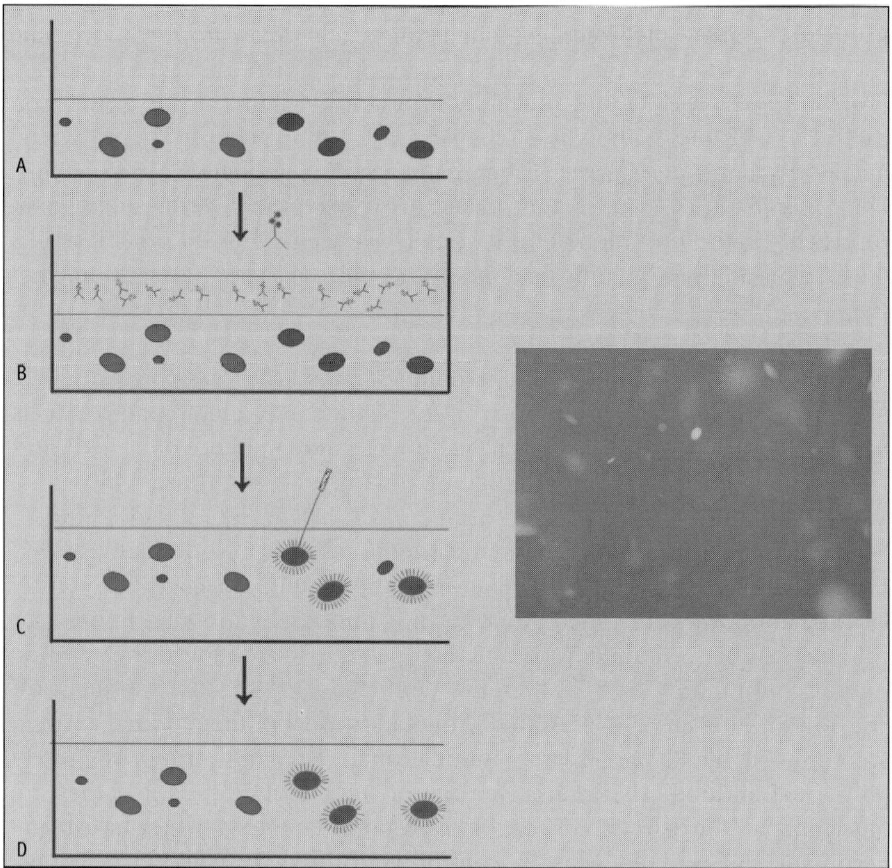

Figure 3. Immunofluorescence colony-staining (IFC) of bacteria. A mixture of bacteria or a complex substrate containing bacteria is grown in a solid medium (A) until colony forming appears, and is then incubated with a FITC-conjugated specific antibody (B). The antibody binds to the target bacteria (C), which results in a bright green fluorescent staining of the colonies containing target bacteria (photograph). The fluorescent colonies can be picked from the medium (D) and can be used for reisolation, detection with molecular detection methods or further characterisation.

4.1.1. Multiplex detection

Although protein platforms are available for affinity research and proteomics, the application of protein-based microarrays for detection of multiple pathogens in a single test has not been exploited extensively. Apart from conventional spotting of glass slides (Suda *et al.*, 2004) rapid and miniaturised methods are developed with microfluidic techniques. The use of flow-through systems (Zhu *et al.*, 2004a) and bead-based conjugation for whole cell detection (Zhu *et al.*, 2004b) are under development. The electric silicon chip principle developed for mRNA analysis in the Bioprocess Chips project (European Commission project QLK3-CT1999-00533) could also be applied for protein analysis (personal communication).

4.2. Nucleic acid-based methods

Since the introduction of amplification methods for DNA and RNA, and the increasing availability of sequence data many methods have been developed for use in detection and identification:

- RNA level: RT-PCR, NASBA or AmpliDet RNA, using primers (and probes, e.g. Molecular Beacons, TaqMan probes)
- DNA level: PCR (e.g. ITS-RFLP, SCAR-PCR, AFLP, multiplex PCR, (nested) PCR, RAPD, DGGE) using primers (and probes, e.g. Molecular Beacons, TaqMan probes)

The amplicons can be detected by traditional gel-electrophoresis or by real-time detection with fluorescent probes. An example of such a probe is the TaqMan probe (Oberst *et al.*, 1998; Schoen *et al.*, 1996; Kalinina *et al.*, 1997; Brandt *et al.*, 1998; Böhm *et al.*, 1999; Zhang *et al.*, 1999; Bonants *et al.*, 2004). This probe consists of two groups: a fluorescent reporter and a quencher. The fluorescence is quenched due to proximity of the two groups. During the annealing step of the polymerase chain reaction the probe is bound on the target and during the elongation step the probe is degraded by the Taq-polymerase resulting in release of the reporter from the probe into solution and increase of fluorescence. The fluorescence can be measured real time during the polymerase chain reaction in a real-time PCR machine. Amplification plots with different amount of template show differences in the number of cycles before an increase in fluorescence above background, the so-called Ct-value. In this way quantification of target DNA is possible. An example of multiplex analysis with TaqMan probes for two potato viruses is shown (Figure 4). PVY and PLRV can be detected simultaneously by using two TaqMan probes labelled with two different fluorescent groups in one amplification reaction.

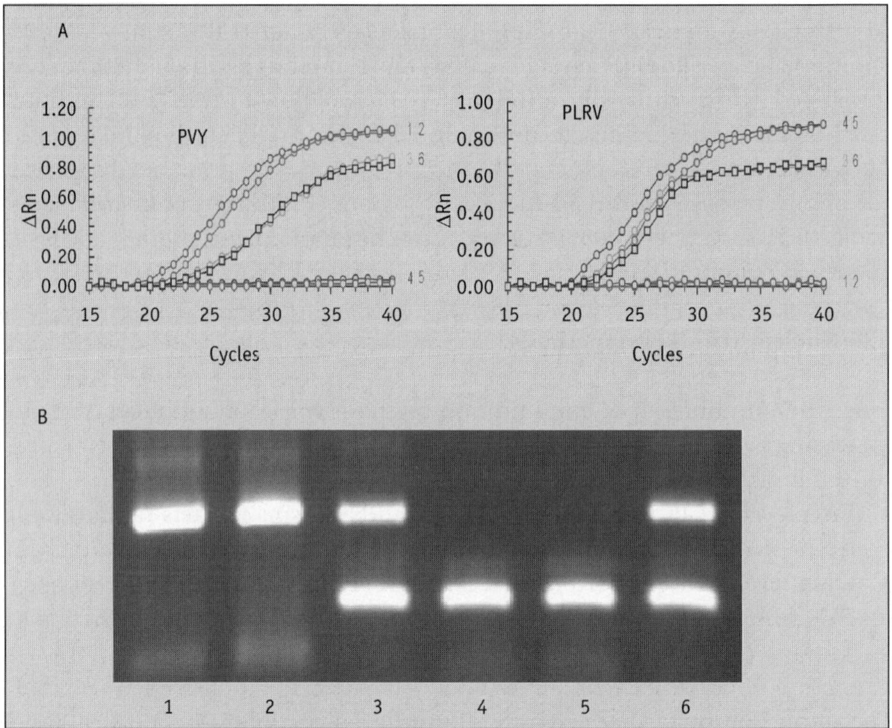

Figure 4. A: Real-time multiplex RT-PCR with TaqMan probes for the detection of potato virus Y (PVY) and potato leaf roll virus (PLRV). Six samples were tested consisting of PVY (1 and 2), PVY and PLRV (3 and 6) or PLRV (4 and 5). Fluorescence is measured using PVY-specific TaqMan probe (left) or PLRV-specific TaqMan probe (right). B: Gel electrophoretic analysis of the amplification products; lanes 1-6 correspond to samples 1-6.

The exquisite sensitivity of this technique is putting a high demand on measures to prevent false positive reactions due to contamination of the laboratory with nucleic acid. Therefore, monitoring of false-positives by the inclusion of negative controls is required. In addition, the reaction performance must be measured. The reliability of the test is increased by inclusion of a general internal amplification control in the reaction mix. When no signals are measured for amplification of the internal control, the test conditions did not allow polymerase chain reaction during the assay and the test results cannot be trusted. This prevents the scoring of false negative results.

To monitor the biosafety of the agrofood production chain, tests are being developed for the detection of *E. coli* O157:H7 and *Salmonella enterica* (Klerks *et al.*, 2004b). In this study, the applicability of the *Salmonella enterica* and *E. coli* O157:H7 TaqMan PCR detection methods was determined with respect to sensitivity, precision and accuracy. Furthermore, a general Internal Amplification Control (IAC) was developed to allow elimination of false negative results. An accurate quantification of initial target DNA was enabled using real-time methods including IAC. Using this TaqMan approach the detection of *Salmonella enterica* in different substrates like soil, manure and compost was found very efficient in combination with a defined optimal DNA extraction method with a priorily added internal procedural control. As few as 1000 CFUs were detectable in these substrates without pre-enrichment. After enrichment even 1 CFU was clearly detected using the described technique. The method provided an efficient and reliable way for evaluating and validating DNA extraction and amplification of each individual sample in one tube, independent of the substrate tested. Moreover, it is applicable for high throughput analysis and routine diagnosis, and it allows a precise quantification of *S. enterica* present in soil, manure or compost (Klerks *et al.*, 2004c).

NASBA (Nucleic Acid Sequence Based Amplification) is another amplification reaction (Compton, 1991). NASBA is an isothermal amplification method suitable for the detection of RNA, which uses three enzymes (reverse transcriptase, RNase H and T7- polymerase) and results in millions of copies of antisense RNA. Within 90 minutes at 41 °C a billion amplification of a specific RNA sequence can be obtained. Because NASBA detects RNA, viability test can be performed (Bentsink *et al.*, 2002). An example with NASBA on *Ralstonia solanacearum* is shown in Figure 5. Bacteria are incubated on metal strips for various hours and the results with NASBA, PCR and plating on medium are shown in Figure 5. In lane 5, 6 and 7 non-viable bacteria can be detected by PCR, but not with NASBA demonstrating that after 24 hours bacteria are dead. To allow real-time amplification and detection, NASBA was combined with molecular beacon probes, called AmpliDet RNA (Leone *et al.*, 1998). A Molecular Beacon (MB) is another fluorescent probe with a reporter and a quencher, which can be used to detect the amplicon (Tyagi and Kramer, 1996). MBs are oligonucleotides with a loop and a stem structure. The loop sequence is complementary to the target sequence. The fluorescent reporter and the quencher are located at the end of the stem sequence. Without target the MB is closed and no fluorescence is recorded. When target is present, the MB is bound to the target and stretched. The reporter and quencher are now distant

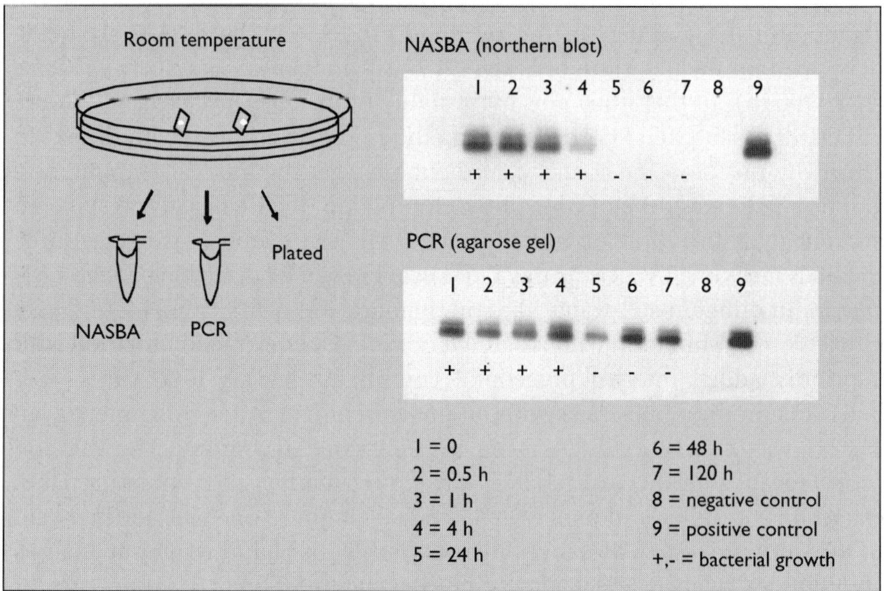

Figure 5: Viability test with NASBA for *Ralstonia solanacearum*. Metal strips were inoculated with *Ralstonia solanacearum* and airdried for 0 to 120 hours. The bacteria were then washed from the strips, plated on a solid medium and tested in NASBA and PCR. The plating results indicate the presence of live, viable cells. NASBA detects 16s RNA and acts and an indicator for viability. PCR detects the presence of genomic target DNA of live and dead cells. Lanes 1-7 indicate the different incubation times of *R. solanacearum* on metal strips; lanes 8 and 9 are controls; + and - indicate bacterial growth on the medium.

from each other and fluorescence will be recorded. Several AmpliDet RNA methods have been developed for detection of ribosomal RNA (van Beckhoven *et al.*, 2002; van der Wolf *et al.*, 2004), messenger RNA (Klerks *et al.*, 2000) and viral RNA (Klerks *et al.*, 2001a, 2001b, 2004a; Vaskova *et al.*, 2004; Szemes *et al.*, 2002). Molecular Beacons have also been used in PCR to detect DNA (Bonants *et al.*, 2004).

The availability of an AmpliDet RNA method for the detection of mRNA of a toxin producing *Fusarium* sp. (Klerks *et al.*, 2000) is of particular interest, since this plant pathogen is recognised as a food safety hazard. Especially on wheat this fungus causes much concern (Waalwijk *et al.*, 1997, 2004). Figure 6 shows

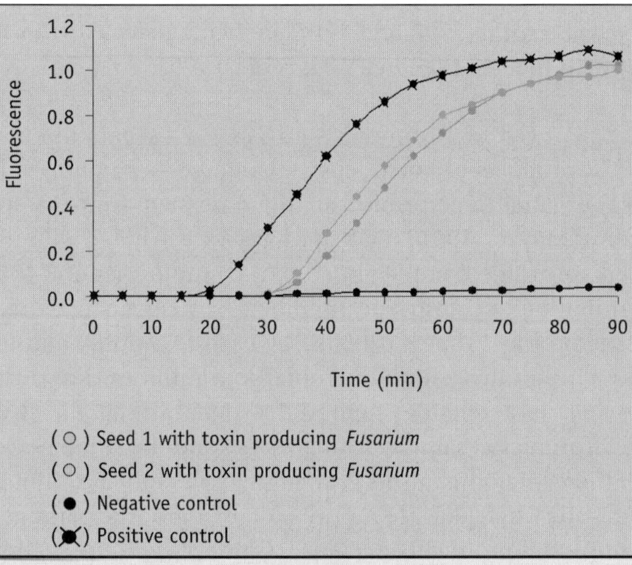

Figure 6. AmpliDet RNA method for the detection of mRNA of a toxin producing *Fusarium* sp.

different seed lots infected with *Fusarium* sp. using the AmpliDet RNA method. An AmpliDet RNA method has also been developed for the detection of two potato viruses (PVY/PLRV) (Klerks *et al.*, 2001a). Both viruses occur as single or mixed infections in potato tubers and are responsible for major economic losses worldwide. Moreover, detection of PLRV and PVY directly on tubers with the currently used enzyme-linked immunosorbent assays (ELISA) is not reliable (Hill and Jackson, 1984; Spiegel and Martin, 1993). The AmpliDet RNA developed enabled the detection of PLRV and PVY directly from tubers in one reaction, already within 2 hours from sampling. Healthy tubers remained negative. The multiplex AmpliDet RNA detected and discriminated the viruses also in mix-infected tubers (results not shown). An AmpliDet RNA method was also developed for several PVY strains with different sequences (Szemes *et al.*, 2002). Different MBs were used to detect the different strains simultaneously in the NASBA reaction (data not shown). The difference between a linear and a MB probe has been studied. Two probes have been developed for *Ralstonia solanacearum*: a linear and a MB with the same sequence. Cross-reaction can be seen with the linear probe, but not with the MB. The closely related bacterial species *Pseudomonas cepatia* 1702 and 1708 are being detected with the linear probe, but not with the MB (Van der Wolf *et al.*, 1999).

Denaturant Gradient Gel Electrophoresis (DGGE) is a separation method for a mixture of PCR products of equal length. Fragment separation is based on melting behaviour and fingerprints can be generated from samples with genetic variability (Muyzer, 1999). PCR products are generated with rRNA gene directed primers to profile complex microbial communities and to infer the phylogenetic affiliation of community members by sequencing or probing the bands within the profile. Primers can also be used to amplify and to analyse variability of specific groups within microbial communities (Kowalchuk *et al.*, 1997). PCR-DGGE is a reliable method for determining the diversity of microbes in environmental samples and goes beyond the small percentage of culturability of these microorganisms Evaluation of microbial communities has been applied to soils where disease suppression of soilborne pathogenic fungi is caused by antibiotics producing bacteria (Garbeva *et al.*, 2004). In Figure 7 an example of a DGGE profile is shown.

AFLP (Amplified Fragment Length Polymorphism) is a powerful technique (Vos *et al.*, 1995) for identification and has often been used to study genetic variation in fungi (Mayer *et al.*, 1997; Bonants *et al.*, 1999, 2000; Baayen *et al.*, 2000, Eikemo *et al.*, 2004; Ivors *et al.*, 2004). *Rhizoctonia solani* is a fungal pathogen on sugar beet. Isolates from different fields were collected and analysed by AFLP DNA fingerprinting (Figure 8) to study genetic variation. Figure 8 shows that

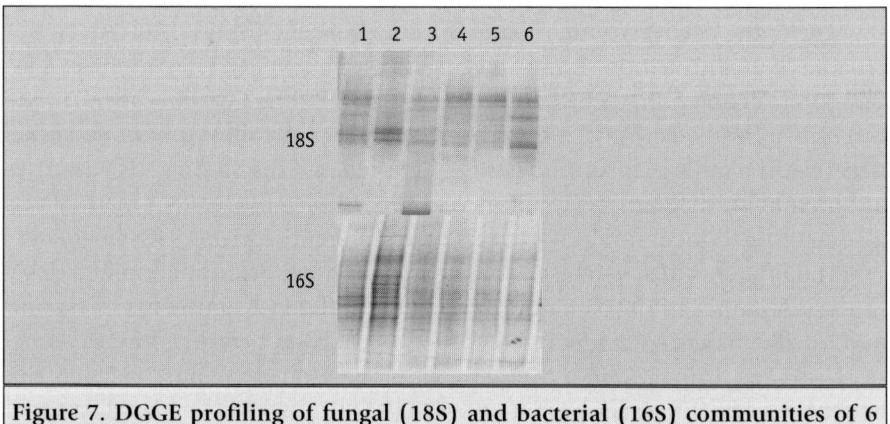

Figure 7. DGGE profiling of fungal (18S) and bacterial (16S) communities of 6 soil samples.

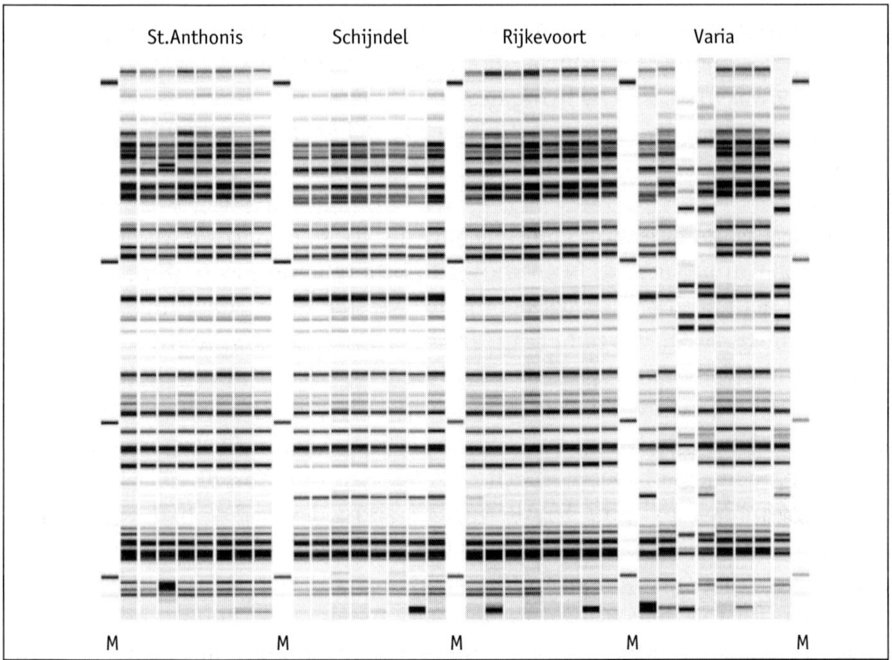

Figure 8. AFLP DNA fingerprinting patterns of *Rhizoctonia solani* isolates from different sugar beet fields.

the genetic variation for isolates from one field is zero, while there is some variation between fields.

For the detection of fungi the ITS (Internal Transcribed Spacer) region of rDNA has often been used. Universal primers (ITS1/ITS4; White *et al.*, 1990) and specific primers were developed to detect a wide range of fungi (Ristaino *et al.*, 1998; Schubert *et al.*, 1999; Stark *et al.*, 1998, Bonants *et al.*, 2003). One example is the pathogen *Phytophthora fragariae* a quarantine-organism for which a sensitive and specific detection method was needed. This Oomycete is pathogenic to strawberry and until now a bait test has been used to detect the fungus in infected material (Duncan, 1980). Sampled strawberry roots are incubated in soil or water with a bait plant for several weeks at 14 °C, after which typical oospores are visible in infected roots of the bait plant. Based on sequence analysis of the ITS regions (Cooke and Duncan, 1997; Cooke *et al.*, 2000) specific primers and probes were developed together with SCRI (Dundee, Scotland) and used for the detection of *P. fragariae* and *P. cactorum* in roots, water and soil (Lacourt *et al.*, 1997; Bonants *et al.*, 1997; 2001; Duncan *et al.*, 2001; Cooke *et al.*, 2001). Sensitivity was increased by using nested PCR (Lacourt *et al.*, 1997; Bonants *et al.*, 1997; Niepold and Schober-Butin, 1997; Niessen *et al.*, 1997; Stark *et al.*, 1998).

Identification and detection methods have also been developed for nematodes. An ITS-RFLP approach has been shown to be a reliable way to distinguish the root-knot nematodes *Meloidogyne chitwoodi, M. fallax, M. hapla* and the species closely related to *M. incognita* (Zijlstra *et al.*, 1995, 1997). Multiplex ITS-PCR is faster and offers a good alternative to estimate species composition in mixtures (Zijlstra, 1997). More recently, species-specific pairs of PCR-primers were developed that amplify Sequence Characterised Amplified Regions (SCARs). Based on sequences of species-specific RAPD (Random Amplified Polymorphic DNA) fragments, pairs of SCAR primers were developed that specifically amplify *M. chitwoodi, M. fallax, M. hapla* (Zijlstra, 2000), *M. incognita, M. javanica* and *M. arenaria* (Zijlstra *et al.*, 2000). Other specific PCR test assays have been developed for the detection of *M. minor* (Karssen *et al.*, 2004) and M. naasi (Zijlstra *et al.*, 2004). More recently, some progress has been made in developing a multiplex TaqMan for the detection of *M. chitwoodi* and/or *M. fallax*.

4.2.1. Multiplex amplification

The newest development in analysis of nucleic acids is the microarray technology, in which different oligonucleotides can be spotted on little more than one square mm. Microarray technology provides the next generation of DNA diagnostics to measure different pathogens on a single chip. For each pathogen many different target DNA sequences can be detected in parallel. This will improve specificity, allowing detection of pathogenic variants of the target pathogen and avoiding laborious confirmation procedures.

Several microarray systems are available. PamGene has developed a revolutionary three-dimensional microarray with a higher capacity to bind oligonucleotides than a two-dimensional glass array, which results in a higher sensitivity. The rate-limiting step in a two-dimensional array planar format is the diffusion of the sample molecules (the target) toward the attached probe during hybridisation. Because diffusion is slow, incubation usually takes place overnight. To bypass this limitation, PamGene has developed a porous aluminium oxide substrate as solid support (Van Beuningen *et al.*, 2001). The substrate, with a thickness of 60 µm, has long branched capillaries, which are interconnected inside (Rigby *et al.*, 1990). The diameter of the individual pores is 200 nm. The reactive surface of this material is increased 500-fold as compared with a flat two-dimensional surface. In addition, the permeable nature of the microarray facilitates the pressurised movement of fluid, such as the sample solution, through its structure. The flow-through microarray substantially reduces hybridisation times and increases signal and signal-to-noise ratios (Schoen *et al.*, 2004). An additional advantage of the system is that the temperature can be regulated and controlled during the hybridisation process.

Amplification of the target DNA of the pathogen to be detected is a prerequisite for sensitive detection. In a simplex PCR assay where one set of PCR primers is used, small samples of DNA can produce sufficient copies to be visualised on microarrays. To amplify multiple targets in one PCR assay, multiple primer sets can be used. However, the drawback of such multiplex PCR assays is the enormous decrease in sensitivity and the limitation in the number of different targets to be amplified in one assay. Moreover, the dynamic range of the targets present in the sample to be tested, is not always reflected in the outcome of the test. Most of the time targets present in very low amounts will not be amplified in contrast to those abundantly present. To overcome some of these problems the pUMA technique was developed at our institute. pUMA stands for "Padlock-based Universal Multiplex detection Array". A schematic representation of the technique is given in Figure 9. DNA extracted from a sample under investigation is mixed with specific padlock probes that bind to the targets to be detected. These specific padlock probes are subsequently ligated, amplified and detected on a microarray. Amplification occurs using one generic PCR primer set. Results with the prototype pUMA enabling simultaneous detection of eight pathogens, look very promising. Combination of microarrays with 96-well plates will enable the analysis of 96 samples at the same time. Application of the pUMA technique is universal and opens possibilities to monitor health hazard problems: in a single test numerous harmful organisms can be detected in a sensitive, reliable and fast manner.

5. Conclusions

In the agrosector methods for detection of plant pathogens have been used for many years. Techniques used for these purposes have rapidly developed during the past years from techniques based on morphological characteristics to

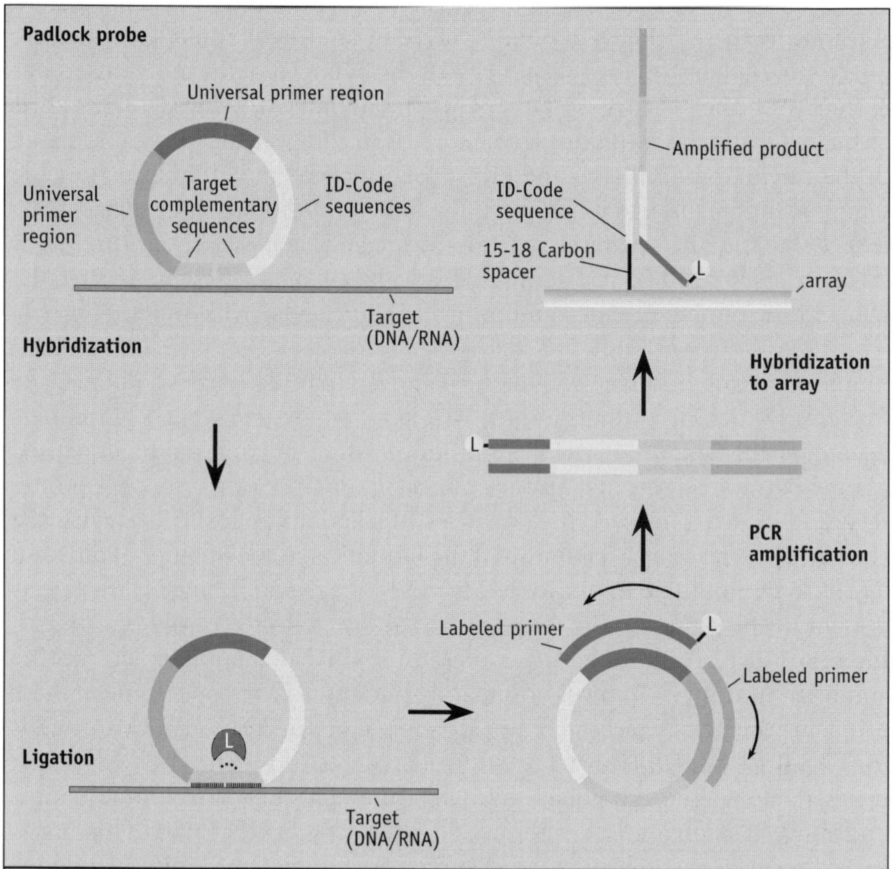

Figure 9. Schematic view for padlock probe hybridisation, ligation and PCR amplification of circularised probes followed by hybridisation to an array system. Padlock probe consists of T1, T2, ID-Code, F and R

T1. Target complementary sequence

T2: Target complementary sequence

ID-Code: unique identifier sequence

F: universal forward primer sequence

R: universal reverse primer sequence

innovative biochemical and molecular tests. Some of the latter techniques are described in this review and are implemented in the plant diagnostic field. Much experience has been gained with extraction of pathogens and/or nucleic acids from complex substrates, complex nucleic acid amplification strategies, application of multiplex detection using microarrays or other platforms, production of recombinant antibodies, etc. The expertise gained by researchers in the agrosector can also be of great value for the development of tests for quality monitoring in agrofood processing systems. Food safety is a relevant issue in society and likewise plant health in general is an important link in the food production chain. An integrated approach to the development of diagnostic tests would be favourable both for the agrosector and the agrofood processing sector.

Acknowledgements

The authors wish to thank José van Beckhoven, Dorine Donkers-Venne, Maudie Egberts, Marga van Gent-Pelzer, Richard van Hoof, Anita Luttikholt, Marjon Krijger, Greet de Raaij, Dianne van der Wal and Marjanne de Weerdt for their contribution to the work presented.

References

Ackers, M.L. and Mahon, B.E., 1998. An outbreak of *Escherichia coli* O157:H7 infections associated with leaf lettuce consumption. Journal of Infectious Diseases 177, p. 1588-1593.

Baayen, R.P., O'Donnell, K., Bonants, P.J.M., Cigelnik, E., Kroon, L.P.N.M., Roebroeck, E.J.A. and Waalwijk, C., 2000. Gene genealogies and AFLP analyses in the *Fusarium oxysporum* complex identifiy monophyletic and nonmonophyletic formae speciales causing wilt and rot disease. Phytopathology 90, p. 891-900.

Bentsink, L., Leone, G., van Beckhoven, J.R.C.M, van Schijndel, H.B., van Gemen, B. and van der Wolf, J.M., 2002. Amplification of RNA by NASBA allows direct detection of viable cells of *Ralstonia solanacearum* in potato. J. Applied Microbiol. 93, p. 647-655.

Böhm, J., Hahn, A., Schubert, R., Bahnweg, G., Adler, N., Nechwatal, J., Oehlmann, R. and Oßwald, W., 1999. Real-time quantitative PCR: DNA determination in isolated spores of the mycorrhizal fungus *Glomus mossae* and monitoring of *Phytophthora infestans* and *Phytophthora citricola* in their respective host plants. J. Phytopathol. 147. p. 409-416.

Bonants, P., Hagenaar-de Weerdt, M., van Gent-Pelzer, M., Lacourt, I., Cooke, D. and Duncan, J., 1997. Detection and identification of *Phytophthora fragariae* Hickman by the polymerase chain reaction. European J. Plant Pathol. 103. p. 345-355.

Bonants, P., Hagenaar-de Weerdt, M., Kema, G., van den Boogert, P., Waalwijk, C. and Baayen, R., 1999. AFLP DNA Fingerprinting of plant pathogenic fungi at IPO-DLO. Proceedings of the COST Action 823, 9-10 July 1998, Faro, Portugal (M. Barba, ed.), Petria 9, p. 97-104.

Bonants, P.J.M., Hagenaar-de Weerdt, M., Man in 't Veld, W.A. and Baayen, R.P., 2000. Molecular characterization of natural hybrids of *Phytophthora nicotianae* and *P. cactorum*. Phytopathology 90, p. 867-874.

Bonants, P.J.M., van Gent-Pelzer, M.P.E. and Hagenaar-de Weerdt, M., 2001. Characterization and detection of *Phytophthora fragariae* in plant and soil by molecular methods. EPPO Conference on Diagnostic Techniques for Plant Pests. Wageningen, 1-4 February 2000. EPPO Bulletin vol. 30 No 3-4, p. 525-531.

Bonants, P.J.M., Carroll, G.C., de Weerdt, M., van Brouwershaven, I.R. and Baayen, R.P., 2003. Development and validation of a fast PCR-based detection method for pathogenic isolates of the citrus black spot fungus, *Guignardia citricarpa*. European Journal of Plant Pathology 109, p. 503-513.

Bonants, P.J.M., van Gent-Pelzer, M.P.E., Hooftman, R., Cooke, D.E.L., Guy, D.C. and Duncan, J.M., 2004. A combination of baiting and different PCR formats, including measurement of real-time quantitative fluorescence, for the detection of *Phytophthora fragariae* in strawberry plants. European Journal of Plant Pathology (accepted).

Brandt, M.E., Padhye, A.A., Mayer, L.W. and Holloway, B.P., 1998. Utility of random amplified polymorphic DNA PCR and TaqMan automated detection in molecular identification of *Aspergillus fumigatus*. J. of Clin. Microbiol. 36. p. 2057-2062.

Compton, J., 1991. Nucleic acid sequence-based amplification. Nature 350, p. 91-92.

Cooke, D.E.L. and Duncan, J.M., 1997. Phylogenetic analysis of *Phytophthora* species based on ITS1 and ITS2 sequences of the ribosomal RNA gene repeat. Mycol. Res. 101. p. 667-677.

Cooke, D.E.L., Drenth, A., Duncan, J.M., Wagels, G. and Brasier, C.M., 2000. A molecular phylogeny of *Phytophthora* and related Oomycetes. Fungal Genetics and Biology 30. p. 17-32.

Cooke, D.E.L., Duncan, J.M., Williams, N.A., Hagenaar-de Weerdt, M. and Bonants, P.J.M., 2001. PCR-based Identification of *Phytophthora* species. EPPO Conference on Diagnostic Techniques for Plant Pests. Wageningen, 1-4 February 2000. EPPO Bulletin vol. 30 No 3-4, p. 519-523.

Duncan, J.M., 1980. A technique for detecting red stele (*Phytophthora fragariae*) infection of strawberry stocks before planting. Plant Disease 64, p. 1023-1025.

Duncan, J.M., Cooke, D.E.L., Seemüller, E., Bonants, P.J.M. and Olsson, C., 2001. European collaboration on eliminating *Phytophthora fragariae* from strawberry propagation. EPPO Conference on Diagnostic Techniques for Plant Pests. Wageningen, 1-4 February 2000. EPPO Bulletin vol. 30 No 3-4, p. 513-517.

Eikemo, H., Klemsdal, S.S., Riisberg, I., Bonants, P.J.M., Stensvand, A. and Tronsmo, A.M., 2004. Genetic variation between *Phytophthora cactorum* isolates differing in their ability to cause crown rot in strawberry. Mycological Research 108, p. 317-324.

Esbenshade, P.R. and Triantaphyllou, A.C., 1985. Use of enzyme phenotypes for identification of *Meloidogyne* species (Nematoda: *Tylenchida*). Journal of Nematology 17, p. 6-20.

Esbenshade, P.R. and Triantaphyllou, A.C., 1990. Isozyme phenotypes for the identification of *Meloidogyne* species. Journal of Nematology 22, p. 10-15.

Garbeva, P., van Veen, J.A. and van Elsas, J.D., 2004. Assessment of the diversity, and antagonism towards *Rhizoctonia solani* AG3, of *Pseudomonas* species in soil from different agricultural regimes. FEMS Microbiology Ecology 47, p. 51-64.

Griep, R.A., Prins, M., van Twisk, C., Keller, H., Kerschbaumer, R.J., Kormelink, R., Goldbach, R.W. and Schots, A., 2000. Application of phage display in selecting tomato spotted wilt virus-specific single-chain antibodies (scFvs) for sensitive diagnosis in ELISA. Phytopathology 90, p. 183-190.

Griep, R.A., van Twisk, C. and Schots, A., 1999. Selection of beet necrotic yellow vein virus specific single-chain Fv antibodies from a semi-synthetic combinatorial antibody library. European Journal of Plant Pathology 105, p. 147-156.

Griep, R.A., van Twisk, C., van Beckhoven, J., van der Wolf, J.M. and Schots, A., 1998. Development of specific recombinant monoclonal antibodies against the lipopolysaccharide of *Ralstonia solanacearum* race 3. Phytopathology 88, p. 795-803.

Griffiths, A.D. and Duncan, A.R., 1998. Strategies for selection of antibodies by phage display. Current Opinion in Biotechnology 9, p. 102-108.

Griffiths, A.D., Williams, S.C., Hartley, O., Tomlinson, I.M., Waterhouse, P., Crosby, W.L., Kontermann, R.E., Jones, P.T., Low, N.M., Allison, T.J., Prospero, T.D., Hoogenboom, H.R., Nissim, A., Cox, J.P.L., Harrison, J.L., Zaccolo, M., Gherardi , E. and Winter, G., 1994. Isolation of high-affinity human-antibodies directly from large synthetic repertoires. EMBO Journal 13, p. 3245-3260.

Guo, X., Chen, J., Brackett, R.E. and Beuchat, L.R., 2001. Survival of salmonellae on and in tomato plants from the time of inoculation at flowering and early stages of fruit development through fruit ripening. Appl. Env. Microbiol. 67, p. 4660-4764.

Guo, X., Chen, J., Brackett, R.E. and Beuchat, L.R., 2002a. Survival of *Salmonella* on tomatoes stored at high relative humidity, in soil, and on tomatoes in contact with soil. J Food Prot. 65: 274-279.

Guo, X., van Iersel, M.W., Chen, J., Brackett, R.E. and Beuchat, L.R., 2002b. Evidence of association of salmonellae with tomato plants grown hydroponically in inoculated nutrient solution. Appl. Env. Microbiol. 68, p. 3639-3643.

Hill, S.A. and Jackson, E.A., 1984. An investigation on the reliability of ELISA as a practical test for detecting potato leafroll virus and potato virus Y in tubers. Plant Pathol. 33, p. 21-26.

Ivors, K., Hayden, K., Bonants, P.J.M., Rizzo, D.M. and Garbelotto, M., 2004. AFLP and phylogenetic analyses of North American and European populations of *Phytophthora ramorum*. Mycol. Res. 108, p. 378-392.

Kalinina, O., Lebedeva, I., Brown, J. and Silver, J., 1997. Nanoliter scale PCR with TaqMan detection. Nucleic Acids Res. 25, p. 1999-2004.

Karssen, G, Bolk, R.J., van Aelst, A.C., van den Beld, I., Kox, L.F.F., Korthals, G., Molendijk, L., Zijlstra, C., van Hoof, R. and Cook, R., 2004. Description of *Meloidogyne minor* n. sp. (Nematoda: *Meloidogynidae*), a root-knot nematode associated with yellow patch disease in golf courses. Nematology (in press).

Klerks, M., Schoen, C. and Waalwijk, C., 2000. The detection of trichodiene synthase mRNA in *Fusarium* on seeds using AmpliDet RNA. Mitteilungen aus der Biologischen Bundesanstalt fur Land- und Forstwirtschaft 377, p. 49.

Klerks, M.M., Leone, G.O.M., Verbeek, M., van den Heuvel, J.F.J.M. and Schoen, C.D., 2001a. Development of a multiplex AmpliDet RNA for the simultaneous detection of Potato leafroll virus and Potato virus Y in potato tubers. J. of Virological Methods 93, p. 115-125.

Klerks, M.M., Leone, G.O., Lindner, J.L., Schoen, C.D. and van den Heuvel, J.F.J.M., 2001b. Rapid and sensitive detection of Apple stem pitting virus in apple trees through RNA amplification and probing with fluorescent molecular beacons. Phytopathology 91, p. 1085-1091.

Klerks, M.M. and Leone, G., 2002. Use of AmpliDet RNA to control the microbiological quality of agro-food products. Agro Food Industry hi-tech 5, p. 37-40.

Klerks, M.M., Lindner, J.L., Vaskova, D., Spak, J., Thompson, J.R., Jelkmann, W. and Schoen, C.D., 2004a. Detection and tentative grouping of Strawberry crinkle virus isolates. Eur. J. Plant Pathol. 110, p. 45-52.

Klerks, M.M., Zijlstra, C. and van Bruggen, A.H.C., 2004b. Comparison of real-time PCR methods for detection of *Salmonella enterica* and *Escherichia coli* O157:H7, and introduction of a general internal amplification control. J. Microbiol. Methods (accepted).

Klerks, M.M., Donnikov, M., Zijlstra, C. and van Bruggen, A.H.C., 2004c. Comparative study of methods to extract *Salmonella enteritidis* from environmental substrates, and complete procedure evaluation by introduction of a general procedural control (in preparation).

Kowalchuk, G.A., Stephen, J.R., De Boer, W., Prosser, J.I., Embley, T.M. and Woldendorp, J.W., 1997. Analysis of ammonia-oxidizing bacteria of the beta subdivision of the class *Proteobacteria* in coastal sand dunes by denaturing gradient gel electrophoresis and sequencing of PCR-amplified 16S ribosomal DNA fragments. Applied and Environmental Microbiology 63, p. 1489-1497.

Lacourt, I., Bonants, P.J.M., van Gent-Pelzer, M.P., Cooke, D.E.L., Hagenaar-de Weerdt, M., Surplus, L. and Duncan, J.M., 1997. The use of nested primers in the polymerase chain reaction for the detection of *Phytophthora fragariae* and *P. cactorum* in strawberry. Proceedings of the third international strawberry symposium, Veldhoven, the Netherlands, 29 April-4 May, 1996. Van der Scheer, H.A.T., Lieten, F. and Dijkstra, J. (eds). Acta Horticulturae 439, p. 829-838.

Leone, G., van Schijndel, H., van Gemen, B., Kramer, F.R. and Schoen, C.D., 1998. Molecular beacon probes combined with amplification by NASBA enable homogeneous, real-time detection of RNA. Nucleic Acids Res. 26, p. 2150-2155.

Mayer, D., Mithen, R., Lewis, B.G., Vos, P. and Oliver, R.P., 1997. The use of AFLP fingerprinting for the detection of genetic variation in fungi. Mycological Research 100, p. 1107-1111.

Muyzer, G., 1999. DGGE/TGGE a method for identifying genes from natural ecosystems Current Opinion in Microbiology 2, p. 317-322.

Nicholson, F.A., Hutchison, M.L., Smith, K.A., Keevil, C.W., Chambers, B.J. and Moore, A., 2000. A study on farm manure applications to agricultural land and an assessment of the risks of pathogen transfer into the food chain. A report to the Ministry of Agriculture Fisheries and Food. London.

Niepold, F. and Schober-Butin, B., 1997. Application of the one-tube PCR technique in combination with a fast DNA extraction procedure for detecting *Phytophthora infestans* in infected potato tubers. Microbiol. Res. 152, p. 345-351.

Niessen, M.L., Klusmann, J. and Vogel, R.F., 1997. Quantitative estimation of *Fusarium graminearum* using a novel solid phase PCR-assay. In: Diagnosis and identification of plant pathogens. Proceedings 4th International Symposium of the European Foundation for Plant Pathology, p. 207-211. Bonn, Germany, 9-12 September 1996. Dehne, H.W., Adam, G., Diekmann, M., Frahm, J., Mauler-Machnik, A. and van Halteren, P. (eds). Kluwer Academic Publishers, Dordrecht, the Netherlands.

Oberst, R.D., Hays, M.P., Bohra, L.K., Phebus, R.K., Yamashiro, C.T., Paszko-Kolva, C., Flood, S.J.A., Sargeant, J.M. and Gillespie, J.R., 1998. PCR-based DNA amplification and presumptive detection of *Escherichia coli* O157:H7 with an internal fluorogenic probe and the 5′ nuclease (TaqMan) assay. Appl. Environ. Microbiol. 64, p. 3389-3396.

Phillips, C.A. 1999. The epidemiology, detection and control of *Escherichia coli* O157. Journal of the Science of Food and Agriculture 79, p. 1367-1381.

Rigby, W.R., Cowieson, D.R., Davies, N.C. and Furneaux, R.C., 1990. An anodizing process for the production of inorganic microfiltration membranes. Trans. Inst. Met. Finish 68: 95-98.

Ristaino, J.B., M. Madritch, C.L. Trout and G. Parra, 1998. PCR amplification of ribosomal DNA for species identification in the plant pathogen genus *Phytophthora*. Appl. Environ. Microbiol. 64, p. 948-954.

Schoen, C.D., Knorr, D. and Leone, G., 1996. Detection of potato leafroll virus in dormant potato tubers by immunocapture and a fluorogenic 5′ nuclease RT-PCR assay. Phytopathology 86, p. 993-999.

Schoen, C.D., de Weerdt, M., Szemes, M., van Beuningen, R., Chan, A., Hilhorst, R., Boender, P. and Bonants, P., 2004. Multiplex detection of plant (Quarantine) pathogens by micro-arrays: an innovative tool for plant health management. Acta Horticultura (in press).

Schubert, R., Bahnweg, G., Nechwatal, T., Jung, T., Cooke, D.E.L., Duncan, J.M., Müller-Starck, G., Langebartels, C., Sandermann Jr., H., and Oßwald, W., 1999. Detection and quantification of *Phytophthora* species which are associated with root-rot diseases in European deciduous forests by species specific polymerase chain reaction. European J. of Forest Pathol. 29, p. 169-187.

Spiegel, S. and Martin, R.R., 1993. Improved detection of potato leafroll virus in dormant potato tubers and microtubers by the polymerase chain reaction and ELISA. Ann. Appl. Biol. 122, p. 493-500.

Stark, K.D., Nicolet, Y. and Frey, Y., 1998. Detection of *Mycoplasma hypopneumoniae* by air sampling with a nested PCR assay. Appl. Environ. Microbiol. 63, p. 543-548.

Suda, K., Rothen-Rutishauser, B., Guenthert, M. and Wunderli-Allenspach, H., 2004. Phenotypic characterization of human umbilical vein endothelial (ECV304) and urinary carcinoma (T24) cells: endothelial versus epithelial features. *In Vitro* Cellular and Developmental Biology - Animal 37, p. 505-514.

Szemes, M., Klerks, M.M., van den Heuvel, J.F.J.M. and Schoen, C.D., 2002. Development of an AmpliDet RNA assay for the specific detection of PVY^{NTN}. J. Virol. Methods 100, p. 83-96.

Toth, R.L., Harper, K., Mayo, M.A., and Torrance, L., 1999. Fusion proteins of single-chain variable fragments derived from phage display libraries are effective reagents for routine diagnosis of potato leafroll virus infection in potato. Phytopathology 89, p. 1015-1021.

Tyagi, S. and Kramer, F.R., 1996. Molecular beacons: probes that fluoresce upon hybridization. Nature Biotechnology 14, p. 303-308.

Uyttendaele, M., De Troy, P. and Debevere, J., 1999. Incidence of *Salmonella, Campylobacter jejuni, Campylobacter coli,* and *Listeria monocytogenes* in poultry carcasses and different types of poultry products for sale on the Belgian retail market. Journal of Food Protection 62, p. 735-740.

Van Beckhoven, J.R.M.C., Stead, D.E. and van der Wolf, J.M., 2002. Detection of *Clavibacter michiganensis* subsp. *Sepedonicus* by AmpliDet RNA, a new technology based or real time monitoring of NASBA amplicons with a molecular beacon. Journal of Applied Microbiology 93, p. 840-849.

Van Beuningen, R., van Damme, H., Boender, P., Bastiaensen, N., Chan, A. and Kievits, T., 2001. Fast and specific hybridization using flow-through microarrays on porous metal oxide. Clinical Chemistry 47, p. 1931-1933.

Van der Wolf, J.M., van Bekkum, P.J., van Elsas, J.D., Nijhuis, E.H., Vriend, S.G.C. and Ruissen. M.A., 1998. Immunofluorescence colony-staining and selective enrichment in liquid medium for studying the population dynamics of *Ralstonia solanacearum* (race 3) in soil. Bulletin OEPP/EPPO Bulletin 28, p. 71-79.

Van der Wolf, J.M., Leone, G.O.M., van Beckhoven, J.C.R.M., Bentsink, L., de Haan, E.G., van den Bovenkamp, G.W., Griep, R.A., van Twisk, C. and Schots, A., 1999. Detection of *Ralstonia solanacearum* biovar 2 in potato by serological methods based on recombinant monoclonal antibodies and AmpliDet RNA. In: Beiträge zur Zürchtungsforschung, Proceedings of the International Symposium on New Aspects of Resistance Research on Cultivated Plants, p. 1-3. November 19, 1998, Aschersleben, Germany.

Van der Wolf, J.M., Vriend, S.G.C., Kastelein, P., Nijhuis, E.H., van Bekkum, P.J., de Vries, Ph.J. and van Vuurde, J.W.L., 2000. Detection of *Ralstonia (Pseudomonas) solanacearum* biovar 2 (race 3) in soil by immunofluorescence colony-staining and PCR-amplification. European Journal of Plant Pathology 106, p. 123-133.

Van der Wolf, J.M. and van Beckhoven, J.R.C.M., 2004. Factors affecting survival of *Clavibacter michiganensis* subsp. *Sepedonicus* in water. J. Phytopathology 152, p. 161-168.

Van der Wolf, J.M., van Beckhoven, J.R.C.M., de Haan, E.G., van den Bovenkamp, G.W. and Leone, G.O.M., 2004. Specific detection of *Ralstonia solanacearum* 16S rRNA sequences by AmpliDet RNA. Eur. J. Plant Pathol. 110, p. 25-33.

Vaskova, D., Spak, J., Klerks, M.M., Schoen, C.D., Thompson, J.R. and Jelkmann, W., 2004. Real time NASBA for detection of strawberry vein banding virus. European Journal of Plant pathology 110, p. 213-221.

Vos, P., Hogers, R., Bleeker, M., Reijans, M., van de Lee, T., Hornes, M., Frijters, A., Pot, J., Peleman, J., Kuiper, M. and Zabeau, M., 1995. AFLP: a new technique for DNA fingerprinting. Nucleic Acids Research 23, p. 4407-4414.

Waalwijk, C., de Koning, J.R.A., Bonants, P.J.M., de Weerdt, M. and Baayen, R.P., 1997. Molecular diagnosis of *Fusarium* spp. Cereal Research Communications 25, p. 251-257.

Waalwijk, C., van der Heide, R., de Vries, I., van der Lee, T., Schoen, C., Costrel-de Corainville, G., Kastelein, P., Kohl, J. and Kema, G., 2004. Quantitative detection of *Fusarium* species in wheat using TaqMan. European Journal of Plant Pathology (in press).

Weissinger, W.R., Chantarapanont, W. and Beuchat, L.R., 2000. Surival and growth of *Salmonella baildon* in shredded lettuce and diced tomatoes, and effectiveness of chlorinated water as a sanitizer. International Journal of Food Microbiology 62, p. 123-131.

White, T.J., Bruns, T., Lee, S.B. and Taylor, J.W., 1990. Amplification and direct sequencing of fungal ribosomal RNA genes for phylogenetics. In: M.A. Innis, D.H. Gelfand, J.J. Sninsky, T.J. White (eds.): PCR protocols: a guide to methods and applications, Academic Press, San Diego, p. 315-322.

Zhang, A.W., Hartman, G.L., Curio-Penny, B., Pedersen, W.L. and Becker, K.B., 1999. Molecular detection of *Diaporthe phaseolorum* and *Phomopsis longicolla* from soybean seeds. Phytopathology 89, p. 796-804.

Zijlstra, C., Lever, A.E.M., Uenk, B.J. and van Silfhout, C.H., 1995. Differences between ITS regions of isolates of the root-knot nematodes *Meloidogyne hapla* and *M. chitwoodi*. Phytopathology 85, p. 1231-1237.

Zijlstra, C., Uenk, B.J. and van Silfhout, C.H., 1997. A reliable, precise method to differentiate species of root-knot nematodes in mixtures on the basis of ITS-RFLPs. Fundamental and Applied Nematology 20, p. 59-63.

Zijlstra, C., 1997. A fast PCR assay to identify *Meloidogyne hapla, M. chitwoodi* and *M. fallax*, and to sensitively differentiate them from each other and from *M. incognita* in mixtures. Fundamental and Applied Nematology 20, p. 505-511.

Zijlstra, C., 2000. Identification of *Meloidogyne* chitwoodi, *M. fallax* and *M. hapla* based on SCAR-PCR: a powerful way of enabling reliable identification of populations or individuals that share common traits. European Journal of Plant Pathology 106, p. 283-290.

Zijlstra, C., Donkers-Venne, T.H.M. and Fargette, M., 2000. Identification of *Meloidogyne incognita, M. javanica* and *M. arenaria* using sequence characterized amplified regions (SCAR) based PCR assays. Nematology 2, p. 847-853.

Zijlstra, C., van Hoof, R.A. and Donkers-Venne, T.H.M., 2004. A PCR test to detect the cereal root-knot nematode *Meloidogyne naasi.* European Journal of Plant Pathology (in press).

Zhu, L., Ang, S., and Liu, W.T., 2004a. Quantum dots as a novel immunofluorescent detection system for *Cryptosporidium parvum* and *Giardia lamblia*. Applied and Environmental Microbiology 70, p. 597-598.

Zhu, L., Zhang, Q. , Feng, H., Ang, S., Chau, F.S. and Liu, W.T., 2004b. Filter-based microfluidic device as a platform for immunofluorescent assay of microbial cells. Lab on a Chip 4: 337-341.

Rapid detection methods for GMOs: Developments and harmonisation in the European Union

C. Paoletti, M. Ermolli, S.R.H. Langrell and G. Van den Eede
Biotechnology and GMOs Unit - Institute for Health and Consumer Protection, European Commission - Joint Research Centre, TP 331, Ispra (VA), 21020 Italy.

1. Introduction

Within Europe European Regulation (EC) No. 1829/2003 on genetically modified (GM) food and feed provides the basis for ensuring a high level of protection of human life and health, animal health and welfare, including environment and consumer interests in relation to GM food and feed, whilst simultaneously ensuring the effective functioning of internal markets. It also lays down Community procedures for the authorisation and supervision, including provision for labelling, of GM food and feed. A key technical element of the European authorisation procedure, is the provision of an event specific quantitative detection method to facilitate eventual control and monitoring of the GM event in the distribution chain, and, more specifically, to assist labelling of such commodities with respect to the 0.9% threshold value for the recognised possibilities of adventitious or technically unavoidable presence of GMOs (which have benefited from a favourable risk assessment) in otherwise non-GM food or feedstuffs. For this purpose sensitive, event specific methods capable of reliably detecting and quantifying the presence of GMOs with limits of quantification within this threshold are necessary. Further, methods should be officially validated to internationally accepted standards, proven fit for regulatory purpose, and their technical deployment and operation harmonised throughout control laboratories of Member States.

2. Definition and requirements

According to Commission Regulation (EC) No. 641/2004, detailing technical measures for the implementation of Regulation 1829/2003, a method shall refer to 'all the methodological steps needed to analyse the relevant material in accordance with Articles 5 (3) (i) and 17 (3) (i) of Regulation 1829/2003'. Technically, event specificity is currently only achievable through a Polymerase Chain Reaction (PCR) approach, and to date, real-time PCR is the only technology that simultaneously satisfies both event specificity and quantification requirements as outlined above, and consequently has received growing attention and consideration. Further, for any given particular material, the method must include a DNA extraction procedure. Together, the entire process, from DNA extraction up to PCR application, constitutes a method. Detailed method acceptance criteria and method performance requirements have been complied through numerous consultations and Working Group deliberations by the European Network of GMO Laboratories (ENGL), a pan-European network of national and/or regional GMO control laboratories originally established with the aim of contributing more effectively to the European harmonisation and standardisation of means and methods for sampling, detection, identification and quantification of GMOs. This information is available through a document entitled 'Method acceptance criteria and method performance requirements' (http://gmo-crl.jrc.it). This document not only outlines the criteria a method should exhibit for acceptance prior to official validation, but also outlines specific criteria against which validation results are evaluated. Currently, protein based methods, although sensitive, lack the ability to satisfy the method requirement for event specificity, but are potentially valuable as versatile cost effective, rapid techniques in GM screening approaches, of which some aspects are briefly addressed below. However, the successful and meaningful application of any method, DNA or protein based, largely depends on the sampling strategy and theoretical assumptions regarding the material under investigation. Here, we highlight some of our current thinking and proposed approaches to sampling raw materials (grains) with respect to European GM control, including research initiatives for evaluating current assumptions.

3. Protein-based GMO detection methods

Protein-based methods are normally developed during the first step in the establishment of a novel GMO and have the potential to serve as cost-efficient, practical and quick tools for screening and for multiplex analysis. These methods can be extremely useful in the wider context of food safety with the same emphasis as PCR methods, and if properly validated, also be used for official control purposes. However, in Europe, due to emphasis on event specificity, protein immunoassays are less used than DNA-based methods, and to date the European Commission has validated only one protein immunoassay ELISA kit.

Although validated detection methods in use for quantitative analysis within the European Union (EU) are based on event specific quantitative PCR protocols, in the USA, protein-based methods are widely used for both qualitative and quantitative approaches. The only one ELISA method validated in the EU is designed for semi-quantitative analysis and therefore mainly utilised for screening approaches that will be further confirmed and quantified by PCR approaches (Lipp and Anklam, 2000). Indeed, one of the major drawbacks of both ELISA and quantitative PCR is that they can only quantify few targets at a time, resulting in a long operational procedure if the presence of a large number of targets has to be assessed. Nevertheless, they are extremely specific and powerful tools in this field.

Currently, a major role of protein-based methods is in support of research and development of novel GM products (Stave, 1999). After the production of cell lines, or genetically transformed plants, selection of lines expressing the appropriate level(s) of protein of the novel trait is determined. Protein-based methods are routinely and extensively applied during selection and bulking of GM transformant plants in fields or greenhouses. To ensure transformed seeds successfully express appreciable levels of novel protein, quality control analysis of the seed production system are performed by ELISA assays. When screening methods are needed directly in field situations or where fast qualitative results are required, protein based lateral flow strips are commonly used. To gain the regulatory approval for novel traits, information on the protein expression level of a novel trait has to be provided as part of the application to the EU competent authority. Therefore, appropriate protein-based methods have to be developed and applied. The protein-based methods are also applicable in other areas within the general context of food safety. As a very wide panorama of new GMO events is in the pipeline for approval for commercialisation in EU markets, both more general screening and specific methods exhibiting efficiency, reliability, and speed of deployment and operation, will be further required. Consequently, work within our laboratory continues to develop and further define protein-based assay for general GM screening purposes.

4. Sampling

The identification and development of appropriate sampling strategies is the first, crucial step for the detection and quantification of GMOs in different market products. By definition, sampling is always a source of error and goal of 'good' sampling practice is to minimise unavoidable sampling error and to maximise sample representativeness. Although this rather simple concept constitutes the basis of sampling theory (Cochran, 1977; Binns *et al.*, 2000), it also represents a real challenge for most surveys. The distribution of GM material within lots of agricultural commodities is generally oversimplified and assumed to be random (Remund *et al.*, 2001). Under this assumption the mean, the standard deviation of the mean and both the producer and consumer's risks can be estimated according to binomial distribution probabilities. However, assuming a random distribution of contaminants in kernel lots is likely to be false (Paoletti *et al.*, 2003).

Indeed, most of the guidelines defining kernel sampling strategies for quality and purity analyses, which are currently adopted for GM contamination detection in kernel lots, acknowledge that the procedures are not effective in the sampling of heterogeneous (non-random) distributions (Kay, 2002). Despite the general consensus within the scientific community that an evaluation of the possibility of non-random distribution of GM materials within lots is a pre-requisite for the definition of effective sampling protocols (Paoletti, 2004; Binns *et al.*, 2000; Kay and Paoletti, 2002; Gilbert, 1999; Coster, 1993; Cochran, 1977), and despite the outcome of simulation studies issuing a clear warning about the possible consequences of heterogeneity when sampling procedures are chosen assuming randomness (Paoletti *et al.*, 2003), neither a specific distribution-free sampling model, nor experimental data investigating directly the distribution of GMOs in large grain or seed lots are available. Our past and on-going research work addresses both these theoretical and experimental sampling open issues.

4.1 Theoretical-research

Following the publication of a new approach to investigate the effects of different levels of heterogeneity on the accuracy of different sampling plans for the detection of GM particles within kernel lots (Paoletti *et al.*, 2003), we have developed a novel statistical model to assess incremental sampling efficiency as function of lot heterogeneity KeSTE (Kernel Sampling Technique Evaluation (http://www.sipeaa.it/ASP/ASP2/KeSTE.asp). In particular, the model allows estimating the sampling error associated to different sampling protocols (in terms of both number and size of samples taken from the lot), applicable to any consignment of particulate material with respect to any kind of contamination, including GMOs. The novelty of our approach is the freedom from any distribution constraints. This new model is at the basis of the sampling protocols for grains recommended in the EU Recommendation "on technical guidance for sampling and detection of GMOs and material produced from GMOs as or in products" in the context of Regulation (EC) 1830/2003.

4.2 Experimental research

From a practical point of view the scenario is even more complex, because studies reporting estimates of GM contamination rates seldom provide details regarding the applied sampling procedures and international standard sampling protocols adopted for GMO testing are often modified depending upon EU Member States contingent needs and facilities. As a result, it is necessary to assess the degree of heterogeneity present in real lots in order to provide the information on lot distribution characteristics, necessary to establish and ensure effective sampling procedures for GMO surveys while minimising sampling related costs.

In collaboration with 10 ENGL designated laboratories from 7 different Member States we have organised a pan-EU research project, KeLDA (Kernel Lot Distribution Assessment), designed to investigate the degree of heterogeneity present in soybean grain lots imported within the European Union. KeLDA, which is almost completed, represents the first study assessing the degree of heterogeneity present in real lots and it is the first study to provide real data on this issue, worldwide. The obtained preliminary results indicate that GMOs distribution can vary among lots and it cannot be assumed *a priori* to be random, as currently done by most internationally adopted sampling protocols. Indeed, excuses to perform incorrect sampling cannot be justified by time or money limitations. A correct sampling is always accurate. If not, no reason exists to carry out any sampling at all.

References

Binns, M.R., J.P. Nyrop and W. van der Werf, 2000. Sampling and monitoring in crop protection. CABI Publishing, CABI International, Wallingford, Oxon, UK.

Cochran, G.W. 1977. Sampling Techniques. 3rd Edition. J. Wiley & Sons Inc., NY, USA.

Coster, R. 1993. Seed lot size limitation as reflected in heterogeneity testing. Seed Science and Technology 21: p. 513-520.

Gilbert, J. 1999. Sampling of raw materials and processed foods for the presence of GMOs. Food Control 10: p. 363-365.

Kay, S. 2002. Comparison of sampling approaches for grain lots European Commission Report, Code EURO20134EN, Joint Research Centre (JRC) Publication Office.

Kay, S. and C. Paoletti, 2002. Sampling strategies for GMO detection and/or quantification. European Commission Report, Code EUR20239EN, Joint Research Centre Publication Office.

Lipp, M. and E. Anklam, 2000. Validation of an immunoassay for detection and quantification of a genetically modified soybean in food and food fractions using reference materials: interlaboratory study. Journal of AOAC International 83: p. 919-927.

Paoletti, C. 2004. Sampling for GMO analysis: the European perspective. In: Testing for genetically modified organisms, Ed. E.A. Farid. The Haworth Press, Inc. London, UK.

Paoletti, C., M. Donatelli, S. Kay and G. van den Eede, 2003. Simulating kernel lot sampling: the effect of heterogeneity on the detection of GMO contaminations. Seed Sci. Technol. 31: p. 629-638.

Remund, K., D. Dixon, D. Wright D. and L. Holden, 2001. Statistical considerations in seed purity testing for transgenic traits. Seed Science Research 11: p. 101-120.

Stave, J.W. 1999. Detection of new or modified proteins in novel foods derived from GMO - future needs. Food Control 10: p. 367-374.

Detection of allergens in food: ELISA, PCR and the multi-allergen screening assay

B. Popping[1], K. Schneede[2], S. Koelln[2], A. Pardigol[3], A. Dan[3] and D. Pinero[4]
[1]Eurofins Scientific, 69a Kilnwick Road, Pocklington, Yorkshire, YO42 2JY, United Kingdom, [2]Eurofins-Wiertz-Eggert-Jörissen GmbH, Stenzelring 14b, D-21107 Hamburg, Germany, [3]Eurofins Scientific Analytics, R. A. Bobierre, F- 44 Nantes, France, [4]Eurofins-Woodson-Tenent Division, Des Moines, USA

Abstract

The problematic of allergens in food gains increasing interest, not least because of the recent directive seeking to protect the consumer from allergens. The analytical situation is still very much focussed on ELISA and (real-time) PCR as tools for detecting allergens or their markers. Recently, a multi-allergen screening assay has been developed and validated to provide cost-efficient and sensitive analysis of allergens in finished products. Other methods, like LC MS/MS are under development, showing promising results but are not yet ready for routine analysis.

Keywords: Food allergens; rapid analysis; ELISA; real-time PCR; multi-allergen screening assay

1. Introduction

In many industrialised countries food allergies are a major concern. Food allergy is a misguided reaction by the immune system, which is not treatable with desensitisation: the body mistakes a food for harmful foreign substance. Allergic symptoms range from mild (itching, burning off lips, watery itchy eyes, running nose, sneezing) to lethal (anaphylactic shock). The time of onset of allergic reaction after consuming the food can range from 3 to 20 minutes, the time to death between 1 1/2 to 3 h. The estimated number of people suffering from food allergies is around 8%. Some data suggest that this number is on the rise. Food allergy is important for two reasons: for the consumer it is a health issue, for the manufacturer or producer a liability issue. For people suffering from food allergies information if the substance they are allergic to is contained in the food they are buying is vital.

2. The labelling history

The European Commission has implemented regulations and directives, which aim to inform and protect the consumer. On the 20th of March 2000 the Directive 2000/13/EC of the European Parliament and the Council has been published. It establishes the general rules for labelling and advertising of foodstuffs including 25% rule for ingredients, i.e. ingredients present on a weight by weight basis of less than 25% do not need to be mentioned. However, this Directive does not cover labelling of allergenic compounds and materials. On the 14 of November 2002 the European Commission published a press release which declares the abolition of the 25% labelling rule as well as new rules also extending to alcoholic beverages. This requires the labelling of allergenic compound even if the ingredient constitutes less than 25% of the total product and, in addition, requires the labelling of alcoholic drinks if allergens are present. This is particularly relevant to wines, where often clearing agents like casein or isinglass, the latter is a protein from fish-bladder, are used. One example of materials which have to be listed under the new directive are compound products like spice-mixes which contain allergenic ingredients like egg, milk, soya or mustard.

On the 20th of November 2003 the European Commission published a common position listing 12 groups of allergenic compounds and their derivatives which will require labelling under the new directive. This list is an amendment to the 2000/13/EC annex IIIa and encompasses the following groups:

- cereals containing gluten and products thereof;
- crustaceans and products thereof;
- eggs and products thereof;
- fish and products thereof;
- peanuts and products thereof;
- soybean's and products thereof;
- milk and dairy products (including lactose);
- nuts and nuts products;
- sesame seed and products thereof;
- celery and products thereof;
- mustard and products thereof;

- sulphite at concentrations of at least 10 mg/kg.

Later on the Commission specified the cereals which require labelling under the new regulation as wheat, rye, barley, oats, spelt, kamut and their hybridised strains and products thereof. The Commission also specified which types of nuts require labelling: almond, hazelnut, walnut, cashew, pecan nut, Brasil nut, pistachio nut, macademia nut, and Queensland and products thereof.

On the10th of November 2003, the European commission adopted Directive 2003/89/EC thereby amending directive 2000/13/EC. The main points of this Directive are the adoption of the 12 long allergen list for labelling, accepting the date for the notification of scientific evidence for products to be exempt from labelling and establishing an exemption list as well as a date for implementation for the Directive (25 August 2004). It also sets a date for the re-examination of the Directive (25 November 2005) and another date for finalising the exemption list (25 November 2007).

3. Hidden allergens

While the new Directive ensures that allergens as part of ingredients are labelled, other aspects are not covered by this directive. Principally there can be three main reasons for hidden allergens: cross-contamination, fraud and wrong assumptions.

Cross-contamination is likely if products, which do not contain allergens, are manufactured in a factory or on a production line where also allergen-containing products are being produced. An example is plain, dark chocolate manufactured on the same line where previously hazelnut-milk chocolate was produced. Even though the line may have been cleaned, traces in sufficient quantities to trigger allergic reactions in highly sensitive individuals may still be present. However, the potential presence of allergens due to cross-contamination is not normally mentioned on the label.

Fraud typically leads to the replacement of higher priced with lower-priced products. If the latter contain allergenic compounds while the former do not, these are for obvious reasons not mentioned on the label. An example is the replacement of the expensive pine nuts for the cheaper peanuts.

Wrong assumptions are often made if products are fermented and the fermentation medium is not part of the final product. Here it is often assumed that there are no leftovers from the medium, which go into the next production step. An example are bacteria grown on soya broth. If the bacteria are then transferred to the next production step, soya proteins may still be present, potentially triggering an allergic reaction in the respective food-intolerant individual.

While hidden allergens are not covered by Directive 2003/89/EC, product liability law requires to appropriately indicating the risk for the consumer. The European Commission has not set a threshold for a quantity above which "hidden" allergens have to be labelled, even if they are in the product as a consequence of cross contamination, but Switzerland has set a level of 1000 ppm for these instances. However, one can imagine that 1000 ppm peanuts would be more than sufficient to trigger a severe allergic reaction to susceptible individuals.

4. The analysis

Principally a wide range of methods is available to analyse, directly or indirectly, for the presence of allergens. Among these methods are ELISA (Enzyme Linked Immuno Sorbent Assay) and PCR (Polymerase Chain Reaction) as the most frequently used ones, followed by Western Blot, Ouchterlonly, Rocket Immuno Electrophoresis and PCR-ELISA.

5. ELISA

ELISA is an immunological test where antibodies, similar to those which are formed in allergic individuals, recognise either the structure (epitope) of the allergenic protein (e.g. Ara h1 protein in peanuts) or recognise the full set of proteins from a particular plant, e.g. peanut proteins. The advantage of recognising a specific protein is the reduced probability for cross-reactivity, i.e. that a similar but not actually allergenic protein is detected and a false-positive result is obtained. The disadvantage of this is that if the target allergen (e.g. Ara h1) is rapidly degraded and therefore no longer detected by the antibody, other allergens (e.g. Ara h2) may still be present. This will result in a false-negative result based on the absence of the individually detected allergen. For the antibodies which recognise the total plant proteins, the opposite is true: chances for an unspecific, false-positive reaction are higher but the sensitivity as well as the probability of still detecting the plant protein, even if one protein type might be degraded faster than the others is higher.

The ELISA works by detecting the target protein and then, catalysed by an enzyme which is bound to the antibody, triggers a colour reaction in the solution. The intensity of the colour is a measure for the amount of target analyte, in this case a specific allergen or total plant protein present.

However, one must not forget that human antibodies can detect epitopes of the allergen, which may differ from the ones detected by the antibodies generated in animals for use in commercial assays. There have been reported cases, where the animal-based sera did not recognise, due to the high degree of processing, the allergen any longer while the human antibody was still capable of doing that. This means that the commercial kit would give a negative result while the allergy-sufferer consuming that food would still show an allergic reaction.

Currently, at least five commercial assays are available for peanut, some of which detect one allergen, others which detect peanut proteins in general. The AOAC Research Institute in the United States has evaluated three of these kits for performance, testing specificity, sensitivity, robustness for a set of matrices. All three kits passed the evaluation study with similar performance, close to 100%. Shortly after, the European Commission Joint Research Centre (EC-JRC) in Geel, Belgium has performed in-house validation of five commercial peanut kits, using complex matrices for the tests. The in-house results showed significant differences between the kits with respect to recovery, specificity, sensitivity, precision, accuracy and robustness. With the aim to validate the kits for acceptance as European Standard by the CEN, the European Standardisation Committee, an international validation study was started, including 31 laboratories from 16 European countries. The initial, so-far unpublished results confirmed the quite different performance of the kits.

Both, AOAC and EC-JRC, data provide a valuable insight into the performance of the kits, assessing different aspects. All data will be considered by the CEN working group in order to evaluate methods for standardisation.

Test results shown in our laboratory showed that also commercially available gluten ELISA kits perform quite differently. Here, gluten levels present in certain matrices were significantly over or underestimated, some kit consistently produced false positive results in specific matrices.

While ELISA remains to be one of the most frequently used test methods for the detection of allergenic organisms like peanuts or hazelnuts, it is absolutely essential to have experience with the particular matrix tested as well as the test-kit performance in order to assess the probability of false-negative or false-positive results.

6. PCR/real-time PCR

An alternative method gaining more and more importance is the indirect detection of the allergenic organism by analysing for the presence of the DNA as a marker for that organism. Several PCR systems have been described in the literature and some systems are available as commercial kits.

The advantage of the PCR-based assays is that they are often capable to detect evidence of the presence even in highly processed products where antibodies do no longer recognise their target (allergen). DNA is in many environments more stable than the proteins but above all, it is easy to generate specific, high-performance detection systems while generating specific antibodies is much more time consuming, labour-intensive and often not possible. As one example: no quantitative commercial ELISA system for soya protein in the ppm-range has so far been commercialised, while a great number of highly sensitive PCR and real-time PCR assays is described in the literature and also available as commercial assay.

The PCR procedure is similar to photocopying. Each cell of an organism has an instruction manual (the nucleus). This manual is made up of several chapters, instructing the cell which proteins to express and how to function. In this book, we find some chapters which are similar in each cell of the same organism (e.g. plant), some of which are specific to a certain tissue in the organism (e.g. root, leaf, stem), and others, which differ between organisms (peanut, hazelnut).

In order to identify the organism present, the chapter (or a part thereof) which differs between the organisms needs to be made visible. However, since the relevant portion of DNA cannot be visualised by means of a microscope, other tools have to be employed. Here, an enzyme, the Taq polymerase copies the relevant section of the manual. This is done with the help of bookmarks (primers) which mark the beginning and the end of the section to be copied by one having the same sequence (same letters) as the first and the other as the last few words of that section. If one looks at any type of (instruction) book, it is easy to spot that the beginning of each section starts and ends with different words, thereby creating specificity. This specificity is used by the primers (the bookmarks) to indicate where the Taq polymerase has to start and where to finish the copying. This process is repeated several times (cycling). After each cycle, the copies made are copied again. Starting with one copy in the first cycle, the original and the copy are put on the photocopy machine (Taq polymerase) for the next round, thereby generating 4 copies, then 8, 16 and so on. The process follows the chain reaction principle 2^n, giving the name to the procedure: polymerase chain reaction or PCR.

After 40 cycles, equalling approximately 3 h, this process has generated more than 1,000,000,000 copies of the sequence which identifies the organism (e.g. peanut). The quantity of that material is so high that it can be made visible by staining, using a substance which specifically stains DNA.

In a standard PCR system, the number of molecules is approximately the same at the end of 40 cycles, irrespective of the starting copy number. The limitations are due to the depletion of the assay components. Therefore the standard PCR can only be qualitative. Using a third element, the probe, which sits between the primers and carries a fluorescent molecule, the progress of the PCR reaction can be monitored by measuring the increase of this fluorescence. This modification of the PCR is called real-time PCR and allows a quantification of the number of starting DNA molecules. If the matrix is well characterised, this technique can be used to semi-quantitate the allergen by relating the number of DNA molecules to the amount of (allergenic) protein present.

The drawback of all of these techniques, ELISA and real-time PCR, is that only one target analyte can be measured, e.g. the Ara h1 protein or generally peanut protein or peanut DNA. However, in most finished products more than one potential allergen can be found. In biscuits it is likely to have peanut or hazelnut, in spices one can find celery, sesame, cereals and other components. In order to cost-effectively, but still sensitively detect these allergens, a multi-allergen screening assays has been developed by Eurofins, allowing to detect up to 8 different allergens in 1 PCR reaction. This was done using common sequences of related plant families, which after PCR were distinguished by melting curve analysis. Combining several of the multiplex assays, the most relevant allergens can be detected by their marker DNA in three PCR assays.

The multi-allergen screening assay has been validated with spiked materials as well as comparing ELISA and PCR results on finished products. Sensitivity of the individual PCR assays varies slightly but has been demonstrated to be in the lower ppm range.

7. New developments

Novel approaches by LC-MS/MS have also shown to be very sensitive, with detection limits in the sub pico-mol range for individual peptides of target proteins (e.g. from Ara h1). The method was capable of detecting peanut traces in vanilla ice-cream. However, the approach is not yet suitable for routine analysis due to complex sample extraction procedures and the lack of automation capabilities.

References

Allmann, M., Candrian, U., Höfelein, C. and Lüthy, J., 1993. Polymerase chain reaction (PCR): a possible alternative to immunochemical methods assuring safety and quality of food. Z Lebensm Unters Forsch., 1993. 196, p. 248-251.

Bock, S., Munoz-Furlong, A. and Sampson, H., 2001. Fatalities due to anaphylactic reactions to foods. J Allergy Clin Immunol 107, p. 191-193.

Burks, W., Bannon, G., Sicherer, S. and Sampson, H., 1999. Peanut-induced anaphylactic reactions. Int Arch Allergy Immunol 119, p. 165-172.

CEN, 2002a. Foodstuffs - Methods of analysis for the detection of genetically modified organisms and derived products - Nucleic acid extraction (ISO/DIS 21571 prEN ISO 21571.

CEN, 2002b. Foodstuffs - Methods of analysis for the detection of genetically modified organisms and derived products - Qualitative nucleic acid based methods (ISO/DIS 21569) prEN ISO 21569:2002.

CEN, 2002c. Foodstuffs - Nucleic acid based methods of analysis for the detection of allergens - General requirements and definitions (ISO/DIS) prEN ISO 24276.

Dahinden, I., von Büren, M. and Lüthy, J., 2001. A quantitative competitive PCR system to detect contamination of wheat, barley or rye in gluten-free food for coeliac patients Eur Food Res Technol 212, p. 228-233.

(DIN)., G.I.f.S., 2001. Part 60: Polymerase chain reaction (PCR), Terminology, general method-specific requirement. In Diagnostics of infectious diseases and diseases of the immune system in serology and molecular biology, DIN 589667-60, Beuth, Berlin.

Hird, H., Lloyd, J., Goodier, R., Brown, J. and Reece, P., 2003. Detection of peanut using real-time polymerase chain reaction Eur Food Res Technol 217, p. 265-68.

Holzhauser, T., Stephan, O. and Vieths, S., 2002. Detection of potentially allergenic hazelnut (Corylus avellana) residues in food: a comparative study with DNA PCR-ELISA and protein sandwich-ELISA J Agric Food Chem, 50, p. 5808-5815.

Holzhauser, T., Wangorsch, A. and Vieths, S., 2000. Polymerase chain reaction (PCR) for detection of potentially allergenic hazelnut residues in complex food matrixes. Eur Food Res Technol, 211, p. 360-365.

Hurst, W.J., Krout, E.R. and Burks, W.R., 2002. A comparison of commercially available peanut ELISA test kits on the analysis of samples of dark and milk chocolate. J Immunoassay Immunochem, 23, p. 451-459.

Hourihane JO'B, Kilburn S, Nordlee J *et al.*, 1997, An evaluation of the sensitivity of subjects with peanut allergy to very low doses of peanut protein: A randomised, double-blind placebo-controlled food challenge study. J Allergy Clin Immunol 100, 596-600.

Koch, P., Schäppi, G.F., Poms, R.E., Wüthrich, B., Anklam, E., and Battaglia, R., 2003. Comparison of commercially available ELISA kits with human sera-based detection methods for peanut allergens in foods. Food Additives and Contaminants, 20, p. 797-803.

Jäger, L. and Wüthrich, B., 1998. Nahrungsmittelallergien und Intoleranzen. Gustav Fischer Verlag, Ulm.

Keck-Gassenmeier B, Benet S, Rosa C, Hischenhuber C., 1999. Determination of peanut traces in food by a commercially-available ELISA test. Food and Agricultural Immunology 11, 243-250.

Koch P, Schäppi G, Poms R *et al.*, 2003. Comparison of commercially available ELISA kits with human sera-based detection methods for peanut allergens in foods. Food Additives and Contaminants 20, p. 797-803.

Meyer, R., Chardonnens, F., Hübner, P., and Lüthy, J., 1996. Polymerase chain reaction (PCR) in the quality and safety assurance of food: detection of soya in processed meat products. Z Lebensm Unters Forsch, 203, p. 339-344.

Lebensmittelkennzeichnungsverordnung (LMKV), Stand Dezember 1999; §3 Kennzeichnungselemente.

Poms R, Klein C, Anklam E., 2004. Methods for allergen analysis in food: a review. Food Additives and Contaminants 21. p. 1-31.

Sandberg, M., Lundberg, L., Ferm, M., and Malmeheden Yman, I., 2003. Real time PCR for the detection and discrimination of cereal contamination in gluten free foods. Eur Food Res Technol, 217, p. 344-349.

SLMB, 2003. Schweizerisches Lebensmittelbuch, Kapitel 52B: Molekularbiologische Methoden, Bern.

Stephan, O. and Vieths, S., 2004. Development of a Real-time PCR and a Sandwich ELISA for Detection of Potentially Allergenic Trace Amounts of Peanut (Arachis hypogaea) in Processed Foods J Agric Food Chem, in press.

Stephan O, Möller N, Lehmann S, Holzhauser T, Vieths S., 2002 Development and validation of two dipstick type immunoassays fort he determination of trace amounts of peanut and hazelnut in processed foods. Eur Food Res Technol 215, p. 431-436.

Taylor, S. and Nordlee, J., 1996. Detection of Food Allergens Food Technol, 50, p. 231-234.

Taylor S, Hefle S, Bindslev-Jensen C *et al.*, 2002. Factors affecting the determination of threshold doses for allergenic foods: How much is too much? J Allergy Clin Immunol 109, p. 24-30.

Richtlinie 2000/13/EG des Europäischen Parlamentes und des Rates vom 20. März 2000 zur Angleichung der Rechtsvorschriften der Mitgliedstaaten über die Etikettierung und Aufmachung von Lebensmitteln sowie die Werbung hierfür.

Richtlinie 2003/89/EG des Europäischen Parlaments und des Rates vom 10. November 2003 zur Änderung der Richtlinie 2000/13/EG hinsichtlich der Angabe der in Lebensmitteln enthaltenen Zutaten. Amtsblatt der Europäischen Union, 25.11.2003.

Wensing M, Penninks A, Hefle S *et al.*, 2002. The range of minimum provoking doses in hazelnut-allergic patients as determined by double-blind, placebo-controlled food challenges. Clin Exp Allergy 32, p. 1757-1762.

Rapid analysis of meat- and bonemeal in feed

R. Margry, S. van den Hoven, C. Gerris and B. Taraba
CCL Research, P.O. Box 107, 5460 AC Veghel, The Netherlands

Abstract

In 2002, the Animal By-Product Regulation (EC) No 1774/2002 was published, imposing a ban on feeding animals with proteins from the same species. The availability of an analytical technique for species identification would facilitate the lift of this ban. The suitability of immunochemical and polymerase chain reaction (PCR) techniques was tested with various reference materials prepared in a dedicated 140 L steriliser under strict conditions and controls (133° and 159°C, pressure 3 bar, 20 minutes). The advantages and disadvantages of immunochemical and polymerase chain reaction assays will be discussed.

Keywords: species identification, cannibalism, PAPs, animal by-products, meat- and bonemeal, MBM, ELISA, PCR, reference materials, feed

1. Introduction

At present, the use of animal by-products in animal feed is prohibited with a few exceptions. The ban on feeding processed animal proteins (PAPs) to any farmed animal, regulated by Regulation (EC) No 999/2001, is an important measure to prevent the spread of Transmissible Spongiform Encephalopathies (TSE). PAPs include, for example, meat- and bonemeal, meatmeal, bonemeal and blood products. Currently, the feed ban is partly lifted for products like milk and egg products, fishmeal for non-ruminants, gelatine originating from pig and poultry, and hydrolysed proteins. Microscopic examination is the official EC technique for detecting violation of the feed ban.

Nearly two years ago, the Animal By-Product Regulation (EC) No 1774/2002 was published, imposing a ban on feeding animals with proteins from the same species. The availability of an analytical technique for species identification would facilitate the lift of this ban. Immunoassays (both dipstick methods and ELISAs) and polymerase chain reaction (PCR) assays are chosen as possible rapid methods.

Immunoassays are commercially available for species identification. These types of techniques are rather easy to perform. The disadvantage is that the proteins, which are tested in ELISAs, are not present in every tissue cell. PCR assays are also commercially available for species identification. DNA is present in every tissue cell with a nucleus. For both types of assays the consequence of heat-treatment of the rendering material on the sensitivity and the detection limit is not clear. For application of the PCR technique there are special requirements for the layout of the laboratory and the skills of the analysts. Therefore, an inventory has been made of published analytical techniques and commercially available kits, being mainly ELISA and PCR kits for food analysis. Some of these are also suitable for the analysis of food processed between 100°C and 120°C or at even higher temperatures. For most of the commercially kits the suppliers indicate that they are not appropriate for species specification of animal material heated at temperatures higher than 133°C.

Initially, the suitability of these ELISA and PCR techniques was tested with samples obtained from various rendering plants. Often false negative and false positive (cross-reactivity) results were observed. Doubts arose about the suitability of the test samples, because the heat-treatments had been much higher than expected. According to Regulation (EC) No 1774/2002 mammalian by-products have to be pressure cooked, which involves heating at temperatures higher than 133°C for at least 20 minutes at a minimum pressure of 3 bar. Both the processing conditions and the composition of the samples (e.g. fat, bone and moisture content) varied for the rendering plants.

Therefore, our laboratory prepared various reference materials in a dedicated 140 L steriliser under strict conditions and controls. Process temperature, pressure and time were chosen and monitored. These reference materials, with a known process history, and some well defined samples from rendering plants have been used for validation of the commercially available kits or assays which are still under development. The advantages and disadvantages of immunochemical and PCR assays will be discussed. The possibilities and limitations (especially the detection limit and the loss of sensitivity) of some of the promising techniques will be shown, including three commercial ELISA and two commercial PCR assays. Also the preliminary results of an ELISA and three PCR techniques, which are under development, will be shown.

Attention will be paid to the consequences of the distinction between non-prohibited and prohibited animal proteins in feed.

2. Materials and methods

2.1 Preparation of reference materials

The pure porcine reference soft material consists of 100% large intestines (empty). The pure porcine reference bone material consists of 5% tails and 95% hind-legs. The pure chicken soft material consists of 100% digestive system (duodenum, pancreas, small intestines, rectum, caecae, cloaca, gall and spleen). The pure chicken bone material consists of 40% heads and 60% shanks.

The materials were processed in a steel 140 L autoclave with a paddle-stirring device for 20 minutes at 133 °C or 159 °C at belonging pressure. Both pre- and post-pressure processed reference materials have been prepared. For the pre-pressure cooking process the steriliser was closed from the beginning until the end. With post-pressure cooking the condensation valve was opened for 30 seconds every five minutes. When the moisture content of the sample was reduced to approx. 15%, the steriliser was closed to continue the sterilisation process. At the end, after cooling down until ambient temperature, the fat was set and removed by hand, followed by centrifugation. The residues from centrifugation were dried at 70 °C for 2 days in a drying stove. After each batch the steriliser was cleaned with a solution of sodium hydroxide at a temperature beyond 100 °C. Thereafter, the debris was removed and the steriliser was rinsed from the top with tap water.

The reference materials mentioned as bovine, ovine, porcine and avian bone material, and bovine muscle heated at 133 °C, 137 °C, 141 °C and 145 °C, respectively, were prepared by PDM Ltd. (S. Woodgate, Doncaster, UK).

Some of the above mentioned pure materials were also mixed. The bovine and avian bone materials of PDM Ltd. were also mixed with compound feed material.

2.2 The following commercially available test kits were used

Food Expert ID, DNA chip technique (BioMérieux); SureFood Animal-ID, PCR-ELISA (Congen Biotechnologie GmbH); MELISA-TEK, ELISA (ELISA Technologies Inc.); AgriScreen, dipstick immunoassay (Neogen Corporation) and FeedChek, lateral flow test (Strategic Diagnostics, Inc.).

3. Results

As an example of the preparation of the reference materials, the temperature profiles of eight of the reference materials are presented in Figure 1a and 1b. The process temperatures, 133°C and 159°C, for the pre-pressure cooked materials are reached in shorter time periods than for de post-pressure cooked materials. Regarding the post-pressure cooked materials, it took approx. 110 minutes for the soft materials and approx. 200 minutes for the bone materials to evaporate the moisture through the open evaporation valve until a moisture

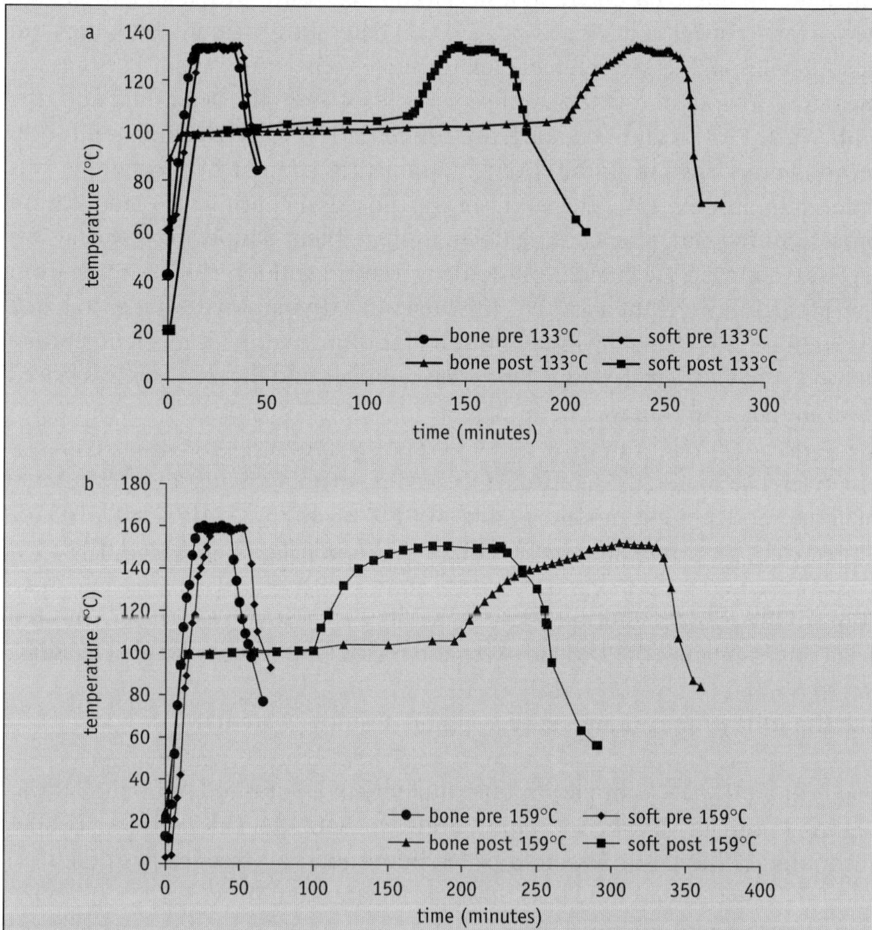

Figure 1. Preparation of chicken reference material at (a) 133 °C and (b) at 159 °C.

content of approx. 15% was reached. These profiles are comparable with the conditions of the rendering process under practical circumstances. The processing of all these reference materials is according to the requirements of Regulation (EC) No 1774/2002 (heating for at least 20 minutes at temperatures higher than 133°C and at a pressure higher than 3 bar).

In view of the legislation, it is important to have a rapid method for identification of poultry material as a contaminant of porcine meat- and bonemeal and, the other way round, for identification of porcine material contaminating poultry meat- and bonemeal (see Table 1a). It is not to be expected that animal proteins in general will be allowed in ruminant feed within the next few years. The most convenient way for control could be the official EC microscopic method (Directive 2003/126/EC of 23 December 2003) or a multi-species DNA chip with which the species of interest could be detected with one single assay (see Table 1b). Ruminant material is also not authorised for use in any feeding stuff other than feed for fur animals or petfood.

The suitability of techniques under development - ELISA (from institute 1) and real-time PCR assays (from institutes 2, 3 and 4) - for species identification was tested with the ruminant, bovine, porcine and avian reference materials (see Table 2).

The ruminant ELISA assay from institute 1 correctly identified the bovine bone and muscle, and ovine bone samples, even at the higher temperature of 145°C. However, false positive results were obtained with all porcine materials. With the porcine assay porcine materials (even the high temperature treated) were all correctly identified, but false positives were obtained for the pure chicken materials. The avian assay gave good results for both the porcine and avian materials, but false positives were found for the porcine bone reference materials. Some contamination of these reference materials prepared by PDM might originate from the rendering plant.

No false positive or negative results were found with the bovine PCR assay from institute 2. However, this assay is not suitable for detection of other ruminants. With the porcine assay false positive results are obtained with most of the pure chicken materials. The avian assay gives some false positive results with some of the pure porcine reference materials.

Table 1a. Techniques for control of meat- and bonemeal (MBM) on contamination with material of prohibited species.

MBM	prohibited material[a] in MBM	detection technique
poultry	ruminant	PCR[b] or immunoassay
poultry (for pig feed)	porcine	PCR[b] or immunoassay
porcine	ruminant	PCR[b] or immunoassay
porcine (for poultry feed)	poultry	PCR[b] or immunoassay

[a]except DCP (dicalcium phospate) and TCP (tricalcium phospate), milk and egg products, hydrolysed proteins and gelatine from non-ruminant origin
[b]PCR, incl. DNA chip (multi-species method)

Table 1b. Techniques for control of feed on contamination with material of prohibited species.

feed	prohibited material in feed	detection technique
ruminant	all animal[a]	microscopy multi-species PCR[c] or immunoassay
poultry	ruminant[b]	PCR[c] or immunoassays
	poultry[b]	PCR[c] or immunoassays
pig	ruminant[b]	PCR[c] or immunoassays
	porcine[b]	PCR[c] or immunoassays

[a]except milk and egg products, hydrolysed proteins and gelatine from non-ruminant origin and fat
[b]except DCP and TCP, milk and egg products, hydrolysed proteins and gelatine from non-ruminant origin
[c]PCR, incl. DNA chip (multi-species method)

Only a few materials were tested with the PCR assay from institute 3. False negatives were found for some porcine and ovine bone materials.

The results obtained with the PCR assay from institute 4 are promising.

Table 2. Analysis of reference materials with tests under development.

				ELISA						Real-time PCR					
				institute 1			institute 2			institute 3			institute 4		
				Ru.	Por.	Av.	Bo.	Por.	Av.	Bo.	Por.	Ov.	Ru.	Por.	Av.
100% Pork	soft	pre-pressure	133 °C	+	+	-	-	+	-					+	-
		cooked	159 °C	+	+	-	-	+	+ ?					+	-
		post-pressure	133 °C	+	+	-	-	+	-					+	-
		cooked	159 °C	+	+	?	-	+	+ ?					+	-
	bone	pre-pressure	133 °C	+	+	-	-	+	-					+	-
		cooked	159 °C	+	+	?	-	+	-					+	-
		post-pressure	133 °C	+	+	-	-	+	-					+	-
		cooked	159 °C	+	+	?	-	+	- ?					+	-
100% Chicken	soft	pre-pressure	133 °C	-	+	+	-	+	+					-	+
		cooked	159 °C	-	+	+	-	+	+					-	+
		post-pressure	133 °C	-	+	+	-	+ ?	+						
		cooked	159 °C	-	+	+	-	+ ?	+						
	bone	pre-pressure	133 °C											-	+
		cooked	159 °C	?	+	+	-	+ ?	+					+	+
		post-pressure	133 °C	?	+	+	-	- ?	+						
		cooked	159 °C	?	+	+	-	+ ?	+						
Bovine bone 133 °C				+			+	-		+					
Bovine bone 137 °C				+			+	-		+					
Bovine bone 141 °C				+	-	-	+	-		+					
Bovine bone 145 °C				+	-	-	+	-		+			+		
Ovine bone 133 °C				+	-	-	-	-				+	+		
Ovine bone 137 °C				+	-	-	-	+ ?				+			
Ovine bone 141 °C				+	-	-	-	+ ?				-			
Ovine bone 145 °C				+	-	-	-	+ ?				-			
Porcine bone 133 °C				+	+	+	-	+			-				
Porcine bone 137 °C				+	+	+	-	+			-				
Porcine bone 141 °C				+	+	+	-	+			-				
Porcine bone 145 °C				+	+	+	-	+ ?			-				

Table 2. Continued.

	ELISA						Real-time PCR					
	institute 1			institute 2			institute 3			institute 4		
	Ru.	Por.	Av.	Bo.	Por.	Av.	Bo.	Por.	Ov.	Ru.	Por.	Av.
Avian bone 133 °C			+	-	-							
Avian bone 137 °C				-	-							
Avian bone 141 °C	-	-	+	-	-							
Avian bone 145 °C	-	-	+	-	+ ?							
Bovine muscle 133 °C	+	-	-							+		
Bovine muscle 137 °C	+	-	-									
Bovine muscle 141 °C	+	-	-									
Bovine muscle 145 °C	+	-	-									

Result okay Wrong / no result

Ru. = ruminant; Por. = porcine; Av. = avian; Bo. = bovine

In addition, some commercially available kits were tested with the ruminant, bovine, porcine and avian reference materials and some mixtures thereof (see Table 3). The Food Expert ID (DNA chip) correctly identified most of the reference materials. A false positive result was found for one chicken bone material. False negative results are obtained with some of the mixtures with contaminant concentrations of 2% or lower.

With the Surefood PCR false negative results are obtained with all the tested bovine, porcine and chicken materials. These kits are not suitable for animal material heated at temperatures beyond 133 °C.

Even though only available as ruminant test kit and only a few materials tested, the MELISA-TEK and Agriscreen kits look rather promising as screening methods. More materials and especially mixtures of animal materials have to be tested.

Table 3. Analysis of reference materials with commercially available tests.

				Food Expert ID (BioMérieux)				Surefood (PCR Congen)			MELISA -TEK (ELISA Techn.)	Agri-Screen (Neogen)	FeedChek (SDI)	
				Bo.	Por.	Av.	Ov.	Bo.	Por.	Av.	Ru.	Ru.	MBM	Mam.
100% Pork	soft	pre-pressure	133 °C	-	+	-	-		-/Inh.				+	-
		cooked	159 °C	-	+	-	-		-				+	-
		post-pressure	133 °C	-	+	-	-		Inh.				+	-
		cooked	159 °C	-	+	-	-		-				+	-
	bone	pre-pressure	133 °C	-	+	-	-		-				+	+
		cooked	159 °C	-	+	-	-		-				+	+
		post-pressure	133 °C	-	+	-	-		-				+	+
		cooked	159 °C	-	+	-	-		-				+	+
100% Chicken	soft	pre-pressure	133 °C	-	-	+	-		-	-	-	-	+	-
		cooked	159 °C	-	-	+	-		-	-			+	-
		post-pressure	133 °C	-	-	+	-		-	+			+	-
		cooked	159 °C	-	-	+	-		-	-			+	-
	bone	pre-pressure	133 °C	-	-	+	+						+	+
		cooked	159 °C	-	-	+	-		-	-			+	+
		post-pressure	133 °C	-	-	+	-		-	-			+	+
		cooked	159 °C	-	-	+	-		-	-			+	+
Chicken 95%+Pork 5% (soft+bone)				-	+	+	-						+	+
Chicken 98%+Pork 2% (soft+bone)				-	+	+	-						+	+
Chicken 99.5%+Pork 0.5% (soft+bone)				-	-	+	-						+	+
Pork 95%+Chicken 5% (soft+bone)				-	+	+	-						+	+
Pork 98%+Chicken 2% (soft+bone)				-	+	-	-						+	+
Pork 99.5%+Chicken 0.5% (soft+bone)				-	+	-	-						+	+
Bovine bone 133 °C				+	-	-	-	-		-	+	+		
Bovine bone 137 °C				+	-	-	-	-		-	+	+		

Table 3. Continued.

	Food Expert ID (BioMérieux)				Surefood (PCR Congen)			MELISA -TEK (ELISA Techn.)	Agri-Screen (Neogen)	FeedChek (SDI)	
	Bo.	Por.	Av.	Ov.	Bo.	Por.	Av.	Ru.	Ru.	MBM	Mam.
Bovine bone 141 °C	+	-	-	-							
Bovine bone 145 °C	+	-	-	-							
Ovine bone 133 °C	-	-	-	+							
Ovine bone 137 °C	-	-	-	+							
Ovine bone 141 °C	-	-	-	-							
Ovine bone 145 °C	-	-	-	-							
Porcine bone 133 °C	-	-	-	-							
Porcine bone 137 °C	-	+	-	-							
Porcine bone 141 °C	-	+	-	-							
Porcine bone 145 °C	-	+	-	-							
Avian bone 133 °C	-	-	+	-	-		-	-	-		
Avian bone 137 °C	-	-	+	-	-		-				
Avian bone 141 °C	-	-	+	-				-	-		
Avian bone 145 °C	-	-	-	-							
Bovine muscle 133 °C	+	-	-	-							
Bovine muscle 137 °C	-	-	-	-							
Bovine muscle 141 °C	+	-	-	-							
Bovine muscle 145 °C	+	-	-	-							
Bo. 50% + Por. 50% (bone)	+	+	-	-							

Result okay Wrong / no result

Bo. = bovine; Por. = porcine; Av. = avian; Ov. = ovine; Ru. = ruminant; MBM = meat- and bonemeal; Mam. = mammalian; Inh. = inhibition of PCR rest

The lateral flow test FeedChek for meat- and bonemeal (MBM) and mammalian material was also tested. The MBM results showed a good correlation. This assay has potential as a control method for determining the absence of processed animal proteins in ruminant feed. The included mammal test showed false negative results for porcine material and false positive results for chicken material. Actually, this kit is not suitable for the detection of mammalian material.

In Table 4 the first results are presented for the analysis of meat- and bonemeal in mixed feed. Until now, only reference bovine and avian bone material were tested with de ELISA assays from institute 1. However, these preliminary results are rather promising. The reference materials tested were correctly identified.

Table 4. Analysis of mixed feed with techniques under development.

	ELISA institute 1			
	Ruminant (%)	Bovine (%)	Porcine (%)	Avian (%)
10% Bovine bone 133 °C in mixed feed	+++++ (10)	+++++ (10)		
1% Bovine bone 133 °C in mixed feed	+ (0.83)	+ (0.65)		
0.5% Bovine bone 133 °C in mixed feed	+ (0.57)	+ (0.5)		
0.1% Bovine bone 133 °C in mixed feed	+ (0.07)	+ (0.04)		
10% Bovine bone 145 °C in mixed feed	++++ (9.1)	+++++ (>10)		
1% Bovine bone 145 °C in mixed feed	+ (0.5)	+ (0.5)		
0.5% Bovine bone 145 °C in mixed feed	+ (0.27)	+ (0.1)		
0.1% Bovine bone 145 °C in mixed feed	+ (0.2)	+ (0.06)		
10% Avian bone 133 °C in mixed feed				++++ (8.2)
1% Avian bone 133 °C in mixed feed				++ (1.0)
0.5% Avian bone 133 °C in mixed feed				+ (0.56)
Avian bone 141 °C	-		-	+++++ (71)
10% Avian bone 141 °C in mixed feed				+++ (4.1)
1% Avian bone 141 °C in mixed feed				++ (1.3)
0.5% Avian bone 141 °C in mixed feed				+ (0.075)

According to the results of Table 2 false positive results are expected with porcine material in mixed feed. The false positive results obtained with the porcine kit for chicken bone are no longer observed after dilution with mixed feed. This ELISA has also potential for semi-quantitative analysis.

4. Discussion

Most of the assays described are able to characterise the species identity of reference materials, even after heating for 20 minutes at the highest temperatures. However, often false positive and sometimes false negative results are found. In order to make a fair comparison between the methods more reference materials have to be analysed.

The DNA chip technique of BioMérieux has a rather good score with the reference materials tested. With a maximum number of eight samples which can be analysed in one day with the standard equipment, it is not a really a rapid method. However, 39 different species can be identified simultaneously in one sample.

Some animal products are excluded from the intra-species recycling ban (1774/2002/EC), the so-called non-prohibited proteins. Their use is authorised for feed for farmed animals. These products include hydrolysed proteins, milk (products), including colostrum, skim and whey, egg (products), fat, gelatine (non-ruminant origin), fishmeal, and di-and tricalciumphosphate. The last three products are not authorised for ruminant feed. It is to be expected that the current ban on the use of blood (products) will be lifted in a later stage. The use of these non-prohibited products in feed will hamper the correct identification (detection) of the other tissues of the species in question, which are not allowed to be present. This dilemma could be raised when a control regime is established with analytical control of prohibited species material present in meat- and bonemeal, in combination with administrative monitoring of compound feed (tracking and tracing).

5. Conclusions

The immunochemical FeedChek kit (Strategic Diagnostics, Inc.) offers good opportunities to detect meat- and bonemeal as an alternative to the microscopic method monitoring the presence of animal material in cattle feed. The ruminant, bovine and avian ELISA assays from institute 1 show good preliminary results for the detection of ruminant and avian material in compound feed. The bovine PCR assay of institute 2 shows unequivocal results for all the reference materials tested. However, the latter kit has the disadvantage that it cannot identify other ruminants than cattle. The DNA chip technique (BioMérieux) has a rather good score for the reference materials. At lower contamination levels some materials are not detected. Both the DNA chip technique and the PCR kit from institute 4 show a good score for the identification of porcine and avian material.

A total scheme with the results of all the reference materials tested with the available kits has to be completed. Although the results are promising, there is no method giving unequivocal results available yet.

Acknowledgements

This study is a collaboration of the European Fat Processors and Renderers Association (EFPRA). It is funded by the Natural Organic Products group of Sobel (Son, the Netherlands).

References

Regulation (EC) No 999/2001 of the European Parliament and of the Council of 22 May 2001 laying down rules for the prevention, control and eradication of certain transmissible spongiform encephalopathies.

Regulation (EC) No 1774/2002 of the European Parliament and of the Council of 3 October 2002 laying down health rules concerning animal by-products not intended for human consumption.

Commission Directive 2003/126/EC of 23 December 2003 on the analytical method for the determination of constituents of animal origin for the official control of feeding stuffs.

Policy on characterisation of antibodies used in immunochemical methods of analysis for mycotoxins and phycotoxins

J.M. Fremy[1] *and E. Usleber*[2]
[1]*French Agency for Food Safety, Research Laboratory for Food Quality and Process, 23 avenue du General de Gaulle, 94706 Maisons Alfort, France,* [2]*Dairy Science, Institute of Veterinary Food Science, Justus-Liebig-University, Ludwigstrasse 21, 35390 Giessen, Germany*

Abstract

The purpose of this document is to provide a policy on antibody characterisation for conducting AOAC INTERNATIONAL® (AOACI) immunochemical method Collaborative Studies for those methods submitted for AOACI® Official Methods℠ Program (OMA) status. The policy defines recommended information and characteristics to be provided by the Study Director, in the protocol of the collaborative study, for approval by AOACI®. The document specifies parameters for characterisation of antibodies used as biological reagent in the protocol of validation of immunochemical methods for the determination of mycotoxins and phycotoxins. These parameters consider the types of antibodies, avidity-affinity, and specificity (cross-reactivity). For each of these parameters, the following is indicated or suggested: (i) the definition or general principle, (ii) the procedure for providing information and/or specification: measurement of the concentration of immunoglobulin G, calculation of affinity constant, calculation of the percentages of relative cross-reactivity, and (iii) recommendation for requested information. This document does not only consider immunoassays, but also immunoaffinity chromatography (IAC). Regarding this technique, based on the use of IAC columns mainly for clean-up purposes, particular parameters such as selectivity and capacity are described and indicated. A table presented as a "Checklist for the description of - and information on - antibodies used in immunochemical methods" is included. The document was published in the Journal of AOACI, Vol. 86, No 4, 2003.

Keywords: Antibodies, affinity, cross-reactivity, immunoaffinity chromatography, selectivity, capacity

1. Introduction

The members of the Method Committee on Natural Toxins and Food Allergens (Committee D) of AOAC INTERNATIONAL® (AOACI) were faced with problems when reviewing protocols and reports of some collaborative studies submitted for AOACI® Official Methods^{SM} Program (OMA) status. These methods were based on the use of antibodies and designed for the quantitative determination of "total toxins" in the field of mycotoxins and phycotoxins. In reality, antibodies do not have a 100% cross-reactivity for other toxin analogues which are intended to be measured. Most mycotoxins and phycotoxins belong to a family of compounds involving several analogues. A lack of information on specificity (cross-reactivity) of the antibodies used as biological reagent in the Collaborative Study Protocol, was noticed several times. As a result, the need to determine criteria for cross-reactivity was identified to assist the Study Director in the protocol description of their methods and for the members of the Methods Committee to have a guide in their reviews.

During the AOACI® Official Method Board (OMB) meeting in September 2000, the development of a task force on "Policy for the Acceptance of Antibody Characteristics Used in Qualitative and Quantitative Methods" (TFAC) was suggested. The settlement and co-ordination of activities of the task force were charged to J. Marc Fremy and Ewald Usleber, chairman and secretary of Committee D respectively.

A final document was produced by the task force, approved by the members of Committee D members and OMB in November 2002, and submitted to the Board of Directors which suggested to publish the document in the Journal of AOAC INTERNATIONAL. After reformatting the document was submitted on 10 February 2003 and accepted for publication. on 26 March 2003 (Fremy, J.M. and E. Usleber, 2003. Journal of AOAC INTERNATIONAL 86: 868-871. Copyright, 2003, by AOAC INTERNATIONAL).

2. General

Immunochemical methods are based on the ability of antibodies to bind specifically different substances. The reversible association between antibodies and their corresponding antigens is called the immunological reaction. The binding forces involve hydrogen bonds and hydrophobic binding as well as weak molecular interactions like Coulomb and Van der Waals forces. Antigens are characterised by their immunogenicity as well as their antigenicity, which resides in their ability to initiate the production of antibodies in animals and subsequently interact or bind with them. Small peptides and small non-peptidic molecules such as mycotoxins and phycotoxins are not immunogenic and are called haptens. Once hapten is conjugated to a protein, it becomes immunogenic. Sensitive immunological methods were developed involving labelling techniques to measure the complex formation when the analyte i.e. the phycotoxin is present at concentrations less than µg-mg/ml. Generally, the terms *'immunochemical methods'* or *'immunoassays'* are used to refer to such methods. Among different labelling techniques used in immunoassay, the best known are radioimmunoassay (RIA), enzyme immunoassay (EIA) and enzyme-linked immunosorbent assay (ELISA). Immunoadsorption methods are based on temporary immobilisation of immunocomplexes on a solid support, immunoaffinity chromatography (IAC) is a means of retaining a given antigen from a complex matrix by using the anti-antigen antibodies. In this case, antibodies are immobilised on support and packed in a small column. This type of immunochemical method is considered as more a cleanup step of a specific analytical method than an immunoassay by itself. However, modification of IAC can be made to achieve an analytical purpose. The overall performance of any immunoassay is a function not only of the immunological principle, but also of the properties of the reagents, the properties of the sample matrix, the assay design, and the experimental errors. These basic principles determine the sensitivity, specificity, precision, and accuracy of the assay. The general protocol for the development of immunochemical methods on mycotoxins and phycotoxins includes in four areas: (a) preparation of immunogen: this includes the formation of a toxin derivative (if necessary) and conjugation of it to a protein; (b) production of antibodies: this includes immunisation, collection of the immune serum and purification of antibodies; (c) characterisation of antibodies: this includes determination of antibody titre and antibody specificity; (d) development and application of immunochemical techniques.

This document is focused mainly on the third area to specify parameters for antibody characterisation.

3. Types of antibodies

3.1. Polyclonal antibodies

Animals can produce several different types of immunoglobulins which bind, at a specific site (epitope), the antigen or hapten. These immunoglobulins are synthesised by several cell types of lymphocytes. Such immune serum contains polyclonal antibodies which generally are not monospecific for a the given antigen/hapten. Thus, these types of purified antibody can react or 'cross-react' with some structurally related antigen/hapten/analyte. The source of these antibodies stops with the death of the immunised animals.

3.2. Monoclonal antibodies

It is very difficult to obtain antisera with identical properties from different animals. Even from the same animal, the antisera successively collected at different times can have different properties. These limitations of polyclonal antibodies (pAb) justify attempts to produce monoclonal antibodies (mAb) which have identical physical, biochemical and immunological properties. The antibody-producing lymphocytes are isolated and then fused with myeloma cells. The hybrid cells are selected in a particular culture medium and then screened for their specific antibody produced. The result is a cell line clone, which only produces a specific monoclonal antibody (mAb). These cells can be maintained *in vitro* over long periods of time and can generate large quantities of identical antibody. They are - theoretically - immortal.

3.3. Recommendation

It is recommended that the type of antibodies - monoclonal or polyclonal - should be clearly mentioned in the collaborative study protocol for the immunoassay validation. It is also recommended that poly- and monoclonal antibodies have to be purified Consequently, the purification method should be stated or reference of the method should be given. In the case of mAb, when the producing cell line was referenced, the reference clone number should be indicated.

The type of immunoglobulins (Igs) should be determined. In the case of IgG, the concentration can be measured using the following equation:

$$IgG\ (mg/mL) = A / \varepsilon$$

A : UV absorption at 280 nm
ε : 1.45 for pure IgG

4. Avidity-Affinity

4.1 Principle

Affinity and avidity are often used synonymously and both relate to the energy of binding of a particular antigen-antibody combination. The term avidity refers specifically to the properties of an antibody, and the term affinity to those of the antigen. Numerically, avidity of the antibody for the antigen is equal to affinity of the antigen for the antibody. The affinity or avidity of the antibody has a strong influence on the further performance of the developed immunoassay or immunopurification chromatography and may be calculated, if the equilibrium constant is known. The antigen-antibody reaction may be described using reaction kinetics as well as thermodynamic equations. However, also time and temperature have influence on assay performance parameters.

In addition, the apparent performance characteristic of competitive immunoassay formats is dependent on the avidity for the labelled antigen, which is usually different from that of the free antigen. In immunoaffinity chromatography applications, avidity and affinity are identical due to the non-competitive conditions. However, for multi-analyte systems, different affinities for different antigens may interfere with each other.

4.2 Recommendation

It is recommended to include data on avidity of the antibodies more adequately the equilibrium constant for the antibody-hapten interaction used in the immunoassay as complementary information in the collaborative study protocol for the immunoassay validation. The method of determination (or estimation) of the avidity should be should be stated or reference of the method should be given.

5. Specificity

5.1. Principle

Next to avidity, the specificity of an antibody is paramount for the performance of the assay. In principle, a specific reaction in immunology may be defined as follows: In the presence of different molecules the specific antibodies must complex only one kind of molecules. The probability of forming a 'wrong' complex determines the specificity of the reaction. In other words, specificity is determined by the steric (tridimensional) match of antigen and antibody as well as by the number of molecular interactions taking place between both molecules. Discussion of specificity requires that both the structure of the antigen and the homogeneity or heterogeneity of the antibodies to be considered. An antibody preparation is homogenic, if all the antibodies bind only the one and the same epitope, although with different affinity. This condition is fulfilled by monoclonal antibodies, but may also be by polyclonal antibodies against compounds of low molecular weight (haptens). On the other hand an antibody preparation is heterogenic, if it contains different antibody populations specific for different epitopes.

On a molecular basis "specific" cross reactivity describes the case in which at least two different antigens compete for the same antibody-binding site. Almost exclusively true cross-reactivity is observed in competitive assays using either monoclonal or polyclonal antibodies against low molecular weight compounds. In practice this means that a high enough concentration of a truly cross-reacting substance gives the same result as the 'right' substances in a competitive assay.

Also, non-specific influence on many instrumental assays can occur. Non-specific cross-reactivity includes many factors that influence the assay including unknown compounds with similar epitopes as the target compound (e.g. matrix effects) and environmental effects (steric effects, pH, temperature, etc.). Such interference occurrence often cannot be distinguished from the specific influence of cross-reacting substances. In both cases, specific and non-specific cross reactivities, the assay response is the same and a false positive or a false negative result may be obtained. Due to the immunoassay principle, however, false negatives are less likely found in competitive assays providing the assay is well optimised.

5.2 Procedure

Test specificity describes the extent to which presence of substances with chemical structure or structural parts similar to that of the original analyte (test substance) in a sample will also result in competitive binding and thus give a positive result in the test system (cross-reaction). Test specificity is a result of antigen-antibody reaction, which is determined by the immunoreagents used in the test (antibody, labelled antigen), and must therefore not be commingled with non-specific influences on the test system (e.g., sample matrix interference; sample solution pH).

The specificity study should be performed under the conditions, which closely match the 'real' analyte situation during sample analysis. To avoid the introduction of anomalous effects due, for example, to solvation effects on the analyte, the solvent concentration for the preparation of the standard curve should be identical with those of analysis of the sample.

In a competitive immunoassay such as those used in the analysis of mycotoxins and phycotoxins, specific cross reactivity is calculated after determination of the concentrations of the test substance [TS] and the cross-reacting substance [CS] required for 50% reduction of the absorbance reading compared to that of the zero control standard using the following equation:
% relative cross-reactivity = [TS]÷[CS] × 100

However, in the non-competitive immunoassays, such as the sandwich immunoassay for the analysis of large molecular weight compounds, e.g. allergens, an increase in absorbance reading is used in the calculations. The identification of non-specific reactivity for some matrices provides valuable information regarding the extent of applicability of the immunochemical method. Non specific cross-reactivity can be determined by a technique described in "Immunoassays in Food Analysis" (Morris and Clifford, see Bibliography). This technique quantifies non specific cross reactivity by comparing a standard curve of varying concentration of specific antigen prepared in buffer, with identical curves prepared in varying and increasing amount of sample matrix. If the curves superimpose at all points, then one can assume that the IgG being examined is specific for the antigen in question and is not susceptible to interference from other related compounds (known or unknown), which may be present in the sample matrix.

5.3. Recommendation

5.3.1. Non specific cross-reactivity

When it has been identified, the occurrence of interferences due to the effect of some matrices has to be indicated in the scope or applicability of the method. Presence of false positive and/or false negative results for some matrices has to be clearly mentioned. Consequently, it should be stated under the scope or applicability section in what matrices (commodities, foods, etc.) the antibody and/or system is intended to be used. For example: "This assay or system is intended for use in corn and wheat only. The effect of other sample matrices on the assay has not been determined. The effect of other sample matrices can give interferences and consequently indicate false positive or negative results", or another example: "This assay or system is intended for use in corn and wheat only. The effect of other sample matrices such as corn gluten, citrus pulp...(list) give interferences and consequently indicate false positive (or negative) results".

5.3.2. Specific cross-reactivity

All significant specific cross-reactions should be indicated. In case the analyte belongs to a group of compounds which are structurally closely related, every group members should be tested. For example, for an antibody used in an assay for one or total aflatoxin(s) in cereals, at least AFB1, AFB2, AFG1 and AFG2 have to be tested. However, for some other groups involving many compounds which can be divided in two or several sub-groups or families, specificity study should be conducted on the major compound or the family of compounds considered as analyte and its most appropriate closely related congeners. Particular recommendation may be addressed according to both particular group of substances and particular assay design

Tests should be grouped according to their specificity pattern as:

- Cross-reactivity ≤ 10%. Substance specific; test may be used for quantitative determinations so long as the cross reacting substance is not present within the sample matrix at levels greatly above those of the test substance (example: even if target compound A and cross-reacting compounds B, C, and D are present in one sample at equal concentrations, the result will be 1.4 times overestimation of A at a maximum, which can be considered to be within the general limits of purpose of enzyme immunoassays).

- Cross-reactivity ranging from > 10% to 100%. Qualitative test; it may be used for quantification only if the identity of the compound is known (example: if compounds A, B, C and D are present at equal concentrations, overestimation of A may range from 1.1 to >4, which cannot be accepted as quantitative).
- Cross-reactivity ranging from ≥ 50% to ≤ 150%. Semi-quantitative group-specific test; may be used for quantification only if the identity of the compound is known (example: compounds A, B, C, and D have cross-reactions of 100%, 150%, 50%, and 100%, respectively. There is a fourfold overestimation of A at equal concentrations of A-D, but since the detection of the group may be desirable the quantification is still acceptable). Also, this must accommodate assays for toxin groups or families in which the differing congeners have similar toxicity (example: when a family of target compounds of similar structure and toxicity are bound by the antibodies within the assay, and cross reaction are within ≥ 50% to ≤ 150% range, the additive or integrative nature of the immunoassay is a major benefit, as the assay can give a measure of 'total X toxin family').

One aspect, which may have to be considered with natural toxins, is the purity of the test standard solutions available. While some substances, e.g. the major aflatoxins, are available with purities >99%, other toxin standards may contain other with structurally related toxins. In the latter case, the information given by specificity tests has to be taken with care.

5.3.3 Immuno Affinity Chromatography (IAC) columns

When antibodies are immobilised in an IAC column, two special criteria have to be mentioned:

- Selectivity. All compounds, which are specifically bound by the antibodies used in IAC under the test protocol, should be listed.
- Capacity. The maximum amount of specific antigen-analyte per column bound under working conditions should be determined and indicated. In the case of multi-analyte column indicate whether or not and to what extent, the presence and binding of analyte A may interfere with binding of analyte B.

Acknowledgements

Table 1. Checklist for the description of antibodies used in immunochemical methods.

Parameter	Specific information required	
Type of antibody	Polyclonal (host)	Monoclonal (cell line reference number)
Antibody purification	e.g., Raw serum fraction, or purified by immunoaffinity chromatography or other procedure (purification method stated).	Purification method stated, when it has been used.
Avidity	Mention the Association Constant (calculation method stated or referenced; e.g., 50% inhibition value of standard curve, if applicable, for example ELISA-type test systems).	
Specificity (cross-reactivity) of related antigens (applicable for ELISA-type test systems)	Mention ALL relevant compounds with % cross-reactivity. Substances with cross-reactivities of less than 10% may be summarised (for example "other aflatoxins tested with cross reactivities of less than 10% were aflatoxin Bx, Cx..."). The basis of calculation of cross-reactivities should be mentioned (e.g., comparison of 50% inhibition values for analytes A, B, C...). According to the specificity pattern, indicate whether the assay is 1) quantitative, 2) semi-quantitative, or 3) qualitative.	
Matrix effects (non-specific binding inhibition) applicable for use of antibodies in ELISA-type and IAC systems	Define whether or not matrix effects (false-positive, false-negative) may occur for certain sample matrices using the respective method protocol. Mention in the scope or applicability section in what matrices (commodities, foods, etc.) the antibody and/or system is intended to be used.	
Selectivity (binding) of antigens in IAC-type extract purification systems	List all compounds which are specifically bound by the antibodies used in IAC under the test protocol.	
Specific capacity of IAC columns	Indicate the maximum amount of antigen per column bound under working conditions. If applicable (multi-analyte columns), indicate whether or not, and to which extent, presence and binding of analyte A may interfere with binding of analyte B	

The authors as co-ordinators of this collective project developed under the auspices of AOACI® sincerely thank firstly the members of Task Force: Fun Chu (USA), Ian Garthwaite (New Zealand), Alice Lee (Australia), Chris Maragos (USA), and George Ware (USA) for their contribution. And secondly the members of AOAC INTERNATIONAL Method Committee on Natural Toxins and Food Allergens: Pedro Burdaspal (Spain); Tetsuhisa Goto (Japan); Winston Hagler (USA), Sue Hefle (USA); Douglas Hite (USA); Susan MacDonald (UK), Tom Nowicki (Canada); Gordon Shephard (South Africa); Hans van Egmond (the Netherlands) for their comments.

References

Fremy, J.M. and Chu, F.S, 1989. Immunological methods for Aflatoxin M1 Determination. Chapter in "Mycotoxins in Dairy Products", H. van Egmond, editor. Elsevier App. Sc., Barking (UK), p. 97-125.

Fremy, J.M., Puech, L., Dragacci, S. and Krys, S., 1999. Immunological Methods for the Detection of Diarrhetic Shellfish Poisoning Toxins. Report for the Symposium on Analytical Chemistry of Phycotoxins in Seafood and Drinking Water, M. Quilliam coord., AOAC Int'l Annual Meeting, Houston, Sept. 26-30, 1999.

Märtlbauer, E., 1993. Basic principles which determine the quality of immunoassays. Int'l Dairy Federation, Editions, Brussels (Belgium). Special issue no. 9302, p. 137.

Morris, B.A. and Clifford, M.N. editors, Immunoassays in Food Analysis. Elsevier Applied Science Publishers. p. 37

Usleber, E., Dietrich, R., Bürk, C., Schneider, E., and Märtlbauer, E., 1999. Immunoassay Methods for Paralytic Shellfish Poisoning Toxins. Report for the Symposium on Analytical Chemistry of Phycotoxins in Seafood and Drinking Water, M. Quilliam coord., AOAC Int'l Annual Meeting, Houston, Sept. 26-30, 1999.

Rapid detection methods for marine toxins: Needs and limits

S. Krys
The French Food Safety Agency, Lerqap, TOP unit, 23 avenue du Général de Gaulle, 94706 Maisons-Alfort cedex

Abstract

The accumulation by filter feeding shellfish of toxins from phytoplanktonic origin represents a threat for human health. Despite progress, scientists are still unable to prevent the natural blooms of phytoplankton and to understand the mechanisms of toxin production. The only way to protect consumers' health is to control the level of contamination of shellfish before harvesting, as prescribed in the EU Directive (91/492/EEC amended by the Directive 97/61/EC and the Decision 2002/225/EC). It has been a challenge for analysts because of the co-occurrence of several analogues of the same family of toxins (PSP toxins) or of several families of toxins (DSP and related toxins). For these reasons, the tests are based on the assessment of the global toxicity with the mouse bioassays (MBA), as the reference methods recommended at the EU level. Despite the efficiency of these tests, the MBA are of great concern because of the scientific and ethical problems they raise. Thus, the EU commission and its network of Community and National Reference Laboratories have been considering as a high priority the replacement of the animal assays (Directive 86/609/EEC). On another side, biotechnology companies have made numerous efforts last few decades in order to develop alternative non-animal, rapid and low cost methods. A large variety of approaches have been described, based either on the mode of action of toxins (functional assays) or on their molecular structures (immunochemical assays). Unfortunately, some of these promising methods did not succeed in the international validation procedures and/or could not be considered as giving an equivalent level of protection of public health in comparison to the MBA, stopping their potential recognition by the regulatory authorities.

Keywords: Phycotoxins, consumers' health protection, official control, mouse bioassays, alternative assays, functional assays, immunochemical assays, validation

1. Introduction

The phytoplankton (vegetal plankton) constitutes the primary producer of the marine food chains and several phytoplanktonic species coexist in seawater. In some favourable combination of ecological and climatic conditions, some species have the ability to proliferate up to gain a predominant occurrence, a natural phenomenon so-called "algal bloom" or "red tides" because of the visible colour of the seawater. Some of the blooming species are able to produce secondary metabolites, the phycotoxins, responsible for specific bioactivity. During a toxic event, their accumulation in filter-feeding shellfish constitutes a potential risk of direct transmission of these toxins to the consumers. Historically, several forms of sickness have been defined according to the symptoms developed in humans; diarrhetic shellfish poisoning (DSP) and azaspiracid poisoning (AZP), both being diarrhetic syndromes, paralytic shellfish poisoning (PSP), amnesic shellfish poisoning (ASP) and neurological shellfish poisoning (NSP) (Yasumoto, 2000). The toxins induce an acute and reversible toxicity (Table 1), and lead mostly from moderate to severe cases, sometimes lethal.

Despite large progress in the knowledge, the scientists are still unable to predict and prevent the natural blooms of some phytoplanktonic species, as well as to understand the mechanisms of toxin production. Since it is not visually possible to distinguish contaminated shellfish from non contaminated ones, and considering the particularly high stability of the toxins (temperature, acidic gastric conditions), the only way to protect the consumers' health is to control the level of contamination of shellfish before harvesting, as prescribed in the EU Directive laying down the health conditions for the production and the placing on the market of bivalve molluscs (Directive 91/492/EEC amended by the Directive 97/61/EC and the Decision 2002/225/EC).

Controlling the level of contamination of shellfish has been a challenge for the analysts because the contamination is not related to the presence of a unique compound but to the co-occurrence of several analogues of the same family of toxins (e.g., PSP toxins) or of several families of toxins (e.g., DSP toxins and related toxins), the various analogues exhibiting different degrees of toxicity and each family having a different mode of action. Rodent assays, particularly mouse bioassays (MBA), constitute the most immediate answer to assess the toxicity but they raise scientific and ethical problems. As a consequence, a crucial need for equivalent alternative non-animal, rapid and low cost methods has

Table 1. Clinical signs of the EU-relevant marine phycotoxins poisonings (from Krys and Lavieille, 2002).

Syndromes		Onset	Major symptoms	Signs of severity	Evolution
• Diarrhetic syndrome	– Diarrhetic phycotoxins	Short, mean: 4 h (30 min - 12 h)	Diarrhoea (92%) Nausea (80%) Vomiting (79%) Abdominal pain (53%)		No after-effect, not lethal, Recovery in 3 days
	– Azaspiracids	Variable: 2.5 h - 18 h	Nausea Vomiting Severe diarrhoea Gastric pain Headache Moderate fever		No after-effect, not lethal, Recovery in 2 - 5 days
• Paralytic syndrome		Very short, 5 - 30 min	– Gastro-intestinal symptoms Nausea, vomiting.	Headache, Confusion	Death possible by respiratory paralysis (2h to 12h after ingestion)
			– Neurological symptoms Paresthesia and numbness of the lips and mouth, might spread to the arms and legs	Ataxia, Respiratory troubles	Recovery (2-3 days), no after-effect
• Amnesic syndrome		Variable: 15 min - 36 h	– Gastro-intestinal symptoms (first 24 h): Nausea, vomiting, headache, diarrhoea, abdominal pain – Neurological symptoms (24-48 h) : Confusion, Mnesic troubles	Increase of the confusion, disorientation, memory loss, coma	Death possible, Slow recovery (days to months), definitive after-effects possible (loss of memory)

appeared for several years, particularly to be used as own-checks by the producers or also to complement or replace the MBA as part of the official controls. A large variety of approaches have been described, based either on the mode of action of the toxins (functional assays as receptor-ligand binding, receptor-mediated cytotoxicity, protein phosphatase inhibition assays) or on the molecular structures of the toxins (immunochemical assays). Unfortunately, for the last few years, some of these promising methods did not succeed in the international validation procedures (ruggedness parameter not satisfying) and/or could not be considered as giving an equivalent level of protection of public health in comparison to the MBA, stopping their potential recognition by the sanitary authorities.

2. The Mouse Bioassay (MBA): a consumer-protective method

As it detects the global toxic bioactivity resulting from several co-occurring toxins, the MBA is the EU-reference test for shellfish in monitoring programs for DSP and PSP toxins since 1996. Despite the proved efficiency of the MBA tests, as shown by the decrease of outbreaks since the preventing systems are in operation, the MBAs are of great concern because of the scientific and ethical problems they raise. Indeed, the tests do not enable the individual identification of the toxins and the determination of their concentrations and have occasionally shown to produce false positive results due to human non-toxic interferences. The MBA methods have never been submitted to a validation procedure mainly because of the lack of standards of the various analogues and the lack of certified reference materials (CRM) leading to insufficient determination of their performance criteria. However the major concern is related to the conflict it creates between the European directives for shellfish toxin testing and the European legislation related to animal welfare (Directive 86/609/EEC) that prescribes the "3R" principles; for Refinement, Reduction and Replacement of animal testing.

In regard to the scientific questioning arising from atypical cases of MBA, consequent conflicts with the industry and in order to increase the compliance to the directive on animal welfare, the EU commission and its network of Community and National Reference Laboratories have been considering as a high priority to move from the MBA to equivalent validated non-animal rapid methods. It is important to notice that this willing has been clearly specified in the new decision (2002/225/EC) of the commission that lays down detailed rules for the implementation of Council Directive 91/492/EEC as regards to the maximum levels and the methods of analysis of certain marine biotoxins in live bivalves molluscs, echinoderms, tunicates and marine gastropods.

3. Another toxicological approach: the functional assays

These tests are based on the interaction between the toxins and their specific receptor leading in some cases to cytotoxicity.

This approach has been used by developers for the detection of PSP toxins and DSP toxins.

3.1 PSP *in vitro* tissue culture bioassays

This test is based on the measurement of the survival of neuroblastoma cells (Neuro 2A), according to the principle described by Kogure *et al.* (1988). It was later developed in a microtiter plate format to become a routine quantitative test (Jellet *et al.*, 1992). Unfortunately, the international interlaboratory study conducted in 1998 on behalf of the AOAC for validation (14 laboratories and 18 duplicate blind samples) failed. Several reasons were identified (damages of the cells during transportation, problems with frozen samples, proficiency of people...). In parallel, a study conducted in Alaska showed very good results, with over 90% of agreement between the results from the MBA and the kit. This was further explained by the fact that the kit was mainly developed and optimised with contaminated shellfish from Alaska, resulting in a very good response for the toxin profiles of that geographic area. This example shows that the kit was not satisfactory for ruggedness, in other words, its transferability was not possible.

3.2 DSP in vitro tissue culture bioassays

Several tests were developed on different cell lines for the detection of the diarrhetic toxins okadaic acid (OA) and dinophysistoxins (DTXs), following morphology changes or cytotoxicity (Amzil *et al.*, 1992; Diogene *et al.*, 1995). A DSP morphology-based test was also developed to detect the 3 classes of toxins of the DSP complex (AO and DTXs, pectenotoxins [PTXs] and yessotoxins [YTXs]), (Aune *et al.*, 1991). However, none of these tests was developed on a routine format, neither submitted to validation. Furthermore, these types of tests have a low specificity and have never been used extensively for regulatory purposes.

3.3 AO/DTXs *in vitro* protein phosphatase inhibition assays

The enzymes protein phosphatases (PP) 2A and to a lesser extent PP1 are inhibited by OA. Based on this inhibition, several assays with different detection modes have been developed, the fluorimetric detection (Vieytes et *al.*, 1997) and the colorimetric detection (Tubaro *et al.*, 1996, Simon and Vernoux, 1994) being the main ones. These assays seem to be very promising to offer an adequate alternative to bioassays for the detection of the OA and DTXs toxin family since they are rapid, sensitive and specific. However it is important to determine their performances on a large scale and especially their ruggedness. At the moment, one interlaboratory study is being organised by the Community Reference Laboratory for biotoxins in order to validate one kit on a microplate format with fluorimetric detection.

3.4 YTXs *in vitro* activation of phosphodiesterases assays

The mode of action of YTXs is not completely elucidated but it is nevertheless known that their presence is responsible for activation of cellular phosphodiesterases (PDEs). An *in vitro* test is being developed based on these intracellular targets, on a microtiter plate format, with fluorimetric detection as an endpoint (Alfonso *et al.*, to be published). Similarly to the PP2A test for OA/DTXs, the validation is required to have a definitive conclusion of the usefulness of this type of test on a large regulatory scale but as it is specific, sensitive and easy to use, it seems to be promising.

4. A structural approach: the immunochemical assays

The common base for these tests is the recognition of a structure by antibodies.

4.1 ELISAs

Two ELISA kits were produced to detect mainly OA and to some extent DTX1 and DTX2. Tested against the MBA, the main problem with these kits was related to the low cross-reactivity with DTX1 leading to an underestimation of the total toxicity. Theses kits are no longer available, but in any case, any new kit developed needs now to integrate a step of hydrolysis in order to detect the acyl-esters of the native toxins, called DTX3, that are known to be metabolised in the shellfish and to exhibit a toxic activity.

Similarly, a commercial kit was available for the detection of STX, one of the major toxins of the PSP toxins family. This ELISA kit used in France led to conflicting results between official control with the MBA and field own-checks performed with the ELISA kit; the latter ones being alternatively overestimated or underestimated. A systematic study carried out in our laboratory showed technical shortcomings and low cross-reactivity to the toxins present in the toxic event, mainly gonyautoxins 1 to 4.

A new kit, based on an improved technology, is now commercialised and is being characterised in our laboratory in order to assess its potentiality to be used on regulatory purposes. The first results are promising but the problem of cross-reactivity with the major analogues might still occur because the kit is mainly developed for STX.

4.2 Lateral flow Immunochromatography

A screening tool for the qualitative detection of the family of PSP toxins is commercialised as a dipstick with visual distinction between positives and negatives (Jellet *et al.*, 2002). It is interesting because it has been designed not only for STX but also for all the analogues, as it uses a mixture of antibodies. Another interesting point is that is was submitted to validation for the extraction step as well as for the detection. It may be used in the field, nevertheless it is important to raise that before its use the test needs to be characterised in relation with the local toxin profiles.

5. Conclusion

It is noticeable that numerous efforts have been made by developers and scientists involved in control of shellfish in order to produce rapid methods for the detection of marine biotoxins. Unfortunately, we are facing the evidence that a very small number of them are in use, especially for regulatory purposes. This latter use implies that the rapid methods must comply with the necessity of offering an equivalent level of protection of the consumer (detection of at least the major analogues) and need to have succeeded in a validation process. It is important to emphasise to the developers that the selection of a rapid method by a regulatory organisation to be an official method might not be possible unless these two requirements are fulfilled. One major parameter of failure in the validation process of promising rapid method has been identified to be the ruggedness, this means their ability to be used in a large geographic scale and to be transferred in a large number of laboratories. An additional major obstacle to the development and to the validation of new rapid methods is directly related to the crucial lack of standards of many analogues of toxins, underlining the high priority that should be given for their production at the EU-level.

References

Alfonso, A., Vieytes, M.R., Yasumoto, T. and Botana, L.M., to be published. A rapid microplate fluorescent method to detect yessotoxins based on their capacity to activate phosphodiesterases. Biochemical Pharmacology.

Amzil, Z., Pouchus, Y.F., Le Boterff, J., Roussakis, C., Verbist, J.F., Marcaillou-Le Baut, C. and Masselin P., 1992. Short-time cytotoxicity of mussel extracts: a new assay for okadaic acid detection. Toxicon 30, p. 1419-1425.

Aune, T., Yasumoto, T. and Engeland, E., 1991. Light and scanning electron microscopic studies on effects of marine algal toxins toward freshly prepared hepatocytes. J. Toxicol. Environ. Health 34. p. 1-9.

Diogene, G., Fessard, V., Ammar, M., Dubreuil, A. and Puiseux Dao, S., 1995. Cytotoxic responses of maitotoxin extracts and okadaic acid. In: P. Lassus, G. Arzul, E. Erard, P. Gentien, C. Marcaillou, eds. Harmful Marine Algal Blooms, p. 285-289. Lavoisier, Paris.

Directive 91/492/CEE du Conseil du 15 juillet 1991, fixant les règles sanitaires régissant la production et la mise sur le marché de mollusques bivalves vivants. Journal officiel L 268 du 24/09/1991, p. 1-14.

Directive 97/61/CE du Conseil du 20 octobre 1997, modifiant l'annexe de la directive 91/492/CEE fixant les règles sanitaires régissant la roduction et la mise sur le marché des mollusques bivalves vivants. Journal officiel L 295 du 29/10/1997, p. 35-36.

Décision 2002/225/CE de la Commission du 15 mars 2002 fixant les modalités d'application de la directive 91/492/CEE du Conseil en ce qui concerne les limites maximales et les méthodes d'analyse de certaines biotoxines marines dans les mollusques bivalves, les échinodermes, les tuniciers et les gastéropodes marins. Journal officiel L 075 du 16/03/2002, p. 62-64.

Jellett, J.F., Marks, L.J., Stewart, J.E., Dorey, M.L., Watson-Wright, W. and Lawrence J.F., 1992. Paralytic shellfish poison (saxitoxin family) bioassays: automated endpoint determination of the in vitro tissue culture bioassay, and comparison with the standard mouse bioassay. Toxicon 30, p. 1143-1156.

Jellett, J.F., Roberts, R.L., Laylock, M.V., Quilliam, M.A. and Barrett, R.E., 2002. Detection of paralytic shellfish poisoning (PSP) toxins in shellfish tissue using MIST Alert™, a new rapid test, in parallel with the regulatory AOAC® mouse assay. Toxicon 40: 1407-1425.

Kogure, K., Tamplin, M.L., Simidu, U. and Colwell, R.R., 1988. A tissue culture assay for tetrodotoxin, saxitoxin and related toxin. Toxicon 26. p. 191-197.

Krys, S. and Lavieille, S., 2002. Intoxications alimentaires par les phycotoxines. La Revue du Praticien 52, p. 2209-2211.

Simon, J.F. and Vernoux, J.P., 1994. Highly sensitive assay of okadaic acid using protein phosphatase and paranitrophenyl phosphate. Nat. Toxins 2, p. 293-301.

Tubaro, A., Florio, C., Luxich, E., Sosa, S., Della Loggia, R. and Yasumoto, T., 1996. A protein phosphatase 2A inhibition assay for a fast and sensitive assessment of okadaic acid contamination in mussels. Toxicon 34, p. 743-752.

Vieytes, M.R., Fontal, O.I., Leira, F., Baptista de Souza, J.M.V. and Botana, L.M., 1997. A fluorescent microplate assay for diarrheic shellfish toxins. Anal. Biochem. 248. p. 258-264.

Yasumoto T., 2000. Historic considerations regarding seafood safety. In: Luis M. Botana, ed. Seafood and freshwater toxins: pharmacology, physiology and detection, p.1-17. Marcel Dekker, New York.

Rapid detection of bacterial food-poisoning toxins: A personal view

G.M. Wyatt
Food Safety & Computational Microbiology Group, Institute of Food Research, Norwich Research Park, Norwich NR4 7UA, United Kingdom

Abstract

Illnesses caused by toxins produced by foodborne bacteria can be divided into those in which the toxin is formed *in vitro* by growth of the bacterium in the food prior to consumption, and those in which the toxin is produced *in vivo* after human consumption of the bacterium. The characteristics of the illness for the two groups are different, and the problems posed for the analytical microbiologist also differ. Classical testing for the presence of toxins often involves the use of animals; in addition to the obvious technical difficulties, it is also highly desirable to replace such tests with molecular methods for animal welfare reasons. In this article, selected developments in rapid molecular methods will be described, and illustrated with examples familiar to the author.

Keywords: bacterial toxins, food, immunochemistry, PCR, dipsticks

1. General considerations

Many types of human illness are caused by toxins produced by foodborne bacteria; broadly, however, such illnesses can be divided into those in which the toxin is formed *in vitro* by growth of cells of the pathogen in the food prior to consumption (foodborne intoxications), and those in which the toxin is produced *in vivo* after consumption of the bacterium in contaminated food (foodborne infections). The latter type may or may not also involve invasion of the host cells. In the context of food-poisoning, the toxins are usually, but not exclusively, enterotoxins, i.e. toxins which affect the human gut, generally producing symptoms such as diarrhoea and vomiting. The principal exception is the neurotoxin produced by *Clostridium botulinum*.

Examples of foodborne intoxications are shown in Table 1, and of foodborne infections in Table 2. Toxin production *in vivo* may also play a major or minor role in the pathogenesis of other foodborne infections, such as those caused by campylobacters and salmonellas, although in many cases neither the structure of the toxin nor its mechanism has been fully elucidated.

The characteristics of the illness for the two groups are different, and the problems posed for the analytical microbiologist also differ.

1.1 Foodborne intoxications

Classical testing for the presence of toxins often involved the use of animals. Examples include the mouse lethality test for *C. botulinum* toxin and rabbit intestinal loops for *E. coli* enterotoxins. In some cases, *in vitro* cell culture has replaced the use of live animals, e.g. the use of Vero cells for detection of shiga-like toxin (verotoxin) from *E. coli*. In addition to the obvious technical difficulties with these tests, it is also highly desirable to replace the tests with molecular methods for animal welfare reasons.

Routine detection of foodborne intoxications, to be effective as a control measure, would probably have to take place immediately prior to consumption, as in many cases it is during the final stages of the food chain that the toxin is formed. For example, toxin production by *Staphylococcus aureus* and (some types of) *Bacillus cereus* often takes place in foods left at room temperature in domestic situations, or uncontrolled catering outlets, particularly during warm summer weather. Toxin testing at point-of-consumption is not routinely done, as far as the author is aware. However, rapid and user-friendly analytical methods might allow this to take place; of course, taking into account the comments above, this would inevitably include testing being carried out by the consumer in their own home or in a restaurant - a challenge indeed to the method developer, and probably also for the social scientist. One can only imagine the work that this might bring to the legal profession, especially if such tests were not 100% specific.

For such situations, completion of the test in a timescale of minutes would be necessary. The test would have to be very user-friendly, with a low skill input; some of the dipstick devices described later have these characteristics. Ideally, the test would have to be 'idiot-proof', if such a thing is possible, but the success of home pregnancy tests in a similar format illustrates that the approach may

Table 1. Characteristics of selected foodborne intoxications (adapted from Johnson, 2003).

Organism	Symptoms	Incubation period	Duration of illness	Toxin type	Toxic dose	Common food vehicles
Bacillus cereus (emetic type)	Nausea & vomiting	1-6 hours	6-24 hours	Heat-stable 1.2kDa peptide	500 µg - 2 mg	Almost always rice dishes
Staphylococcus aureus	Nausea, vomiting, diarrhoea	1-6 hours	6-24 hours	Heat-stable 26-29 kDa protein	100 - 200 ng	Salads, ham, cream cakes
Clostridium botulinum	Neurological (breathing, paralysis); may be fatal	12-72 hours	Weeks or months	Heat-labile 150 kDa protein	10 ng?	Home-canned foods, fermented fish

Table 2. Characteristics of selected foodborne infections (adapted from Johnson, 2003).

Organism	Symptoms	Incubation period	Duration of illness	Infectious dose	Toxin type	Common food vehicles
Bacillus cereus (diarrhoeal type)	Intestinal pain, diarrhoea	6-12 hours	12-24 hours	10^5-10^7 cells	Heat-stable, structure not fully known	Bulk-cooked meats and vegetables
Clostridium perfringens	Intestinal pain, nausea, diarrhoea, fever	8-16 hours	16-24 hours	10^7-10^8 cells	Heat-labile 35kDa protein	Bulk-cooked meats and vegetables
Escherichia coli:						
ETEC (entero-toxigenic)	Principally diarrhoea (travellers diarrhoea)	16-48 hours	1-3 days	10^5-10^8 cells	Many, protein, stable and labile types	Salads, unpasteurised dairy products (contaminated water)
VTEC (vero-toxigenic)	Bloody diarrhoea, intestinal pain; systemic conditions (HUS)	1-7 days	May be weeks	10 cells or less	Heat-labile protein, 70kDa	Beefburgers, milk, apple 'cider'. Also . direct contact with animals

be feasible. The sensitivity (minimum level of detection) would have to be at least as low as the toxic dose, which means in the ng - µg range depending on the toxin, and the specificity would need to be very close to 100% for the legal and social reasons mentioned above. There is, however, still the problem of food sampling which would have to be both representative and efficient; such considerations lie outside the scope of this article.

More likely is that a rapid test for a foodborne toxin would be applied post-consumption and post-illness, as a public health or epidemiological measure to trace the source of illness. In this case, the test device would need to be suitable for use by a public health professional, possibly at the patient's home or in a restaurant. Here, although it would still need to be user-friendly, a moderate skill input could be tolerated as it would be used by trained staff. A slightly longer test timescale might be acceptable, maybe up to 30 minutes. However, if the food sample is taken to a laboratory, the timescale could be hours. Sensitivity and specificity of detection similar to the above would still be required.

1.2 Foodborne infections

In the case of foodborne infections, detection of the organism, perhaps coupled with confirmation of the presence of the toxin gene(s), is the usual analytical route. Here, the potential for time-saving with a rapid method is enormous. As there is (usually) a low initial number of the target pathogenic organism in a food sample, traditional microbiological methods often involve selective enrichment culture, i.e. growth in a broth culture medium formulated to increase the number of target cells and to produce a more favourable ratio of target cells:background flora. This is usually followed by culture on a diagnostic agar to provide a provisional identification, and subsequently biochemical confirmation of the identity of the isolated organism and classical toxin testing.

In many countries there is a legislative requirement to ensure that certain foods are free of particular pathogens; in practice, this means a sensitivity of 1 cell per 25 g food, hence the need for the enrichment culture technique. Even if this sensitivity could be achieved by a rapid method used directly on the food, the problem of actually finding a single cell in a 25g sample remains. Therefore, it is sensible to take advantage of the natural 'amplification system' provided by division of bacterial cells during growth to lower this very stringent sensitivity requirement and also increase the chances of locating the target organism(s).

Traditional approaches are, of course, lengthy, often taking several days before the presence of the toxin-producing organism in the food sample is confirmed. A 'rapid' method saving perhaps even one day may offer advantages over the traditional technique. Time saving depends largely on the point of application of such a test and here the use of a PCR method directly on, for example, the initial enrichment culture of the food can reduce the time markedly; the use of such an approach for *C. botulinum* is discussed later. Clearly, testing for pathogenic bacteria in food is likely to be laboratory-based rather than being used by the consumer.

Usually for foodborne pathogenic bacteria it is not necessary to enumerate (quantify) the organism; indeed, the use of enrichment culture obscures the original number of cells present in the food sample, making this difficult. For most purposes, including the legislative reasons mentioned above, it is sufficient merely to have a presence/absence result for the test. Here a test for the presence of a toxin gene in the mixed population of bacteria that results from culture of a food sample, such as the PCR method to be described later, is especially useful.

The specificity of the detection test is obviously important, especially when there are potential legal implications of false-negative results. Again, a new test should be user-friendly, which might encourage wider screening of foods for quality control purposes or HACCP schemes, in addition to the legal requirements. As the method is likely to be laboratory-based, moderate or high skill input can be anticipated.

1.3 Testing 'at the bedside' or 'in the field'

Although testing of food samples in the home or catering situation (in the case of toxins) was discussed above, some rapid method formats also have the potential for medical, agricultural or veterinary use outside the laboratory, mainly for direct detection of pathogenic bacteria rather than their toxins. For example, in many cases of gastrointestinal illness the pathogenic bacterium is present in high numbers in the faeces of the patient, e.g. salmonella, rather than the much lower number likely to be found in a food sample, and so enrichment of the organism by culture before detection is unnecessary. Rapid patient-side testing in this situation may allow faster and/or more appropriate antibiotic therapy to be given. In this case, the timescale for the test must again be in the order of minutes to be useful. In the same context, cow-side testing of milk for the detection of mastitis, for example, may be possible with test devices that are portable, thus reducing the possibility of infected milk entering the food chain.

In these instances, sensitivity should not be a problem, as high numbers of target organisms are present, but a short test timescale and low skill input would be highly desirable.

1.4 Test formats

This article is not intended to be a comprehensive review of rapid methods in food microbiology, so only selected technologies and formats will be described in detail. In general, however, it can be said that rapid methods for detection or quantification of toxins are mainly limited to the use of binding molecules, i.e. antibodies and receptors, and combinations of these molecules. Assay types include ELISAs or equivalent assays with other binding molecules, and precipitation/agglutination assays.

To detect the organism and/or toxin genes, many technologies and formats are in use or under investigation. For example, immunochemical and immunophysical methods include labelling methods as diverse as EIA/ELISA and fluorescence microscopy, and precipitation/complexing methods such as agglutination, immunomagnetic separation and chromatographic techniques. Nucleic acid-based methods include PCR and other amplification technologies (e.g. NASBA for RNA) and hybridisation technologies. Other technologies include those based on the biological activity of bacteriophages, such as the introduction of bioluminescence to the target organism based on the *lux* gene or green fluorescent protein, fluorescent staining of phage DNA or the ice nucleation gene. One problem with phage-based methods is that the host range is often too specific, recognising only a sub-set of a particular bacterial species. This is being overcome by the use of phage cocktails or by genetic engineering techniques.

These technologies have been incorporated into manual formats such as slides, microplates or wells, dipsticks and dot-blots/arrays, and automated and/or instrumental formats such as flow cytometry, time-resolved fluorescence (with europium-labelled antibodies) and biosensors based on binding molecules (SPR and other formats).

For more details of these methods, the special edition of the Journal of Microbiological Methods (volume 53, issue 2, 2003) is a good starting point.

2. Some interesting examples of rapid methods for foodborne toxins

We will now look in detail at a selection of rapid methods for foodborne toxins or toxin genes of which the author has personal experience. These are targeted at *C. botulinum* toxin (where the classical method is the mouse bioassay), *S. aureus* toxin (where the classical method is to challenge a laboratory animal, e.g. cat, dog), and the verocytotoxin produced by a sub-group of *E. coli* (where the standard method is based on cytotoxic effects on Vero cells in culture).

2.1 *Clostridium botulinum* toxin

Here we will look at two very different methods. The first method is a multiplex PCR designed to detect the genes for botulinum toxin types A, B, E and F, the principal serological types toxic to man. This method was developed by Miia Lindström (Lindström *et al.*, 2001) who is currently a visiting worker in our group. It is the first PCR for botulinum toxin specifically designed for multiplex use; previously-published PCR primers for individual toxin types are not mutually compatible, having widely different melting points, and are therefore unsuitable for use in a multiplex format. We have applied the method to detection of *C. botulinum* in a survey of the incidence of spores of the organism in a wide range of food samples. An enrichment culture of the food sample is made using TPGY medium (FDA, 2001) in strictly-anaerobic conditions (i.e. under a headspace gas of 5% H_2, 10% CO_2 and 85% N_2). Prior to incubation, the food samples are heat-treated to reduce the competing natural flora. The enrichment culture is followed by a simple extraction of 1 ml culture fluid by boiling or, more efficiently, by using 'PrepMan' reagent (Applied BioSystems). The PCR is then carried out using 1μl of the extract in a 25 μl reaction mix. Figure 1 shows agarose gel electrophoresis of the products of the reaction for each toxin type separately, and in combination in the same sample.

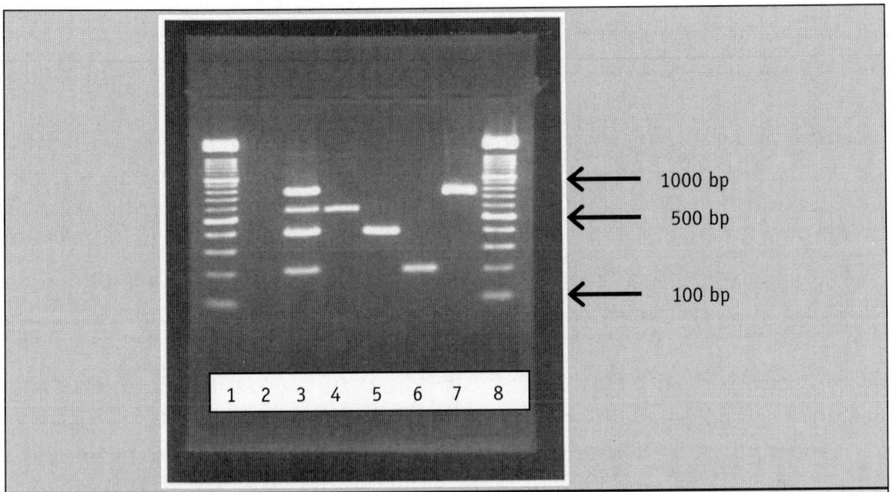

Figure 1. Agarose gel electrophoresis of products of *C. botulinum* toxin gene PCR (lanes 1 and 8, size marker; lane 2, negative control; lane 3, sample containing extracts of cells of *C. botulinum* toxin types A, B, E and F; lane 4, toxin type F only; lane 5, toxin type E only; lane 6, toxin type B only; lane 7, toxin type A only).

The second method for *C. botulinum*, also in use in our laboratory, involves indirect detection of the toxin by monitoring its biological activity. Botulinum neurotoxins mediate their intracellular action via highly specific zinc endopeptidases which cleave synaptic proteins at different sites, e.g. types A and E cleave SNAP-25 (a synaptosome-associated protein), and types B, D, F, and G cleave VAMP (a vesicle-associated membrane protein). Hallis *et al.* (1996) developed an assay for the toxin type B based on this endopeptidase activity. Synthetic polypeptide containing the VAMP sequence which is the target for the enzyme is immobilised and exposed to samples containing the toxin. The enzyme cleaves the polypeptide, exposing a new terminus which is then recognised by a specific enzyme-labelled antibody in the second stage of the assay. Figure 2 shows the principle for this assay.

The sensitivity of this assay for type B toxin was approximately 1.9 ng/ml with purified neurotoxin. Addition of an enzymic amplification system increased the sensitivity 10-fold to approximately 0.17 ng/ml. However, the mouse toxicity test is 100-fold more sensitive (10 to 20 pg/ml), but the authors regarded their endopeptidase method as a first-generation assay, not fully optimised.

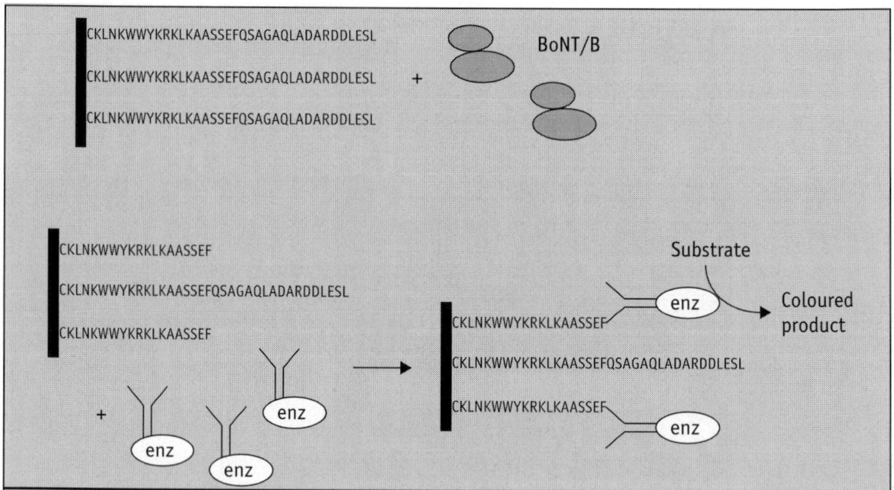

Figure 2. Endopeptidase assay for *C. botulinum* type B toxin (adapted from Hallis *et al.*, 1996).

Subsequently, a modified assay was produced by the same group (Wictome *et al.*, 1999); the main modification was to use immobilised antibody to the toxin, and then present the peptide substrate in solution rather than as a solid-phase substrate. With an incubation time of 2.5 h, a sensitivity of 10 pg/ml was achieved, i.e. equivalent to the mouse bioassay. A similar assay has been developed for *C. botulinum* toxin type A (Ekong *et al.*, 1995), using a polypeptide substrates derived from the SNAP-25 protein; this assay was also of similar sensitivity to the mouse assay. Thus, for at least two botulinum toxins, a potential replacement for the animal test is available.

2.2 *Staphylococcus aureus* enterotoxin

Here a 'home-made' ELISA is described. This was developed (de Oliveira *et al.*, 1994) in our laboratory in collaboration with Tereza Oliveira, a visiting scientist from Brazil. *S. aureus* enterotoxin A is a particular problem in Brazil and only limited funding is available for diagnosis. Commercial ELISA kits are very expensive, and thus this assay proved to be very useful and economical. For antibody production, rabbits were immunised with increasing amounts of the toxin (six amounts from 5-200 μg) over a 10-week period to avoid adverse reactions in the animals. The antiserum was used in a 'sandwich' ELISA format, and sensitivity with purified toxin was found to be about 0.25 ng/ml. A simple aqueous extraction method for food samples was also developed, and sensitivities for spiked samples were in the range <0.125-0.5 ng/ml, depending on food type (Table 3). These detection levels are very adequate for foods containing the known toxic dose of the toxin (Table 1).

Table 3. Sensitivity of *S. aureus* enterotoxin A microplate ELISA with various foods (adapted from de Oliveira *et al.*, 1994).

Matrix	Limit of detection (ng/ml extract[a])	
	80 min incubation	140 min incubation
PBST	<0.125	<0.125
Lemon tart	0.125	<0.125
Milk	0.50[b]	0.50[b]
Sausage	0.50	<0.125
Potato mayonnaise	<0.125	<0.125
Cheese	0.125	<0.125
Ham and turkey paté	0.25	<0.125

[a]aqueous extract (50 g food + 100 ml water)
[b]without extraction

2.3 Verotoxin

Verotoxin (sometimes also called 'shiga-like toxin' because of its close similarity to the shiga toxin of *Shigella* species) is produced by several strains of *E. coli*, generally designated as VTEC (verotoxigenic *E. coli*). The most common serogroup to produce the toxin was, until recently, thought to be *E. coli* O157, but many other serogroups are now known to be toxigenic. Illness is caused when the toxin is produced by infection *in vivo* (Table 1), rather than by growth of the organism in food, but because the toxigenic strains are otherwise very similar to non-toxigenic *E. coli*, toxin-production is used in laboratory examination of suspected food vehicles as a distinguishing feature. To assess toxin production in laboratory cultures, the standard method is to look for effects of the culture fluid on Vero cells *in vitro*. This test requires skilled staff, and also is rather subjective. Here I will describe two types of rapid method for verotoxin, the first based on antibodies in two different dipstick formats, and the second on a combination of the natural toxin receptor with an antibody.

First, in collaboration with Agrotechnology & Food Innovations, Wageningen, the Netherlands, we have developed two types of dipstick for rapid and simple detection of the two principal forms of the toxin, designated VT1 and VT2 (Aldus *et al.*, 2003). These are in a multi-analyte format, giving separate signals for the two toxins, together with an indication of the serogroup of *E. coli* responsible. The multi-target capability of these dipsticks is an important feature.

The first dipstick type uses antibodies in a 'sandwich' ELISA format, with the capture antibody immobilised on a plastic paddle (Figure 3). End-point detection uses an HRP-labelled detector antibody, via an insoluble colorimetric substrate which produces a red deposit on the reaction surface. Using different capture antibodies in spots at different positions on the paddle allows a multi-analyte format. After an incubation period in the sample, the paddle is washed, incubated with the labelled detector antibody(ies), washed again and immersed in the substrate solution (Aldus *et al.*, 2003).

The second type of dipstick is based on lateral-flow chromatography, with immobilised capture antibodies and carbon-labelled detector antibodies (van Amerongen *et al.*, 1994). Again, it is a multi-analyte format, with individual analytes distinguished as separate lines in a window of the cassette device (Figure 4).

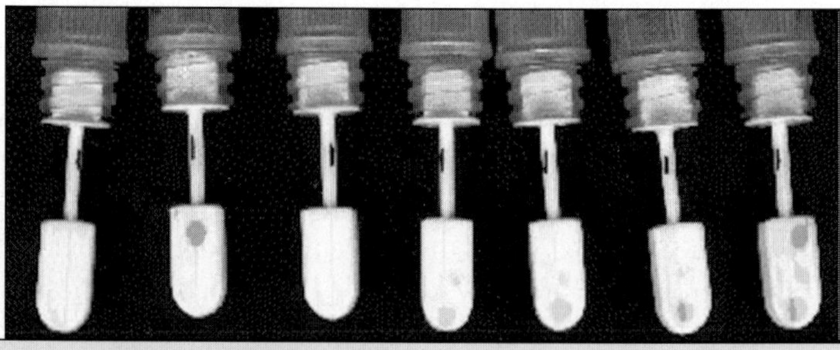

Figure 3. An example of a multi-analyte paddle-dipstick, showing a maximum of three spots for different analytes (e.g. verotoxins, *E. coli* serogroups).

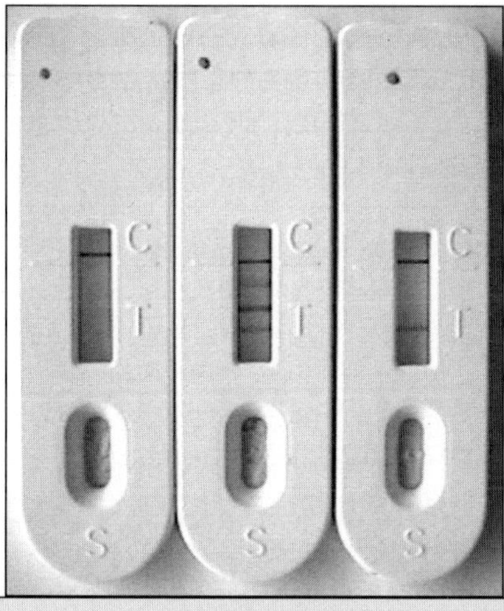

Figure 4. Mutli-analyte lateral-flow dipstick device for detection of verotoxigenic *E. coli* and the toxins. Each device includes a control (upper line) and a maximum of three positive test lines (as illustrated in the middle example).

The lateral-flow dipsticks are much more user-friendly than the paddle-type, being self-contained and requiring no reagent additions or washing stages. The sample (commonly 100 µl) is applied via the window marked 'S' in Figure 4, and the test is completed when the control line is clearly visible (usually 15-30 min). We tested a lateral-flow dipstick for detection of verotoxin in enrichment cultures of foods in an interlaboratory trial. The samples were spiked with several verotoxigenic *E. coli* strains at levels mainly in a range below 100 cells per 25 g food sample. Results (Capps *et al.*, 2004) showed (Table 4) very good identification of negative samples (mean 94.0% for three foods across 13 laboratories) but less good detection of positive samples (mean of 61.3%), indicating that the minimum detectable level was not sufficiently low. However, this device was a prototype, built using antibodies available to us at the time; we are confident that this could be improved with antibodies of higher affinity.

The final assay for verotoxin to be described here is based on capture of the toxin in a microplate format using the natural cell receptor molecule, a specific glycolipid known as globotriosyl ceramide (Gb_3; Figure 5), which binds to the B-subunit of the toxin. Each toxin molecule has five copies of the B-

Table 4. Results of an interlaboratory trial for detection of verotoxin in enrichment cultures of artificially-inoculated foods using a lateral flow dipstick.

	Matrix:		
	Apple juice	Beef	Milk
Number of testing labs	12	13	13
Total number of samples included in analyses	360	390	455
Specificity[a], %	95.0	92.3	94.6
(Lower-upper)	(89.5-97.7)	(83.2-96.7)	(89.3-97.4)
Sensitivity[b], %	66.2	50.8	66.8
(Lower-upper)	(60.1-71.9)	(45.4-56.2)	(61.5-71.7)

[a]Specificity: percentage of samples correctly identified as negative
[b]Sensitivity: percentage of samples correctly identified as positive

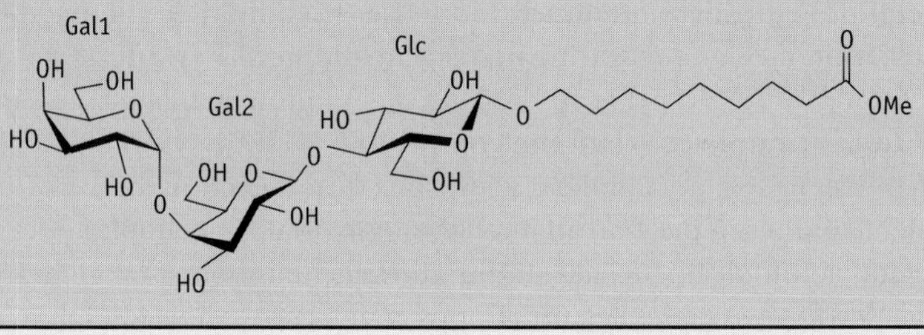

Figure 5. Structure of globotriosyl ceramide, Gb_3 (adapted from Ling *et al.*, 1998).

subunit, with three Gb_3 binding sites per subunit, thus considerably enhancing the binding potential. Bound toxin is then detected with enzyme-labelled monoclonal antibodies.

This assay (Ashkenazi and Cleary, 1989) has a sensitivity for purified verotoxin of 2 ng/ml. The assay was positive with sonicated extracts of cultures of *Shigella dysenteriae* serotype 1, *E. coli* O157:H7 and *E. coli* O26:H11, and negative with purified *E. coli* enterotoxins, *C. difficile* cytotoxin, and sonicated extracts of cultures of other shigellae, *E. coli*, *Salmonella* and *Campylobacter*. Thus it appears to be specific for shiga toxin and verotoxin.

Recently, Gb_3 was cloned into the surface lipopolysaccharide of an *E. coli* host strain which was suggested for use as a therapeutic agent in cases of human verotoxin poisoning (Paton *et al.*, 2000). However, this development also offers exciting possibilities for new diagnostics for the toxin, based on use of the cloned receptor as both capture and detector ligand.

3. Conclusions

Originally, most rapid methods were antibody-based, and in a 96-well microplate format. However, as this article has tried to show, a much greater variety of technologies and, increasingly, of formats for the tests is now being employed. Many of these formats, particularly some of the dipstick types, are very user-friendly and this may, in the future, allow analysis to take place outside the laboratory. However, if the consumer is to test food prior to consumption at home in a catering establishment, then there may be considerable social and legal implications.

Acknowledgements

I would like to thank Clare Aldus, Miia Lindstrom and Mike Peck, my colleagues in the IFR Food Safety & Computational Microbiology Group, and also Aart van Amerongen, Jan Wichers and Renata Ariëns, our collaborators at Agrotechnology & Food Innovations. Part of the work was supported by the UK Food Standards Agency.

References

Aldus, C.F., van Amerongen, A., Ariens, R.M.C., Peck, M.W., Wichers, J.H. and Wyatt, G. M., 2003. Principles of some novel rapid dipstick methods for detection and characterization of verotoxigenic *Escherichia coli*. Journal of Applied Microbiology 95, p. 380-389.

van Amerongen, A., Wichers, J.H., Berendsen, L.B.J.M. and van Doorn, A.W.J., 1994. Immunoassays using a carbon sol label. European Patent Application no. 94201817.7

Ashkenazi, S. and Cleary, T. G. 1989. Rapid method to detect shiga toxin and shiga-like toxin-1 based on binding to globotriosyl ceramide (Gb3), their natural receptor. Journal of Clinical Microbiology 27, p. 1145-1150.

Capps, K.L., Mclaughlin, E.M., Murray, A.W.A., Aldus, C.F., Wyatt, G.M., Peck, M.W., van Amerongen, A., Ariëns, R.M.C., Wichers, J.H., Baylis, C.B., Wareing, D.R.A. and Bolton, F.J., 2004. Validation of three rapid screening methods for detection of verotoxin-producing *Escherichia coli* in foods: interlaboratory study. Journal of AOAC International 87, p. 68-77.

Ekong, T., Mclellan, K. and Sesardic, D., 1997. Immunological detection of *Clostridium botulinum* toxin type-A in therapeutic preparations. Journal of Immunological Methods 180, p. 181-191.

FDA, 2001. U.S. Food & Drug Administration, Center for Food Safety & Applied Nutrition , Bacteriological Analytical Manual Online www.cfsan.fda.gov/~ebam/bam-toc.html

Hallis, B., James, B.A.F. and Shone, C.C., 1996. Development of novel assays for botulinum type A and B neurotoxins based on their endopeptidase activities. Journal of Clinical Microbiology 34, p. 1934-1938.

Johnson, E.A., 2003. Bacterial Pathogens and Toxins in Foodborne Disease. In: Food Safety: Contaminants and Toxins, Chapter 2, edited by J.P.F. D'Mello, CABI Publishing, Cambridge MA, USA.

Lindström, M., Keto, R., Markkula, A., Nevas, M., Hielm, S. and Korkeala, H. 2001. Multiplex PCR assay for detection and identification of *Clostridium botulinum* types A, B, E, and F in food and fecal material. Applied and Environmental Microbiology 67, p. 5694-5699.

Ling, H., Boodhoo, A., Hazes, B., Cummings, M.D., Armstrong, G.D., Brunton, J.L., and Read, R.J., 1998. Structure of the shiga-like toxin I B-pentamer complexed with an analogue of its receptor Gb_3. Biochemistry 37, p. 1777-1788.

de Oliveira, T.C.R.M., Lee, H.A., Wyatt, G.M., Hirooka, E.Y. and Morgan, M.R.A., 1994. A simple and rapid antibody-capture ELISA for the detection of staphylococcal enterotoxin A in food including a simple extraction step. International Journal of Food Science and Technology 29, p. 563-573.

Paton, A.W., Morona, R., Paton, J.C., 2000. A new biological agent for treatment of shiga toxigenic *Escherichia coli* infections and dysentery in humans. Nature Medicine 6, p. 265-270.

Wictome, M., Newton, K. Jameson, K., Hallis, B., Dunnigan, P., Mackay, E. Clarke, S. Taylor, R. Gaze, J. Foster, K. and Shone, C., 1999. Development of an *in vitro* bioassay for *Clostridium botulinum* type B neurotoxin in foods that is more sensitive than the mouse bioassay. Applied and Environmental Microbiology 65, p. 3787-3792.

Enzyme sensor array for the determination of biogenic amines

C. Wittmann
Neubrandenburg University of Applied Sciences, Department of Technology, Brodaer Strasse 2, 17033 Neubrandenburg, Germany

Abstract

An enzyme sensor array for the simultaneous determination of three biogenic amines (histamine, tyramine and putrescine) using pattern recognition by an artificial neural network was applied to the measurement of sauerkraut samples. The sensor array took 20 minutes to complete. The measuring range for histamine and tyramine reached from 10 mg/kg to 200 mg/kg, and for putrescine from 5 mg/kg to 100 mg/kg. A mean correlation coefficient of 0.928 was determined by measuring the sauerkraut samples and comparing the results with liquid chromatography data. Biogenic amine concentrations in sauerkraut were higher for spontaneous fermentations as compared to the use of starter cultures. During lactic fermentation amine concentrations were highest in the mid-period (i.e. 11 days) of total time (21 days). In commercial canned sauerkraut higher amine contents were measured (sum of amines ranging from 103 to 235 mg/kg) than in the samples produced in a pilot plant (from 12 to 37 mg/kg).

Keywords: biogenic amines, histamine, tyramine, enzyme sensor array, sauerkraut

1. Introduction

The determination of biogenic amines is of great interest today. Biogenic amines (e.g. histamine, tyramine, putrescine) are formed by decarboxylation of amino acids and are degraded during metabolism in animals, plants and microorganisms (Figure 1).

Foods, which may contain particularly high levels of biogenic amines, are cheese, fish, beer, wine, chocolate, sauerkraut and further fermented foods such as sausage. Spoiled foods are also rich in biogenic amines and usually contain

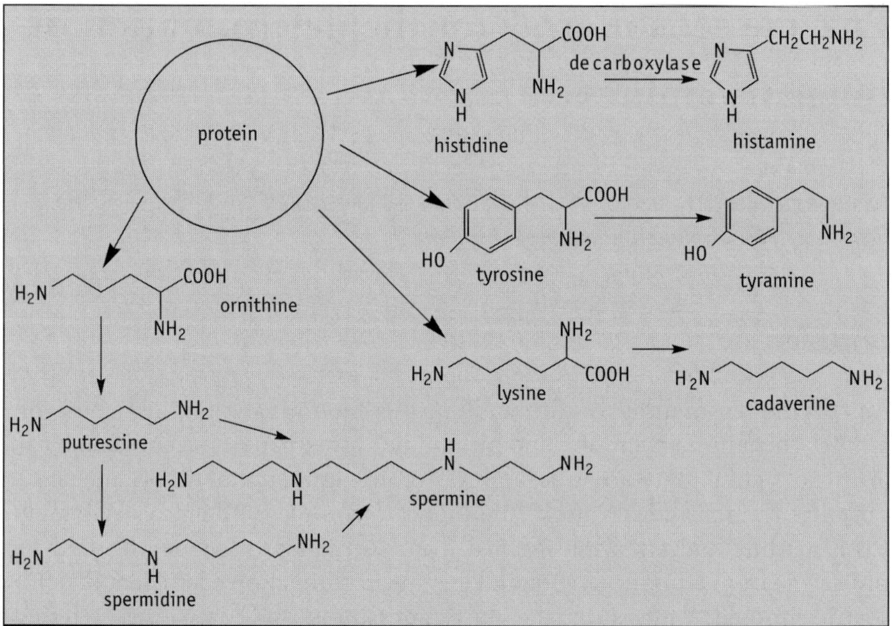

Figure 1. Formation of biogenic amines by decarboxylation of amino acids.

high levels of putrescine and cadaverine. However, food contaminated with bacteria often includes high levels of bacterial toxins and metabolites that would render the food unsafe for consumption regardless of any toxic effects of the biogenic amines present. Histamine, putrescine, and cadaverine have been implicated in human poisoning, when present at considerable concentrations. Symptoms such as headaches, nausea, hypo- or hypertension, cardiac palpitations and even anaphylactic shock have been reported as a result of excessive intake of biogenic amines, especially for histamine. Therefore, in most countries a maximum permissible concentration for histamine in fish and fish products is prescribed, which is in Germany according to the "Fischhygieneverordnung" 200 mg/kg, whereas in other countries a limit of 100 mg/kg has been set.

Traditionally, histamine has been measured by derivatisation with fluorescent reagents followed by chromatographic separation (liquid chromatography, in most cases). Liquid chromatography (LC) analysis of biogenic amines is tedious with regard to sample clean-up prior to analysis and requires trained personnel in combination with quite expensive equipment. Other analysis methods have also been developed, such as capillary electrophoresis (Lange *et al.*, 2002), immunochemical methods as ELISA (Aygün *et al.*, 1999), and gas chromatographic procedures (Rogers and Staruszkiewicz, 1997). To reduce the time needed for analysis and to offer a rapid screening method for industrial food quality testing, some enzymatic methods and several enzyme sensors have been described so far. A major drawback of these enzyme sensors is that the majority of amine oxidases used do not only react with a single biogenic amine, but also with different amines to various extents. Therefore, an enzyme sensor array for the simultaneous determination of three biogenic amines (histamine, tyramine and putrescine) was developed using pattern recognition by an artificial neural network (Lange and Wittmann, 2002). The aim of this investigation was to evaluate the applicability of the enzyme sensor array to bioprocess control. Lactic fermentation of white cabbage (producing sauerkraut in a small pilot plant) was chosen as a representative bioprocess. The influence of different fermentation process parameters (spontaneous fermentation vs. the use of starter cultures, fermentation temperature and time, and storage conditions of white cabbage) on the formation of biogenic amines was studied. Process optimisation to reduce biogenic amine contents in the final food product sauerkraut will be a future challenge. To this end, commercially available canned sauerkraut samples were also analysed. To get information on any other biogenic amines present, LC analyses were performed for method comparison.

2. Materials and methods

The enzyme sensor array is prepared by the separate immobilisation of a monoamine oxidase (tyramine oxidase from *Arthrobacter* spp., EC 1.4.3.4, Sigma and Aldrich, Deisenhofen, Germany), a plasma amine oxidase (EC 1.4.3.6, Sigma), and a diamine oxidase (EC 1.4.3.6 from pea seedlings, as obtained following the procedure of Güvenilir and Deveci (1996)) via glutaraldehyde onto the surface of screen-printed thick-film electrodes. Immobilisation details are indicated by Lange and Wittmann (2002). Different formats of screen-printed electrodes were obtained from Forschungszentrum Sensorik e. V. (Greifswald, Germany). Figure 2 shows the layout of one screen-printing electrode format to detect hydrogen peroxide generated from the enzymatic reaction (Figure 3) at a potential of +700 mV.

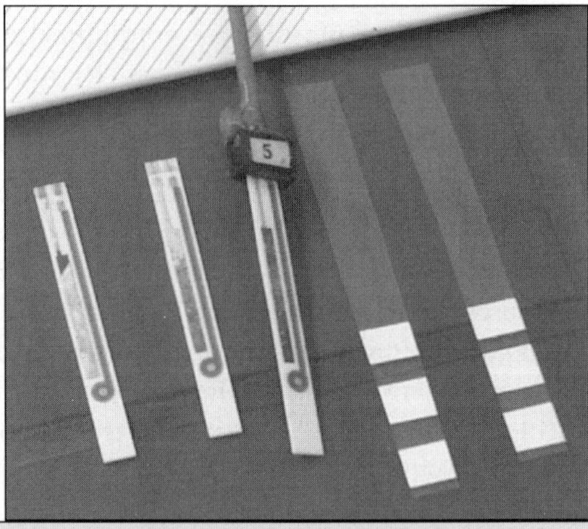

Figure 2. Layout of screen-printed thick-film electrodes.

(1) $$R{-}CH_2{-}NH_2 + O_2 + H_2O \xrightarrow{\text{amine oxidase}} R{-}CHO + NH_3 + H_2O_2$$

(2) Sensor: Electrode reaction

$$H_2O_2 \longrightarrow O_2 + 2H^+ + 2e^-$$

Figure 3. Detection principle of the enzyme sensor.

The sensor depicted consisted of a round platinum working electrode and a Ag/AgCl counter electrode. The working electrode and the connecting wire were coated with an insulating layer. After enzyme immobilisation a cellulose acetate layer was formed on the sensor surface to achieve an improved selectivity to hydrogen peroxide. The electrodes were contacted by a clip to the electrode cable.

Electrochemical measurements were carried out with an Autolab PGSTAT 10 and GPES software (both Deutsche Metrohm, Filderstadt, Germany). A four-electrode system was employed. For amperometric measurements the four electrodes (three with a separate enzyme layer and an additional one with a reagent blank) were connected to the Autolab PGSTAT 10 system. Every working and counter electrode was connected via an adapter to the Autolab system in such a way that every working electrode could be addressed separately. The working electrodes were all polarised at +0.7 V against the counter electrodes. All electrochemical measurements were carried out at room temperature. Although for chronoamperometric measurements the response time was less than 5 seconds, it appeared more suitable to collect current data after reaching the steady-state. The original signals (t vs. A) were transferred to Excel as ASCII files reducing the data amount. To calculate the amount of a specific biogenic amine the raw data from the multichannel software were transferred to a neural network. The files could then be read as such and analysed with a modified SPSS (SPSS Neural Connection 2.1, SPSS Inc., Chicago, USA) neural network software. A feedforward net with three input and three output units was the basis to find out a suitable network topology. For a variety of food samples (e.g. wine, beer, dairy products, sausages, fish samples, yeast, sauerkraut) one or more hidden layers were studied together with additional shortcut connections. Comparing the mean and maximum deviation for the backward propagated output values, the best topology turned out to be the use of only one hidden layer. With this topology different learning methods - backpropagation, variations of backpropagation (with momentum term, flash-spot elimination and weight decay) and Rprop - were investigated in comparing different learning parameters. Finally, it was decided to restrict the learning method to counterpropagation with the exception of cheese samples, where standard backpropagation worked best.

For sample measurements the enzyme electrodes were first equilibrated in 0.1 M potassium phosphate buffer. After the output current had reached a stable baseline, the sensors were calibrated by using standard series for histamine, putrescine and tyramine prepared in a brine solution (of a similar composition as the one used later in sauerkraut sample measurements). For food sample measurements aqueous sauerkraut extracts were used (requiring no previous neutralisation step). The sauerkraut sample was only homogenised in distilled water with the aid of a laboratory mixer. An aqueous extract turned out to be superior to routine extraction with trichloroacetic acid and subsequent neutralisation.

In general, sauerkraut is produced from white cabbage as the raw material by adding 1-2% salt on the basis of a spontaneous lactic fermentation or by using defined starter cultures. Whole fermentation takes 2-6 weeks to be completed depending mainly on the process parameters, i.e. spontaneous fermentation vs. starter cultures, temperature during fermentation, storage conditions for white cabbage and time-period for lactic fermentation. In our study, sauerkraut was prepared in a small pilot plant by spontaneous fermentation and by using different starter cultures. For a better control of the fermentation process the use of starter cultures is recommendable. The following starter cultures were used for sauerkraut preparation: *Lactobacillus casei, Lactobacillus paracasei, Lactobacillus plantarum* 1 and *Lactobacillus plantarum* 2, *Lactobacillus sakeim* and *Pediococcus pentosaceus*. In addition, the fermentation process should be stopped when a pH value of 4.0 is reached. Therefore, samples were collected after defined time periods within 21 days (days 1, 2, 7, 11, 14, and 21) carrying out a routine pH check. Fermentation temperature chosen depends on the microorganisms present and should, in principle be low. With regard to GMP regulations of the sauerkraut producing industry the following threshold values for biogenic amines should not be exceeded: 10 mg/kg for histamine, 20 mg/kg for tyramine, 5 mg/kg for β-phenylethylamine, 50 mg/kg for putrescine, and 25 mg/kg for cadaverine with a sum of 90 mg/kg biogenic amines. By high performance liquid chromatography (method description cf. Lange *et al.*, 2002) five biogenic amines (histamine, tyramine, putrescine, β-phenylethylamine, and cadaverine) could be detected in the sauerkraut samples prepared in the pilot plant. Industrially produced sauerkraut (six commercially available sauerkraut samples) was also analysed and biogenic amine concentrations were compared with the ones in the sauerkraut produced by the pilot plant.

3. Results

Previously, the enzyme sensor array applied here to the measurement of sauerkraut samples for biogenic amine contents, was evaluated for a variety of foods (Lange and Wittmann, 2002). The main array characteristics are listed in Table 1 and compared with the respective HPLC method (Lange *et al.*, 2002).

With food measurements a recalibration was necessary every five samples, carried out with at least one standard concentration (in most cases 50 mg/kg). Due to the detection limits of the enzyme sensor array, food samples containing only small amounts of histamine and/or tyramine gave a false negative response. On the other hand, histamine and tyramine were detected in the canned sauerkraut sample, which could not be confirmed by HPLC analysis. Experiments performed with a blank food matrix spiked with increasing amounts of ascorbic acid, showed that a signal was obtained, i.e. the

Table 1. Enzyme sensor array characteristics as compared to HPLC method.

Aspect	Enzyme sensor array	HPLC
• Food samples analysed	Beer, wine, cheese, ham, salami, fish, yeast, sauerkraut	Beer, wine, cheese, ham, salami, sauerkraut, fish, chocolate
• Sample pre-treatment	Extraction and neutralisation required	Extraction, neutralisation, liquid-liquid or solid phase extraction necessary (with membrane filtration of the extracts)
• Correlation with HPLC	Coefficient: 0.854	-
• Analysis time	20 min + additional time for data calculation (external: 3 min per sample)	Analysis time (excluding sample pre-treatment and clean-up): 30 min
• Lower detection limit	10 mg/kg for histamine and tyramine, 5 mg/kg for putrescine	0.5-2 mg/kg depending on the amine or amino acid
• Linear range	Up to 200 mg/kg for histamine and tyramine, up to 100 mg/kg for putrescine	Up to 1000 mg/kg
• Analytes	Histamine, putrescine, tyramine	Histamine, tyramine, cadaverine, β-phenylethylamine, spermine, spermidine, agmatine, putrescine, tryptamine and other amines,and several amino acids
• Validity of analysis data	12% false positive results, 25% false negative results	(< 1% false positive results as registered by LC/MS confirmatory analyses)
• Reproducibility	High if the same sensor is used	Very high
• Application area	Limited if used as online sensor because of sample pretreatment needed prior to analysis	Dependent on the availability of equipment for online studies

determination might be disturbed if the sample contained ascorbic acid in a concentration higher than 120 mg/kg. In our case, the sauerkraut had a vitamin C content of 20 mg/100 g as measured by a redox titration. As the aim of our study was to follow the lactic fermentation process of white cabbage to sauerkraut by applying the enzyme sensor array, calibration for histamine, tyramine, and putrescine was performed in a matrix similar to the sauerkraut sample by adding salt and 200 mg/kg ascorbic acid to the histamine, tyramine, and putrescine standard series. In addition, preliminary experiments for biogenic amine extraction have showed that the use of aqueous extracts turned out to be favourable. Therefore, the samples were homogenised in a laboratory mixer upon the addition of a defined amount of distilled water.

At first, the sauerkraut produced in the pilot plant was analysed by HPLC for biogenic amines formed during lactic fermentation within a time-period of 21 days. Five biogenic amines. i.e. histamine, tyramine, putrescine, β-phenylethylamine, and cadaverine were detected as shown in Figure 4. Biogenic amines were formed in higher concentrations during a spontaneous fermentation as compared to the use of defined starter cultures. During the fermentation process biogenic amine concentrations were highest in the mid (i.e. after 11 days) of total time period (21 days) as can be seen in Figures 5a and 5b. This

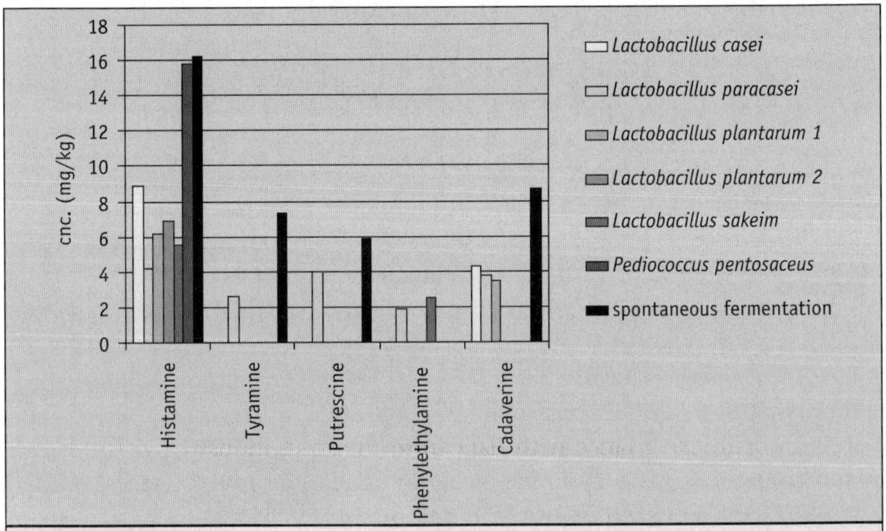

Figure 4. HPLC analysis of biogenic amine concentrations in sauerkraut produced by using different starter cultures in comparison to a spontaneous fermentation.

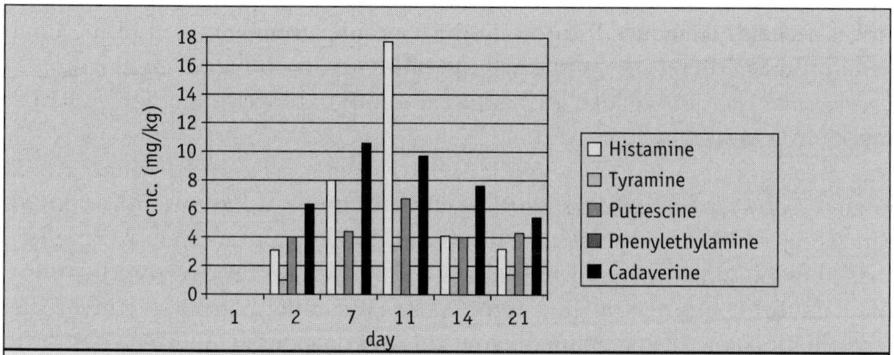

Figure 5a. Course of biogenic amine formation during spontaneous fermentation. Histamine, tyramine and putrescine concentrations were measured with the enzyme sensor array. β-Phenylethylamine and cadaverine concentrations were determined by HPLC analyses.

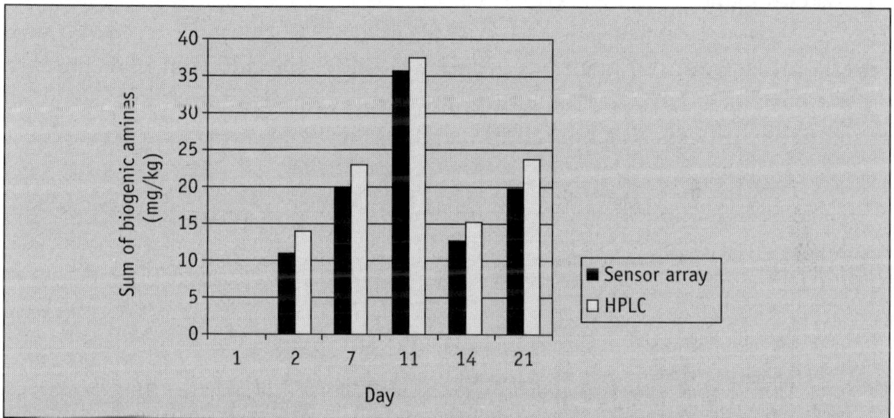

Figure 5b. Comparison of the results obtained by the enzyme sensor array and HPLC for the sum of biogenic amines generated in spontaneous sauerkraut fermentation.

result is not in accordance with data from the food industry, where amine concentrations were observed to be highest at the end of the fermentation process. It is also obvious from Figure 5b that the enzyme sensor slightly underestimated the biogenic amine contents due to the presence of cadaverine, which is not detectable by the enzyme sensor array.

It was our aim to reduce the amount of biogenic amines present in the final food product. Therefore, we studied the influence of different starter cultures. *Lactobacillus paracasei* yielded the highest amounts of biogenic amines with 15.2 mg/kg in total (Figure 6).

Figure 7 shows that the biogenic amine concentrations of industrially produced sauerkraut were generally higher (sum of amines ranging from 103 to 235 mg/kg) than the amounts measured in the sauerkraut samples produced in the pilot plant (sum of amines ranging from 12 to 37 mg/kg), independent of the fermentation type (using defined starter cultures or spontaneous fermentation). This may be explained by the fact that in some cases the commercial canned sauerkraut contained white wine, which may be an additional source of biogenic amines. Especially for putrescine the threshold value set by the industry (50 mg/kg) was exceeded in all six samples, reaching a maximum of 161 mg/kg.

4. Discussion

Sauerkraut is produced from white cabbage under the addition of 1-2% salt on the basis of a lactic fermentation, either by using starter cultures or in a spontaneous process. Fermentation usually takes 2 to 6 weeks to complete.

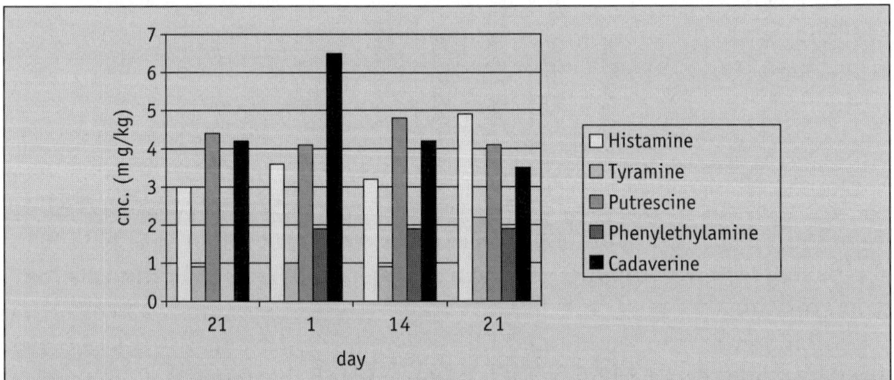

Figure 6. Course of biogenic amine formation using *Lactobacillus paracasei* as a starter culture. The amounts of histamine, tyramine, and putrescine were determined by the enzyme sensor array, whereas β-phenylethylamine and cadaverine concentrations were analysed by HPLC.

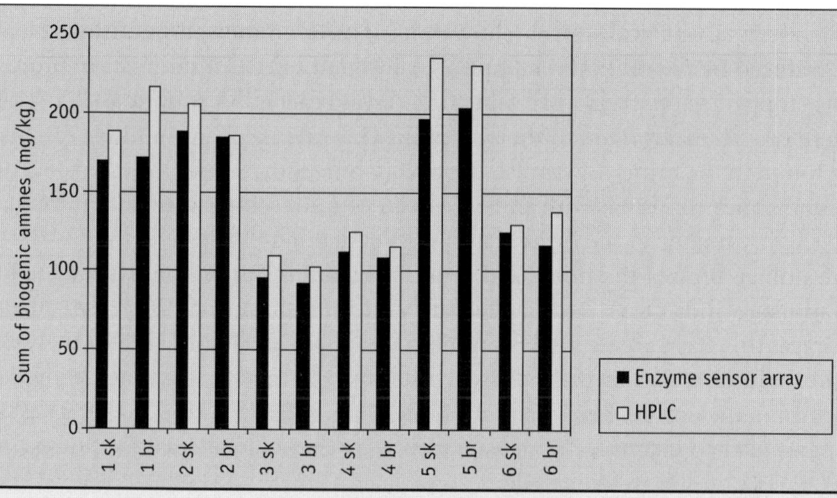

Figure 7. Comparison of the results obtained for biogenic amine concentrations in commercial canned sauerkraut samples by enzyme sensor array and HPLC. Sk: sauerkraut; br: brine.

According to information given by food industry it is described that sauerkraut may contain high amounts of tyramine and histamine depending on process parameters. Three possible pathways for the formation of biogenic amines exist. The main process is a proteolysis and subsequent decarboxylation of amino acids. In addition, biogenic amines can be generated by amination of aldehydes and ketones present or by the degradation of nitrogen compounds by hydrolytic cleavage of amines (e.g. ethanolamine from phosphatides). In the main decarboxylation reaction pyridoxalphosphate acts as a cofactor for decarboxylases, which are specific for certain amino acids. The activity of decarboxylases depends on temperature, pH and surrounding media. The pH optimum for decarboxylases lies between pH 6.0 and 9.5. In the case of a spontaneous fermentation heterofermentative *Leuconostoc mesenteroides* predominates at the beginning, followed by homofermentative *Lactobacillus plantarum* and *Lactobacillus brevis*. At the end of the fermentation process *Pediococcus cerevisiae* predominates. The lower the pH value during fermentation the better *Pediococcus cerevisiae* could grow and multiply. For a better control of the fermentation process the use of starter cultures is recommendable.

The results of our study clearly show that biogenic amine concentrations could be reduced by fulfilling several prerequisites. At first, the fermentation process was stopped as soon as a pH value of 4.0 was reached. A routine pH check is therefore necessary. Due to their solubility in water biogenic amines can also be found in the brine, as was also the case in our study. Sometimes the biogenic amine concentrations (with the exception of putrescine) were determined to be slightly higher in the brine than in the sauerkraut itself. A further aspect is the storage time of the final food product. Especially tyramine, putrescine (as a spoilage indicator) and cadaverine concentrations may increase during storage. This was the case for all commercial canned sauerkraut samples analysed. Threshold values set by the industry (histamine, 10 mg/kg; tyramine, 20 mg/kg; β-phenylethylamine, 5 mg/kg; putrescine, 50 mg/kg; cadaverine, 25 mg/kg) as a parameter to guarantee a high quality, were exceeded at least for putrescine by all six samples investigated.

For sauerkraut samples, a mean correlation coefficient of 0.928 was reached for the measurement of the biogenic amines by the enzyme sensor array as compared with the respective HPLC data. The main reason for this improvement as compared with the coefficient for all food samples studied so far (0.854, see also Lange and Wittmann, 2002), is considered to be the preparation of the biogenic amine standard series in a solution of brine containing 200 mg/kg ascorbic acid.

5. Conclusions

The enzyme sensor array developed for the analysis of the three biogenic amines histamine, putrescine and tyramine could be applied to study the formation of biogenic amines during lactic fermentation of white cabbage to sauerkraut. As the biogenic amines cadaverine and, only to a minor extent, β-phenylethylamine were also present in the sauerkraut samples, an underestimation was observed with the enzyme sensor array analysis. Therefore, improvement of the biological compound of the biosensor system still remains an important aspect. In our case this concerns the supply of enzymes exhibiting high specific activities and, additionally, either a single substrate specificity or nearly the same specificity for a broad range of different substrates. This can be achieved, at least in theory, by tailor-made enzymes using site-directed mutagenesis. However, the course of biogenic amine formation could be followed during fermentation with the advantage of avoiding tedious sample pre-treatment, expensive equipment, and the requirement of skilled personnel. As an outlook to reach the stage of an online sensor, several critical points have to be solved. A disadvantage of biosensor applications in the area of food and feed analysis is that at least some sample pre-treatment has to be carried out (i.e. in general an extraction step), which prevents the system from being applicable as an online sensor. In principle, the sample pre-treatment step could be embedded in the overall analysis procedure. For this purpose, there is a high demand for a precise and easy liquid handling, e.g., flow injection analysis systems becoming automated and eventually directly integrated in biosensor assembly as a kind of a lab-on-a-chip.

Acknowledgements

This project was funded by BMBF (grant 1700498). We thank Dr. Peter Abel and his colleagues (Forschungszentrum Sensorik Greifswald e.V. and BCS GmbH) for providing us with the screen-printed thick-film electrodes. We gratefully acknowledge technical support by Mr. Friedrich from Hochschule Wismar in modifying the SPSS Neural Network software for data transfer.

References

Aygün, O., Schneider, E., Scheuer, R., Usleber, E., Gareis, M. and Märtlbauer, E., 1999. Comparison of ELISA and HPLC for the determination of histamine in cheese. J. Agric. Food Chem. 47, p. 1961-1964.

Güvenilir, Y. and Deveci, N., 1996. The isolation and purification of diamine oxidase of pea seedlings and pig liver. Appl. Biochem. Biotechnol. 56, p. 235-241.

Lange, J., Thomas, K. and Wittmann, C., 2002. Comparison of a capillary electrophoresis method with high performance liquid chromatography for the determination of biogenic amines in various food samples. Journal of Chromatography B 779, p. 229-239.

Lange, J. and Wittmann, C., 2002. Enzyme sensor array for the analysis of biogenic amines in food samples. Analytical and Bioanalytical Chemistry 372, p. 276-283.

Rogers, P.L. and Staruszkiewicz, W., 1997. Gas chromatographic method for putrescine and cadaverine in canned tuna and Mahimahi and fluorometric method for histamine (minor modification of AOAC Official method 977.13). J. AOAC Int. 80, p. 591-602.

Rapid assays for detection of residues of veterinary drugs

A.A. Bergwerff
Department of Public Health and Food Safety, Faculty of Veterinary Medicine, Utrecht University, P.O. Box 80175, 3508 TD Utrecht, the Netherlands

Abstract

The use of veterinary drugs, such as antibiotics, hormones, anthelmintics and tranquillisers, is an integral part of an intensive agricultural production involving animals. Screening and/or rapid assays are needed to support effective and affordable regulatory monitoring, and to facilitate self-control programs. An adequate application of such assays will contribute to the safety of food of animal origin and to a relief of public concerns on food safety issues. This chapter comprises a review of rapid diagnostic methods available for the determination of residues of veterinary drugs in the animal production chains. It summarises different principles of analysis involving immuno- and receptor technologies in single and multi- endpoint approaches. Besides microbial growth inhibition and different conventional immuno-, enzyme and receptor assays, new technologies in food safety analysis, such as biosensors, cellular bioassays, transcriptomics and proteomics, are discussed. The choice of the analyst for a certain technique to check the absence or presence of a certain residue should, however, fit its purpose, and is discussed comprehensively here as well. Finally, it is concluded that rapid assays are available for many situations for many substances and many analytical matrices, but that developments are ongoing and are going swiftly. The ultimate at-line assay, which can perform remotely and non-destructively multiple analyses with instantaneous results, is yet still very far away.

Keywords: receptor assays, immunoaffinity assays, residue analysis, veterinary public health, food analysis

1. Introduction

The turnover in food production is high and is not without the use of veterinary pharmaceuticals in animal husbandry. Without a certified description of the history of a product in the animal production chain, each raw material or product could be unintentionally, but unacceptably, contaminated with residues of veterinary pharmaceutical products. Control of residues of these substances in food of animal origin is of importance for the protection of public health. This control is not effective without applying adequate analysis procedures, which are able to reveal these contaminants when present. As a matter of fact, the consumer is often convinced that all foodstuffs are inspected for all possible contaminants. In practice, this is not feasible as a substantial part of resources in terms of laboratory facilities, personnel and regulatory agency activities, is already needed to execute residue control programs that sample and analyse only a relative small percentage of the total food production for certain targeted compounds.

Besides the regulatory-generated need for residue analyses, integrated chain management or quality assurance systems, HACCP, veterinary control systems or good veterinary practices (GVP) may dictate that the producer, either being feed producer, farmer or food processor, is executing a transparent self-control programme with respect to the possible occurrence of residues of veterinary drugs. Analytical assays are therefore applied:

- to monitor random survey samples for compliance for compounds with regulatory tolerances;
- to targeted survey samples where there may be suspicion of non-compliance with regulatory tolerances;
- to monitor compliance in the use of banned compounds, i.e. which are not approved for use in food animal production and have therefore no tolerated levels in foodstuffs.

This paper reviews the assays either commercially available, underway to commercialisation or currently still at the 'design table of the analytical engineer' enabling fast detection, including in some cases identification and/or quantification, of residues of veterinary drugs in our food.

2. Analytical challenges

For the residue analyst, the detection of an anonymous residue is a major challenge for some, especially technical, reasons (Table 1). Firstly, it should be realised that about 3,000 veterinary pharmaceuticals have been described. These preparations contain several hundreds of active components (Aerts *et al.*, 1995). In addition, another several hundreds of non-registered and banned substances, including those with hormonal activity, have been found or can be anticipated. On top of this gamma of biologically active molecules potentially present, it should be reckoned with their corresponding degradation products and metabolites, and excipients as well. The use of veterinary drugs in animal production may amount a quantity in excess of what is being used to treat humans. The chance that a certain concentration of any residue ends up in any animal product is therefore not negligible and this awareness fuels public

Table 1. Some questions raised by the residue analyst, when requested to investigate a sample or a series of samples. The answers to these questions determine the analysis strategy.

Analyst's question	More specific questions
Which information is required?	Enforcement of a (monitoring) programme? What is degree of accuracy and precision needed? What is the budget?
How many samples do I receive?	Number of samples? One shipment or continuous delivery of samples?
What are the properties of the analyte(s)?	Tolerated or banned? What are the biological and chemical properties? Can I expect metabolites?
Which analysis system can I use?	Which sensitivity is required? Individual or population of animals? Equipment available?
Quality assurance?	What does the task require: GLP or ISO? Is the customer satisfied?

concerns with respect to this type of unintentional contamination of food. The challenge is therefore to determine the correct analyte in a correctly collected and appropriate sample reflecting the entity, whether tissue, animal or group of animals, from which the sample was derived.

Secondly, the final residue is occurring at a relatively low concentration ranging from subnanograms to a few milligrams per kilogram of product. As an exception, a meat product containing a (fresh) injection or implantation site may contain residues at a level of grams per kilogram. Typically, the residue analyst is literally looking for a pin in a haystack and thus needs ultra-sensitive and very specific analytical methods. In addition, the matrix is a heterologous composition of numerous different molecules with different physicochemical characteristics, which further hampers a simple and fast finding of the 'pin'.

Thirdly, the residue analyst is asked to generate the outcome of his analysis promptly with short analysis times, as not only perishable products are involved, but any temporary storage of the product that cannot be released, is an economic loss for the producer. The time pressure can therefore be relatively high. This is in particular the case in outbreak situations, as evidenced by the recent scandals with dioxins in poultry products (1999, Belgium), nitrofen in feed (2002, Germany), medroxyprogesteron acetate in pork (2002, the Netherlands), chloramphenicol and nitrofurans in aquacultural, honey, poultry and rabbit products (since 2000, from outside the EU). In such situations, a so-called standstill, export/ import stop can be ordered and no product is allowed to be traded without a certificate clearing the feed, animal and/or food product.

3. Need for rapid diagnostic assays

In particular, outbreak situations make the need for rapid assays for residue detection and crisis management very understandable. It is clear that rapid assays with high sample throughput capacities are then preferred. Besides incidents, such high throughput assays are also required in the abattoir, by food processors or in the private residue analysis laboratories monitoring residues for their clients, including farms. It should, however, be realised that high throughput is not needed for the veterinarian or farmer. They may wish to check only if for example the milk from a treated cow that suffered from mastitis, can be combined with the tank milk. Only a single or a few samples should then be tested to determine whether the aminoglycoside and/or β-lactam antibiotic concentrations are below their corresponding regulatory tolerances.

On-site methods should thus generate rapid 'yes/no' results. Here, the term rapid is rather elastic. In a production line, rapid might mean having analytical results in sufficient time to remove non-compliant material prior to processing. In a dairy plant it might mean having results for a tanker or silo of milk prior to product manufacture, while only a few minutes lag time would be admissible for removal of non-compliant carcasses to meet veterinary inspection requirements. The rapid assay should therefore provide real-time results, i.e. results obtained in a timescale that allows for a response suited to the needs of the user (Andersen, 1994).

In self-control programmes, the analyst may have access to multiple similar samples to improve the precision of the result. In many situations, the self-controller is not a trained chemist. This person, therefore, requests not only reliable, rapid assays, but also user-friendly procedures. Although not always the same, user-friendliness can here be understood as easy to perform as well.

It is comprehensible that the applied analysis method should be cost-effective. The user-friendliness of a method is generally correlated to costs for analysis. In addition, speed of the method is often reciprocal to the final analysis costs. It should be noted that multiple analyses to improve the reliability of the result, is not very cost-effective. So, the assay should be accurate and precise. Not performing any analysis, however, can be very expensive, in particular at the moment that the presence of foreign compounds is discovered by regulatory authorities (recall and destruction of suspected products), customer (retraction of business relationship) or consumer (damaged brand image).

4. Analytical performance criteria

While much effort has gone into the development of physiological and histological tests for rapid identification of potential residue non-compliant animals (Groot *et al.*, 1996; Groot and Biolatti, 2004), they are at best used to help inspectors to identify suspect animals (O'Keeffe, 1996). Phenomena such as metaplastic and hyperplastic lesions of the prostate as indicative of the use of oestrogenic substances (Weijman *et al.*, 1990) or anatomo-pathological alteration of the trachea as indicative of β-agonist drug usage (Macrì *et al.*, 1993) only have limited value (O'Keeffe, 2000). Physicochemical analytical techniques, which are appropriate for the determination of organic molecules like most residues, have therefore to be exploited.

In its simplest form, one applies a single analytical method that, when needed, enables to screen an appropriate number of samples for the occurrence of a variety of residues, and simultaneously enables to identify and to quantify the residues that have been detected. High-end analytical methods for determination of every possible residue in any matrix may be very accurate, sensitive and reliable, but are far too expensive and therefore not realistic to perform in relation to the final price of the product. In fact, the low return of food products stimulates the development of sophisticated sampling models, selection of a number of target analytes and detection of foreign compounds at an early moment in the chain based on risk assessments. Rapid assays have certainly their place in such strategies.

Contemporary analysis strategies involve screening methods, i.e. the first procedure to establish the presence or absence of a residue, as a first step in order to utilise the infrastructure and financial facilities as efficiently as possible (Figure 1). Upon the finding of a suspected sample, the screening can be followed by a confirmatory analytical method. Confirmatory methods are more selective, more specific, more sensitive, more accurate and have an ability to identify the residue. When the sample is considered non-compliant, re-confirmation can be appealed, in particular in an arbitrage situation. For this purpose reference methods with a high performance level are applied. It is clear that with each level in this analytical strategy, the analytical capacity decreases and the costs increase. Suitable methods for confirmation can in most cases not be considered as rapid.

Screening methods should be as simple as possible and when a residue is present, it should give a possibility to test the level of the analyte against a regulatory allowed maximum residue concentration. In this respect, the methodological approach is conditioned by regulations on veterinary drugs. In the case of registered pharmaceuticals, a certain concentration, up to the so-called maximum residue limit (MRL), is allowed to occur in defined food matrices, including muscle (meat), liver, kidney, fat (and skin), milk, honey and egg. In the EU, these tolerated substances are referred to as Group B substances (Table 2). In the case of the banned Group A substances, the method should detect the analyte at a level as low as possible, as in such cases no residues are accepted.

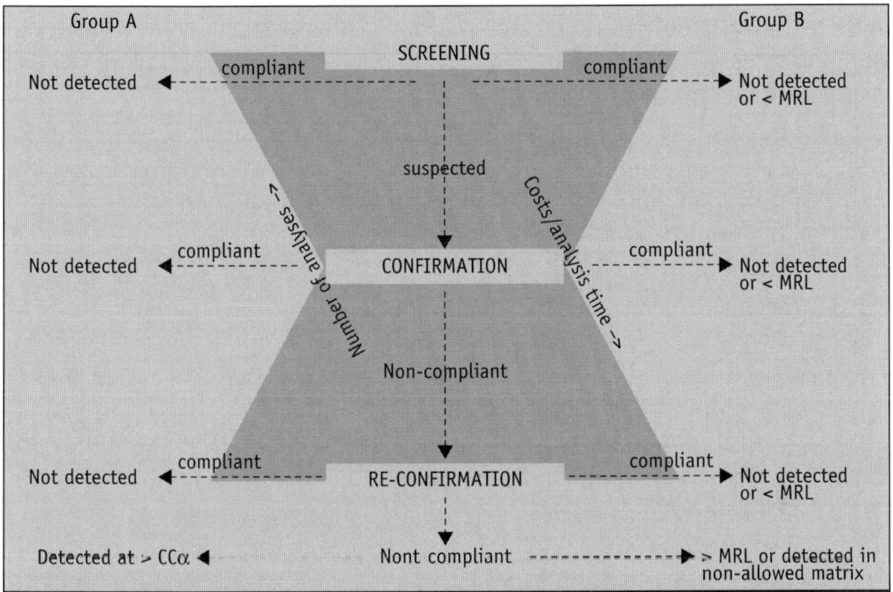

Figure 1. A typical residue analysis strategy. A high number of samples are screened for the presence of residues as a first step. Suspected samples may be confirmed in the second step using physicochemical instrumental techniques, and only in a few cases non-compliant samples are re-confirmed. In this way resources are used as efficiently as the costs for analysis rise with every step. Group A and B refer to acceptable and banned veterinary drugs (see text and Table 2 for more details).

It should be noted that for a series of banned veterinary drugs the European Commission has established so-called minimum required performance limits (MRPL), which an analytical method should fulfil. The MRPL (a method criterion) should not mistakenly be confused with the MRL (a toxicological criterion). In fact, an analytical method should be able to detect residues of banned substances at MRPL and higher. In some cases, however, a validated method may perform detection at a concentration lower than the MRPL (Table 3). Concentrations found below MRPL must be reported as non-compliant in such situations.

Many screening methods can be considered as rapid diagnostics assays. Rapid assays are, however, not screening assays *per se* and *vice versa*. Rapid diagnostic

Table 2. Classification of residues and contaminants according to Council Directive 96/23/EC. Group B3 'Other substances and environmental contaminants' is not included in this Table.

GROUP A - Substances having anabolic effect and unauthorised substances	
A1.	Stilbenes, stilbene derivatives, and their salts and esters
A2.	Antithyroid agents
A3	Steroids
A4.	Resorcylic acid lactones (including zeranol)
A5.	β-agonists
A6.	Substances which have been banned because they pose a risk to human health (annex IV of 2377/90/EEC)
GROUP B - Veterinary drugs and contaminants	
B1.	Antibacterial substances, including sulphonamides, quinolones
B2.	Other veterinary drugs
a.	Anthelmintics
b.	Anticoccidials, including nitroimidazoles
c.	Carbamates and pyrethroids
d.	Sedatives
e.	Non-steroidal anti-inflammatory drugs (NSAIDs)
f.	Other pharmacologically active substances

Table 3. Residues with described MRPLs and analytical methods used to detect these residues at least their corresponding MRPL levels.

Residue	MRPL (μg kg^{-1})	Possible method for screening
Chloramphenicol	0.3	BIA, ELISA
Malachite green	2[a]	HPLC, LC-MS
Medroxyprogesterone	1	ELISA
Nitrofuran metabolites	1	LC-MS

[a] Total sum of malachite green and its metabolite leuco-malachite green.

assays are not in all cases able to process large number of samples and are not always cheap, but do not require - or only need simple - equipment to perform. Screening methods, on the other hand, are in most cases relatively cheap, but not always rapid. Most of the regulatory microbial growth inhibition assays are screening methods hampered by their long incubation times to enable detection of inhibition zones. The New Dutch Kidney Test, European Four-Plate Test and Modified Four-Plate Test used in most EU member states or the 7-plate assay used by the US Department of Agriculture's Food Safety and Inspection Service (FSIS) require at least 16 hours for microbial development. Measurement of impedance or conductance influenced by the metabolic activity of the indicator microorganism improves speed (Myllyniemi *et al.*, 2002). The ready-to-use, all-in-one ampoule system PremiTest (DSM, the Netherlands) generates results in 4 hours and seems to be much faster.

An important quality criterion for rapid analytical methods that are used to filter suspected samples, are their false-compliant and false non-compliant rates (Table 4). A cleared product that is found non-compliant afterwards at an inspection action, is as expensive as not performing analyses at all with respect to lost trust, recall actions, and destruction of product. On the other hand, every false non-compliant result may generate unnecessary costs for confirmatory analysis. Hence, the use of methods with a poor predictive value can be very expensive and thus not cost-effective for the producer.

Commission Decision 2002/657/EC has tried to cover the uncertainty of the result, which had major juridical consequences, by the introduction of the so-called detection capability (CCβ) of a method, which is also applicable to rapid and/or screening methods. It is the lowest concentration of an analyte, which can be detected in a sample with an error probability of β. For rapid and/or screening methods, β is set at 5%. Only when this criterion is met and the screening method is validated according to Commission Decision 2002/657/EC, *i.e.* they must 'fit the purpose', it is allowed in regulatory monitoring as described in Council Directive 96/23/EC.

It has to be realised that, when confirmation is needed in all suspected cases, this would lead to an enormous increase in financial costs (Figure 1). Many countries, but also stakeholders mutually, have agreed to condemn animal products on the result of a screening test. The consequence will be that some false non-compliant suspected products will be destructed; on the other hand the current CC_{β} allows 5% of the non-compliant materials to be judged as false compliant.

Table 4. Important parameters for each analytical test disclosing interference susceptibility, ruggedness, accuracy and applicability of the method.	
Performance characteristic	Description
Selectivity	Ability of a test method to detect truly compliant samples as compliant regardless of sample source
Selectivity ratio	Percentage of truly compliant samples that are found to be compliant by the test method
False compliant rate	Percentage of truly compliant samples that are wrongly detected as non-compliant
Specificity	Ability of the test method to distinguish between the analyte of interest and any other substances that may be present
Specificity ratio	Percentage of truly non-compliant samples that are found non-compliant by the test method
False non-compliant rate	Percentage of truly non-compliant samples that are wrongly detected as compliant
Sensitivity	Decision limit (CCα), which is the limit at and above which it can be concluded with an error probability of α that a sample is non-compliant. Detection capability (CCβ) is the smallest concentration of the substance that may be detected, identified and/or quantified in a sample with an error probability of β.

The study described by (Kroll *et al.*, 2000) in which the performance of rapid assays was compared, is an example of many published. In that study, the performance of assays for the detection of β-lactam antibiotics in raw milk was investigated, namely β-STAR, Charm MRL (Charm Sciences, USA), Delvo-X-Press βL-II (DSM, the Netherlands), Penzyme 50 (UCB Bioproducts, Belgium) and SNAP (IDEXX, USA).

5. Sample preparation

Despite the many advances in analytical technology, sample preparation remains the most difficult stage in residue analysis and remains most prone to errors. The best sample preparation is of course no sample preparation at all, but sample preparation is generally needed to:

- disclose the analyte;
- remove the matrix, including interfering substances;
- cleave conjugates or release so-called bound residues;
- concentrate the analyte;
- improve chromatographic and/or detection properties.

Sample preparation is a major challenge for any rapid assay procedure. In fact, sample preparation will claim about 50 to 75% of the total analysis time (Stolker, 2000). Despite that, most technical innovations and improvements are made in the separation and detection procedures. It must be noted that automation of sample preparation procedures can not only reduce the chance for errors, but can also speed up the generation of the results considerably. Although the time gain for a single sample is limited, the fact that many samples are treated simultaneously or at different stages at the same time, will add to the speed of the assays.

Liquid samples, including serum, urine, milk, honey or egg, are generally considered as relatively simple to handle. Such samples may need only dilution or minimal sample work-up for the analysis to take place. Detection of substances in 'solid' matrices, such as meat, fat, kidney, liver, hair or retina, however, requires liberation of the enclosed analytes. It must be noted that irrespective the sample matrix, the analyte may have to be released from conjugates such as glucuronic acid and sulphate, to obtain the analyte in a form that is suitable for analysis. This will need extra handling and several hours of incubation with splitting enzymes. In a similar way, residues may be covalently linked to larger biomolecules, such as proteins. These so-called bound residues have to be chemically released to enable their detection.

Only a few methods, based on the diffusion of small molecules, allow direct detection of the target in the solid matrix, such as in the European Four-Plate Test using slices of meat. However, such assays are generally not considered as fast. Alternatively, automated on-line dialysis HPLC has been suggested to automate and to speed up residue analysis (Bisschop *et al.*, 2000). Simple extraction approaches also include, for example, the generation of tissue drip through a simple garlic press, freeze/ thaw cycles or boiling the meat in a closed plastic bag. The drip is then analysed. Although described by the producer this extract may not be sufficient to screen effectively for antibiotics using the commercially available PremiTest, as a generic solvent extraction to enhance its performance was proposed (Stead *et al.*, 2004a; Stead *et al.*, 2004b). It is noteworthy that this study used p-aminobenzoic acid and β-lactamase to reveal the identity of residue in a suspected sample, namely sulphonamide or β-lactam compound, respectively.

As a first step, extraction of the target residues from tissue needs homogenisation of the sample, which includes laborious cutting and blending prone to the risk of cross-contamination. Mechanical extraction using disposable filter (Stomacher) bags, such as used in microbial analysis, can be more convenient and has been used earlier for the detection of sulphonamides in pork (Ackermans *et al.*, 1992).

As extraction is in most cases necessary, its procedure should have regard to the gross composition and relevant properties of the analytical matrix and the physico-chemical characteristics, including hydrophobicity, ion density, pK_a and molecular mass of the targeted molecule(s). An overview of typical properties of some relevant matrices, including urine, plasma, milk, animal tissues and eggs are found in Aerts *et al.* (1995).

Substantial advances in sample preparation are achieved using novel procedures such as microwave-assisted extraction (MASE), pressurised liquid extraction (Hogendoom *et al.*, 2001; Pallaroni and von Holst, 2003), supercritical fluid-extraction (Zoontjes *et al.*, 2000) and accelerated solvent extraction (Draisci *et al.*, 2000; Stolker, 2000). These approaches are slowly gaining popularity, but are still used in niche applications. Specific solid sorbents may be used in solid phase extraction (SPE), which enables a certain level of selectivity and concentration in a single step. Combinations of stationary phases within a single cartridge (mixed bed SPE) are available. So-called restricted access materials (RAM) combine gel filtration and solid-phase extraction within a single sorbent. These special SPE-materials allow the extraction of aqueous extracts or liquid samples, like milk, without prior de-proteination. Haasnoot coined the name immunofiltration (IF) for the ultrafiltration of a sample, which was first mixed with free antibodies, over a 30-kDa filter (Haasnoot *et al.*, 2000). Retained complexes were washed, dissociated and filtered to obtain a purified and concentrated analyte-containing filtrate. IF turned out to be a comfortable and fast alternative for the use of immunoaffinity columns, avoiding much of the non-specific bound material.

More recently, discs consisting of membranes of polymers with high, reversed and ion-exchange phase capacity have been introduced. These discs are used as filters, allow fast-flow processing, and also purify and concentrate analytes in small to large sample volumes. Interfaces are available enabling the coupling of the discs to analytical devices such as HPLC-columns.

Another sorbent extraction approach is matrix solid-phase dispersion (MSPD), which allows homogenisation and extraction in a single run omitting liquid extraction of the homogenate as a separate step. Columns prepared of the mixture of sample and sorbent material are eluted in a similar way as SPE cartridges.

To complement fast detection, affinity extraction using antibodies (IAC) or even receptors, is employed to clean up the extract, although solvent incompatibility may be a problem. To perform affinity extraction, a suitable specific antibody or receptor is needed. Its dissociation constant should be within a certain range to allow separation of analytes from matrix components and should allow elution without harsh conditions that could modify the residue (Godfrey, 1998).

To circumvent the generation of suitable antibodies, sorbents called molecularly imprinted polymers (MIPs) can be obtained as highly cross-linked polymers synthesised in the presence of a print molecule, typically a variant of the analyte. Removal of the print molecule leaves an imprinted sorbent for extraction of the analyte (Brüggemann *et al.*, 2000; Haginaka, 2004). Molecular imprint techniques have not yet reached full maturity and only a few examples of their application in residue analysis have been described so far. Clenbuterol, for example, has been isolated successfully from calf urine in this way (Blomgren *et al.*, 2002) and recently the extraction was also reported for other β-agonists from cattle, duck, rabbit and turkey meat (Kootstra *et al.*, 2004). The synthesis of MIPs targeted towards water-soluble molecules has met with limited success. Therefore, synthetic oligo- or polynucleic acid ligands, known as aptamers, have attracted much interest. Aptamers also exhibit binding characteristics comparable to or even better than monoclonal antibodies (Burgstaller *et al.*, 2002). They have been selected against a variety of targets, ranging from small organic molecules up to supramolecular structures, including organisms. An RNA aptamer with high affinity for 6-hydroxy containing tetracyclines has been described for example (Berens *et al.*, 2001). As obvious from a search for information on aptamers on the internet, the potential of these molecules has been recognised in many analytical scientific disciplines.

6. Assay formats

An analyte-containing sample (extract) is rich in many potentially interfering (co-extracted) matrix components, such as fat, proteins, carbohydrates and small organic molecules. A method can only be fast and compatible with multiple matrices when it introduces high selectivity to cope with these potential matrix interferences. Frequently, the required selectivity is obtained by introducing a high degree of specificity as well.

Despite the associated costs, mass spectrometric (MS) methods are increasingly attracting attention as a technology with a potential to design rapid methods, which are very selective and specific and offer multi-analyte detection. As an example, instead of screening kidney samples using test kits, the planned 1,000 pork kidney samples are screened and confirmed simultaneously for β-lactams, macrolides, lincomycin, sulfonamides and tetracyclines in a Canadian surveillance program using MS (Noot *et al.*, 2004). Advanced sample preparation is, however, essential in order to manage matrix-induced ion-suppression and/or deposition of matrix debris in the ionisation source affecting the stability of instrument operation.

To acquire the necessary selectivity, many rapid assays are based on biomolecular affinity detection of the targeted residues with the majority of applications using antibodies. Binding biomolecules, however, does not only comprise antibodies in immuno-assays (IA), but enzymes and receptors in receptor assays (RA) as well. These IA and RA assays, whether dipstick, lateral flow device (LFD) or enzyme immunoassays (EIA)-like configurations, allow detection of low molecular substances including most residues of veterinary drugs, almost exclusively through a competitive assay set-up using labelled antigens (Ag^*) or receptor ligand (D^*):

IA: $$Ag + Ag^* + Ab \Leftrightarrow AgAb + Ag^*Ab \quad (1)$$
RA: $$D + D^* + R \Leftrightarrow DR + D^*R \quad (2)$$

where Ag is the free antigen and Ab is the antibody in the IA, and D is the free drug residue and R is the receptor or enzyme in the RA.

From these formulae it can be recognised that the concentration of the analyte is inversely proportional to the assay response by detection of the label. To discriminate unreacted and complexed reactants, most assays use Ab/R (direct assay) or Ag/D (indirect assay) immobilised to a solid phase, such as in ELISAs, through which isolation of complexes can be obtained (Haasnoot and Schilt, 2000).

The evolution of IAs started with radio IA (RIA) using radioactive isotope labelled antigens. The RIA concept was proposed in the midst of the 1950s following experiments with ^{131}I-labbeled insulin in patients (Berson and Yalow, 1959). In residue analysis, RIA was used in particular in the 1970s and 1980s, ELISAs followed in the 1980s and 1990s, (time-resolved) fluorescent IAs (FIA) came in the 1990s, and in the new millennium successful times for (immuno)affinity biosensors seem to dawn. The detection systems have thus evolved from the use of undesired radioactive tags, to the use of non-radioactive labels, such as chromophoric and later fluorescent conjugates, to the detection of physical intrinsic characteristics of the analyte (Table 5).

It must be noted that many variations on the listed assay concepts are published. Compared to ELISA, the sorbent surface can be increased, and hence sensitivity, by the use of (latex) beads. The latex agglutination inhibition assay (LAIA) observes the agglutination of the beads in the presence of the analyte, as shown for the on-site screening for sulphamethazine in pork (O'Keeffe *et al.*, 2000). Instead of ELISA plates, tubes have been used as the sorbent in a unique competitive assay called tube enzyme immunoassay (Haasnoot *et al.*, 1996). The tubes with pre-coated antibodies are robust and very fast. Using the provided portable differential photometer it gives inspectors a semi-quantitative answer on the presence of clenbuterol residues on-site within 20 minutes using simple tap water (W. Haasnoot, personal communication).

As a result of the introduced specificity by binding molecules, rapid methods can only detect a single drug residue or a limited number of residues showing cross-reactivity with the applied binding protein(s). Some methods therefore attempt to broaden the number of targeted analytes by combining a number of capturing biomolecules showing affinity for a group or groups of residues. This is for example the case in the CHARM Test II (Charm Sciences, USA), which assays for a number of antibiotic classes by including several receptors in the assay (Korsrud *et al.*, 1996).

Much research was performed on steroid receptors as potential binding elements in screening assays. Arts and co-workers used for example the oestrogen receptor to probe calf urine for the abuse of exogenous anabolic

Table 5. Summary of advantages and disadvantages of several IAs in residue analysis. The same holds for comparable RAs.

Technique	Advantage	Disadvantage
RIA	• Adequate sensitivity • Robust • Easy to use and rapid • Low false non-compliant rate • High-throughput possible • Inexpensive	• Use of radiolabels • Radiolabelling by specialised companies
ELISA	• Adequate sensitivity • High-throughput possible • Inexpensive	• Limited stability of some conjugates • High false non-compliant rates in some tests
Affinity biosensors	• Reduced matrix effects • Reduced sample preparation • Low false non-compliant rate • No tracer • Versatile system - multi-analyte detection	• Price of equipment • Limited number of different residue tests available
Dipstick/ LFD	• Robust • Easy to use and very fast • No equipment needed	• Not suitable for high throughput • Limited specificity • Expensive

substance with oestrogenic activity (Arts *et al.*, 1989; Arts and Van den Berg, 1989). The assay never became incorporated into routine testing programmes, as the preparation of a stable receptor was a too high hurdle to take.

Molecular biology and genetic engineering technologies have evolved enormously since then, and in potential, binding proteins, either receptors, enzymes or antibodies, can be engineered on a routine basis to enhance their avidity, to broaden their specificity or to increase their stability and thus suitability in analytical tests. Relatively many dedicated companies already offer this service. An example of this approach is the directed mutation of an anti-sulphonamide antibody that enabled the binding of 18 sulphonamides instead of only a few (Korpimaki *et al.*, 2004). As an other example, generic antibodies binding every potential β-agonist, which are at least 50 different compounds, are hitherto not available. For that reason, the membrane-bound human β_2-adrenergic receptor was expressed in *E. coli*. Although currently under construction, the recombinant DNA receptor is planned to be used in a field test (dipstick assay) and a laboratory method (affinity biosensor assay) (Danyi *et al.*, 2004).

The dipstick system is simple in handling, whereas the technology behind it is advanced and subtle (Figure 2). The extract, urine, diluted honey or milk migrate on a support coated with two zones of receptor ligands and/or antigens (indirect assay) or two zones of receptors and/or immunoglobulins (direct assay). Just before application, the sample is mixed with the appropriate soluble labelled reagent, either a tagged antibody/receptor (indirect assay) or labelled ligand/antigen (direct assay). One of the added binding partners is thus conjugated. In the case of binding proteins gold nanoparticles are the most likely conjugates. Upon the binding at the capturing zones the gold particles are concentrated and a dark, black-reddish band becomes visible (Figure 2). Whether an indirect or a direct assay set-up, upon the development of the stick the first band will indicate the presence or absence of the targeted analyte, while the second will indicate the success of the migration. On the basis of this technology, commercial dipstick kits are offered to test veterinary drug residues, such as β-lactams (β-STAR assay, UCB Bioproducts, Belgium) (Janosi *et al.*, 2000) or tetracyclines (Tetrasensor-assay, Unisensor, Belgium) (Okerman *et al.*, 2004).

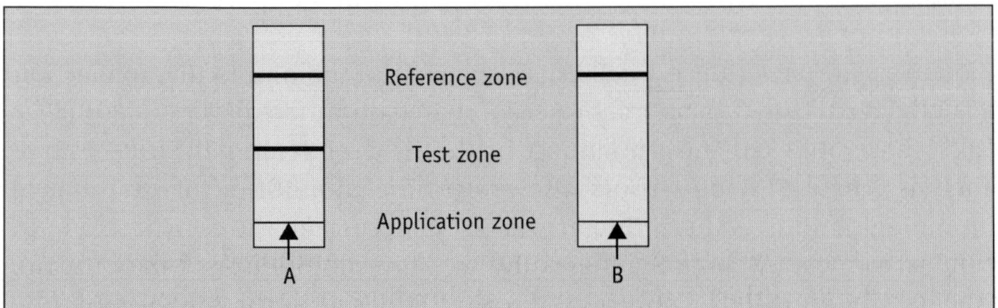

Figure 2. A simple schematic representation of a dipstick or of a lateral flow system. Analyte-containing solutions are applied at the arrow. In an indirect assay set-up: A, compliant sample and B, non-compliant sample.

The one-step strip test, better known as lateral flow device (LFD), derived from the dipstick system is even simpler in handling and is a field test par excellence. The system is very successfully applied as a pregnancy test available at the pharmacy or retail shops. The analyte-containing solution or suspension is not mixed with reagents, but while migrating, matrix components, including the analyte, are mixed with analyte-binding and reference-binding molecules in a reagent pad just after application of the liquid sample (Verheijen and Haasnoot, 2000). Lateral flow devices are available to test for example sulfadimidine (Verheijen *et al.*, 1998), (dihydro)streptomycin (Verheijen *et al.*, 2000), enrofloxacin (Watanabe *et al.*, 2002) or salinomycin (Watanabe *et al.*, 2001), but also to test a naturally occurring toxin microcystin produced by cyanobacteria in fresh water relevant to recognise animal intoxications (Kim *et al.*, 2003).

An interesting development is a biochip used as a small reaction platform (9 mm^2) with multiple specific antibodies or antigens immobilised at 100 defined sites on the surface. Following exposure of the chip to a sample the binding of analytes is examined through an imaging system recording generated chemiluminescence. It is claimed that many different analytes in thousands of samples per hour can be detected (Anonymous, 2004; McConnel *et al.*, 2000).

Another promising analysis format is the one-step all-in-one-well concept exploiting dry reagent chemistry and time-resolved fluoroimmunoassay (TR-FIA) technology (Hagrén *et al.*, 2004; Peippo *et al.*, 2004). In this format extracted samples are added automatically to the dry microtiter well followed by an incubation, wash-step and (automatic) recording of a response.

Biosensors are powerful and well-established tools for biomolecular interaction analysis (BIA). Hitherto, optical biosensors based on surface plasmon resonance (SPR) have attracted most attention in the field of veterinary drug residue analysis. These SPR biosensors allow real-time monitoring of the analyte immunoglobulin/receptor interaction as it occurs at the sensor chip. In contrast to most IAs and RAs, these advanced direct biosensors do not require labelling. In an EC 4th Framework Project (Foodsense) the capability of the technology to perform rapid and reliable analysis on samples with a minimal amount of sample preparation was clearly demonstrated using Biacore technology (Biacore International AB, Sweden). Indeed, a fast growing list of applications appears in the literature demonstrating its user friendliness, flexibility and suitability in food analysis (Cacciatore *et al.*, 2004a; Elliott *et al.*, 1998). Tests are, among many more, reported for the detection of chloramphenicol (Gaudin and Maris, 2001), clenbuterol (Haughey *et al.*, 2001), enrofloxacin and ciprofloxacin (Mellgren and Sternesjo, 1998), β-lactams (Cacciatore *et al.*, 2004b; Gustavsson and Sternesjo, 2004), (dihydro) streptomycin (Baxter *et al.*, 2001; Haasnoot *et al.*, 2002), ractopamine (Shelver and Smith, 2003), gentamycin (Haasnoot and Verheijen, 2001), sulphonamides (Bjurling *et al.*, 2000; Crooks *et al.*, 1998; Elliott *et al.*, 1999). The technology is finding a place in the routine laboratory for chemical contaminant analysis. The Central Science Laboratory (York, UK), for example, is implementing an analytical strategy based on a combination of the PremiTest and SPR biosensor in the screening phase of the routine monitoring plan (Dr. S. Stead, personal communication).

An intriguing concept supportive for development of fast analysis procedures is the detection of predictive indicators reporting the presence or absence of targeted residues. The new Dutch kidney test, which is a microbial growth inhibition assay, relies on this principle assaying the pre-urine for antibiotics while in fact testing for antibiotics in the muscle (meat) of the animal (see Table 6 for more examples). Predictive indicators are already widely used in clinical laboratories. Main advantage of their use in residue analysis lies in the ability to analyse fluid samples without the need for (much) clean-up, thus giving the ability to produce automated testing procedures with high throughput when necessary. This idea was brought to live in two studies involving a surface plasmon resonance biosensor to test the presence of sulphonamides in swine real-time at the slaughter line (Baxter *et al.*, 1999; Situ *et al.*, 2002). The biosensors, which were physically present in the abattoir in close proximity to the kill line, showed sufficient robustness to withstand humidity, bumping and

Table 6. Some examples of studies using indicators predictive for the presence or absence of targeted residues of veterinary drugs.

Analyte	Indicator employed	Reference
Sulphonamides	Bile	(Crooks *et al.*, 1998)
Sulphadimidine	Urine	(Verheijen *et al.*, 1998)
Glucocorticoids	Blood leukocytes	(Brambilla *et al.*, 1998)
Antibiotics	Urine	(Korsrud *et al.*, 1998)
Anabolics	Blood	(Dijkmans *et al.*, 2002)

temperature changes (Dr. C.T. Elliott, personal communication). The biosensors tested for unacceptable concentrations of sulphonamides in pork meat by detecting these substances in easily collectable bile. The false non-compliant rates were very low (between 0.13 and 0.74%) and no false compliant samples were revealed. The biosensor assay facilitated the daily screening of 650 animals, i.e. more than 20% of the total kill (Situ *et al.*, 2002), which is much more than currently in many 'conventional' monitoring programs.

7. Detection

At the end of the analysis procedure a signal that can be interpreted as some residue content, has to be obtained. Final reading of the result in rapid assays is usually an easily observable change in or generation of colour, which can in some cases be correlated to a certain concentration of the residue (Table 7). Assays exploiting fluorescence, chemoluminescence or radioisotopes need equipment to record the assay response.

Manufacturers of rapid assays have to pass some obstacles. Typically, rapid methods do not reveal molecular structure information and binding molecules may capture homologues and/or analogues of the targeted analyte(s). As a consequence antibodies cross-reacting with degradation products or metabolites of the analyte, which are not included in the final MRL, will overestimate the final residue level and may thereby wrongly condemn a product or carcass. This cannot be solved easily by adaptation of the cut-off values, as metabolic profiles are, among many other parameters, tissue and species dependent. In the second

Table 7. Sensitivity of detection principle or methodological approach.

Methodological approach or detection technique	Lowest detectable concentrations
UV absorbance	$\mu g\ kg^{-1}$ to $mg\ kg^{-1}$
Mass spectrometry	$ng\ kg^{-1}$ to $\mu g\ kg^{-1}$
Rapid immunoassays / receptor assays[a]	$\mu g\ kg^{-1}$
Bioassays	$ng\ kg^{-1}$ to $\mu g\ kg^{-1}$

[a] Assays producing fluorescence, chemoluminescence or radioisotope responses may be an order more sensitive.

place, each residue has its own MRL and often this MRL is matrix-specific for fat (and skin), kidney, liver and muscle. As long as the identity of the residue is not disclosed by the assay, the cut-off value or decision level is based on the greatest common denominator for the considered MRLs or based on the lowest MRL of the residues for which the assay was developed. Consequently, this decision level will determine a certain rate of false compliant and non-compliant results (cf. Table 4). It explicates that in the analysis strategy employed, screening should be followed by confirmation analysis of the residues in the suspected samples to establish identity and concentration of the targeted molecules as depicted in Figure 1.

8. Bioassays

Bioassays are based on *in vivo* or *in vitro* biological effects provoked specifically by target compounds. In fact it is a logical exploitation of biological processes, as residue tolerances are per definition based on (adverse) biological effects in animals and humans. Rats and mice, for example, are used routinely to screen bivalves for the presence of diarrhoeic (DST) and paralytic (PST) shellfish toxins, as suitable chemical analytical methods are not available. Assuming that a drug interacts reversibly with a cellular component, the following equation can be derived (Haasnoot and Schilt, 2000):

$$\text{Drug (D)} + \text{Receptor (R)} \Leftrightarrow \text{DR} \Rightarrow \text{effect} \qquad (3)$$

The determination of an effect, which is also per definition observable and likely quantifiable, initiated by some biological active substance(s) is significant different from direct detection of analytes in non-bioassays. The effect does not necessarily correlate with the concentration of the (biologically active) analyte(s) in the sample, as discussed above for the microbial growth inhibition assays.

Reporter gene-based bioassays use mammalian or yeast cells, which may be genetically modified for the expression of a certain analyte-specific receptor and/or for introduction of a reporter gene in the DNA material. Through binding to the receptor a residue will trigger the expression of increased levels of reporter gene product(s), such as luciferase for the production of luminescence (Zacharewski, 1997). The reporter gene-based bioassays can be divided in two categories depending on whether the measurement involves detection of a single endpoint or the simultaneous assessment of complex molecular patterns representing multiple endpoints (Naegeli *et al.*, 2004) (Figure 3).

An example of a single-endpoint bioassay is the chemical-activated luciferase gene expression (CALUX) method, which is very successfully applied to monitor toxic planar PCB and dioxin congeners in a broad range of samples (Hoogenboom *et al.*, 1999; Murk *et al.*, 1996). An oestrogen activity-detecting variant of this assay (ER-CALUX) was obtained by transfection of human breast cancer cells with a responsive luciferase responsive element (Hoogenboom *et al.*, 2001; Legler *et al.*, 1999). The cell system is not only responsive to 17β-estradiol, but also to its derivatives estrone, 17α-estradiol and the synthetic 17-hydroxy benzoate ester of 17β-estradiol and several metabolites with different sensitivities (Hoogenboom *et al.*, 2001).

Recently, a yeast bioassay stably expressing human oestrogen α- and β-receptors, and green fluorescent protein in response to oestrogen was described (Bovee *et al.*, 2004). This so-called RIKILT oestrogen assay (REA) responds relatively fast without cell wall disruption or addition of substrates. Cells are suspended in the sample (extract) and incubated while monitoring the development of bioluminescence. Concentrations of 1 ng ml^{-1} estrone, 17α-estradiol, 17α-estradiol or estriol were easily detected, but substances with oestrogenic activity often present in samples, such as derivatives of bisphenol, nonylphenol and phytoestrogens, were detected as well. This interference necessitated liquid chromatography (LC) fractionation of the sample before applying the bioassay.

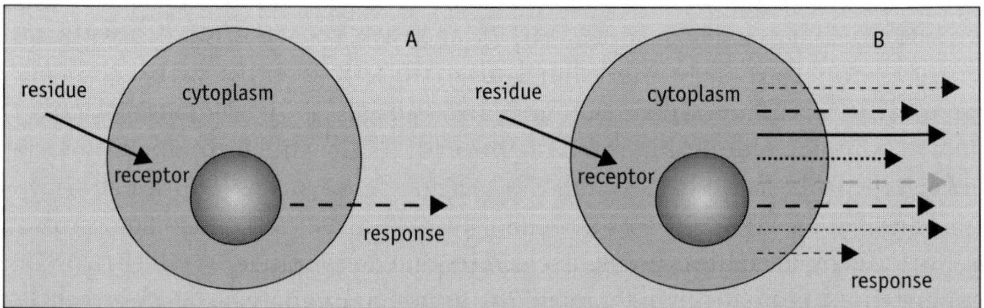

Figure 3. Cellular response to exogenous substances. (Genetically modified) cells produce A) a single measurable effect (single endpoint assay), or B) multiple effects monitored and assessed through some array measurement (multi-endpoint assay).

A dual luciferase reporter screening assay using a transfected sensitised human cell line, has been constructed to produce two different luminescence signals, one of which serves as internal control to adjust for assay variations (Schumacher *et al.*, 2003). Detection of a broad range of glucocorticoids, such as dexamethasone, flumethasone, methylprednisolone and triamcinolone, was demonstrated. The assay seems to be a new tool to detect illicit use of this type of veterinary drugs for growth promotion.

The availability of complex microarrays consisting of synthetic oligonucleotides ('DNA chips') complementary to the sequences of essentially all *e.g.* approx. 33,000 human genes, enabled the analysis of genome-wide changes in transcriptional activity. In other words, multi-endpoint measurement reporting specifically the presence of certain toxic compounds or compound classes assessed through transcriptomics technology, has become available. Through hybridisation of the total messenger RNA (mRNA) isolate from target cells exposed to the food product extracts, it has become possible to investigate the changes in gene expression by monitoring thousands of specific mRNA molecules in a single analysis run. This transcriptomics technology is still in its infancy. It is pioneered by Naegeli and co-workers (Naegeli *et al.*, 2004) in the field of food contaminant and residue analysis. They observed specific genomic fingerprints in human cells exposed to PCB54, PCB153, TCDD and the phytoestrogen genistein (Figure 4).

The procedures including sample preparation and incubation of cells, are not fast, but the derived information is in potential virtually unlimited. Each fingerprint of mRNA is unique for a certain bioactive compound to which the hosting cell was exposed. So, despite the relative long analysis run, the 'high-density' of information in terms of maximum number of detected substances per time unit, outperforms easily 'conventional' rapid assays.

In a similar way changes in protein arrays may reflect the presence of (multiple) foreign bioactive molecules as well. The (post)transcriptional, and the co- and posttranslational changes giving rise to a so-called microheterogeneity of most polypeptides, is a challenge to assess in a way that can be executed easily in a residue laboratory. Nevertheless, the application of proteomics as a multi-endpoint bioassay, may lead to the identification of protein biomarkers indicative of specific exposures that can be detected in new rapid assays (Dijkmans *et al.*, 2002). Such biomarkers can be found in exposed cells, but also in the living animal such as in its blood. The monitoring of biomarker

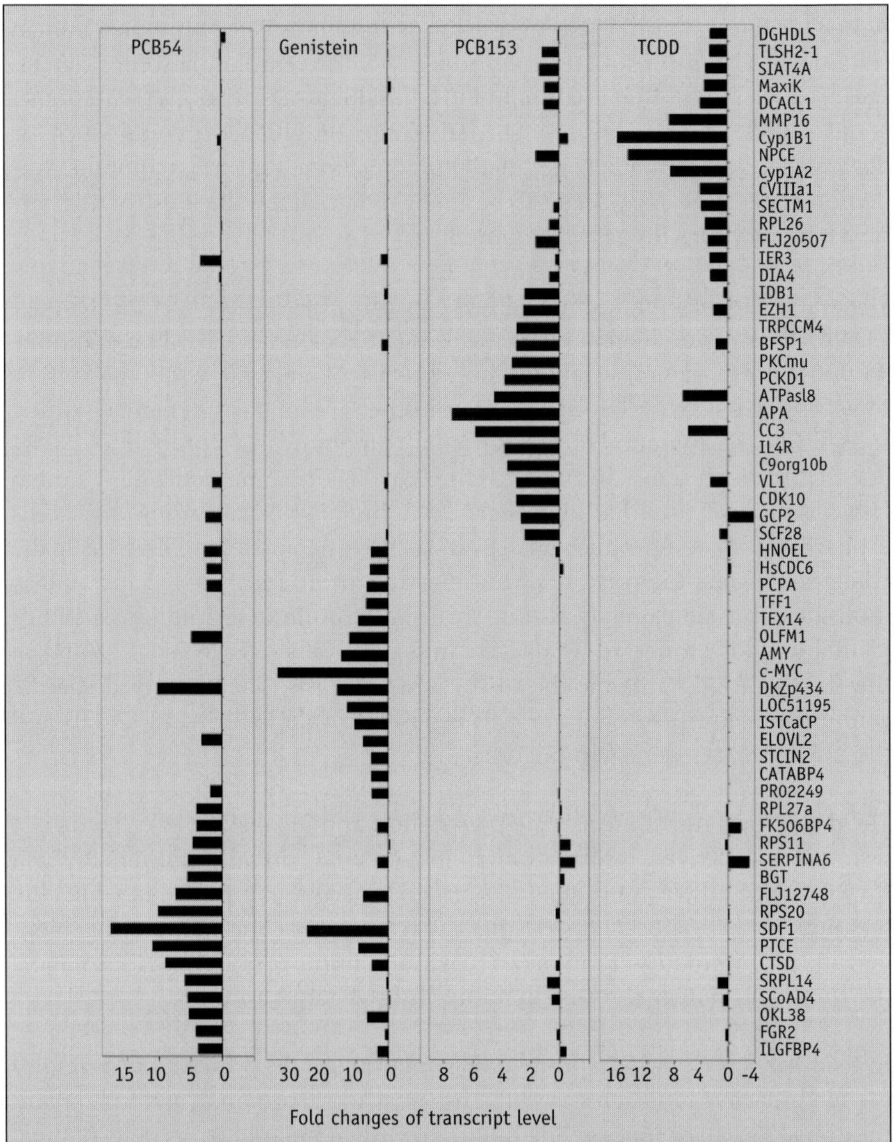

Figure 4. Genomic fingerprints from T47D cells, which were exposed to the polychlorinated biphenyls PCB54 and PCB153, tetrachlorodibenzo-dioxin (TCDD), and the phytoestrogen genistein for 24 h. Changes in mRNA levels were assessed by hybridisation to Affimetrix U133 DNA microarrays. Courtesy of Dr. H. Naegeli, University of Zurich, Switzerland.

signatures may certainly help to achieve the mandatory monitoring for illegal anabolic growth promoters occurring at only a few promille of farm animals. This needs a high number of samples even in the case of very reliable methods to come across (Table 8). When only conventional methods are applied, such high sample numbers are giving resource problems

Effect assays, such as microbial or cell growth inhibition tests or reporter gene-based assays, can thus detect a wide range of substances. In the case that multiple pharmaceuticals have been used, these assays may detect the sum of all biologically active compounds. Especially when cocktails of banned drugs were used, such effect assays may therefore outperform physicochemical methods, as they do not encounter difficulties to detect each individual substance at low concentrations. Notably, bioassays are very suitable for the detection of 'novel' compounds. This is of importance in the race to reveal new substances, which are 'invented' for illegal animal fattening practices. The power of bioassays in this way has been shown recently by the discovery of a novel β-agonist in feed (Nielen *et al.*, 2003).

Table 8. Minimum required number of samples to detect at least a single non-compliant result with pre-defined probability in a population with a known compliance prevalence (Boison, 2004).

Non-compliant prevalence (% in population)	Confidence level		
	90%	95%	99%
20	11	14	21
15	15	19	29
10	22	29	44
5	45	59	90
1	230	299	459
0.5	460	598	919
0.1	2302	2995	4603

9. Conclusion

This review has indicated a clear need for rapid assays in residue analysis. Disaster preparedness, crisis management, increased requirement for self-control, increased degree of liability of each link in the food chain, the appeal to give transparency of production conditions in (international) trade regulations, and public concerns are important drivers for the development of such tests. In addition, the financial penalties imposed on producers who violate residue regulations are increasing. Therefore, there is a growing demand for analyses of an increasing number of possible analytes and proportionally there will be an ever-increasing demand for rapid, reliable and user-friendly assays. This places a greater onus on the residue analyst to guarantee good analysis with sound data supporting the correct release or condemnation of (a batch of) animal products. In addition, the application of rapid assays is not only performed by trained residue analysts, but increasingly also by laymen as well. For this reason, robust and user-friendly methods are needed, which indeed have become commercially available in many formats.

The disadvantage of many biomolecular affinity-based formats is that they are only able to measure a single analyte or a small group of analytes. The 'fishing' biomolecules are being engineered by recombinant DNA techniques for more favourable characteristics in 'renewed' diagnostic assays. Multi-analyte methods using a combination of (engineered) binding proteins with relative high throughput are dawning, but they fail to give molecular information with respect to the co-occurrence of metabolites.

Despite the availability of an increasing number of rapid diagnostic assays, the current bottleneck remains sample preparation consuming in general the major part of the total analysis time. It is necessary that substantial progress is made in this area to keep pace with the developments in detection systems. To circumvent the need for (extensive) sample clean-up, the detection of indicator molecules, biomarkers or effects in an indicator microorganism or cell may be very promising. Single or multi-endpoint bioassays have therefore the potential to become valuable tools for rapid and high throughput screening for residues of veterinary drugs in a variety of samples.

Irrespective the advances in rapid assays, they have to be applied carefully depending on the purpose for which they are used. In other words, the data produced by the assay should enable a user to make technically and administratively correct decisions for a stated objective.

References

Ackermans, M.T., Beckers, J.L., Everaerts, F.M., Hoogland, H. and Tomassen, M.J., 1992. Determination of sulphonamides in pork meat extracts by capillary zone electrophoresis. Journal of Chromatography 596, p. 101-109.

Aerts, M.M., Hogenboom, A.C. and Brinkman, U.A., 1995. Analytical strategies for the screening of veterinary drugs and their residues in edible products. Journal of Chromatography B-Biomedical Applications 667, p. 1-40.

Andersen, J.R., 1994. Development of real time methods for residues of veterinary drugs. In: Residues of veterinary drugs and mycotoxins in animal products - new methods for risk assessment and quality control, edited by G. Enne, H.A. Kuiper and A. Valentini. Wageningen Pers, Wageningen, the Netherlands, p. 41-51.

Anonymous, 2004. [Online]. Available by Randox Laboratories Ltd., www.randox.com (Last visited: 30 August, 2004).

Arts, C.J., Kemperman, P.T. and van den Berg, H., 1989. Oestrogen radioreceptor assay for multi-residue screening of bovine urine for oestrogenic anabolic compounds. Food Additives and Contaminants 6, p. 103-115.

Arts, C.J. and van den Berg, H., 1989. Multi-residue screening of bovine urine on xenobiotic oestrogens with an oestrogen radioreceptor assay. Journal of Chromatography 489, p. 225-234.

Baxter, G.A., O'Connor, M.C., Haughey, S.A., Crooks, S.R. and Elliott, C.T., 1999. Evaluation of an immunobiosensor for the on-site testing of veterinary drug residues at an abattoir. Screening for sulfamethazine in pigs. Analyst 124, p. 1315-1318.

Baxter, G.A., Ferguson, J.P., O'Connor, M.C. and Elliott, C.T., 2001. Detection of streptomycin residues in whole milk using an optical immunobiosensor. Journal of Agricultural and Food Chemistry 49, p. 3204-3207.

Berens, C., Thain, A. and Schroeder, R., 2001. A tetracycline-binding RNA aptamer. Bioorganic and Medicinal Chemistry 9, p. 2549-2556.

Berson, S.A. and Yalow, R.S., 1959. Quantitative aspects of the reaction between insulin and insulin-binding antibody. Journal of Clinical Investigation 38, p. 1996-2016.

Bisschop, H.W.M., Radstake, D.J. and Stern, A.G., 2000. On-line dialysis as a tool in the analysis of residues of veterinary drugs. In: Proceedings of EuroResidue IV, edited by L. Van Ginkel and A. Ruiter, National Institute for Public Health and the Environment, Bilthoven, p. 220-225.

Bjurling, P., Baxter, G.A., Caselunghe, M., Jonson, C., O'Connor, M., Persson, B. and Elliott, C.T., 2000. Biosensor assay of sulfadiazine and sulfamethazine residues in pork. Analyst 125, p. 1771-1774.

Blomgren, A., Berggren, C., Holmberg, A., Larsson, F., Sellergren, B. and Ensing, K., 2002. Extraction of clenbuterol from calf urine using a molecularly imprinted polymer followed by quantitation by high-performance liquid chromatography with UV detection. Journal of Chromatography A 975, p. 157-164.

Boison, J.O., 2004. New trends in analytical methodology - new developments in the validation of screening methods. In: Proceedings of EuroResidue V, edited by R. Stephany and A. Bergwerff, National Institute for Public Health and the Environment & Utrecht University, Utrecht, the Netherlands, p. 45-50.

Bovee, T.F., Helsdingen, R.J., Rietjens, I.M., Keijer, J. and Hoogenboom, R.L., 2004. Rapid yeast oestrogen bioassays stably expressing human oestrogen receptors alpha and beta, and green fluorescent protein: a comparison of different compounds with both receptor types. Journal of Steroid Biochemistry and Molecular Biology 91, p. 99-109.

Brambilla, G., Fiori, M., Pierdominici, E., Antonucci, G., Giorgi, P., Ramazza, V. and Zucchi, M., 1998. A possible correlation between the blood leukocyte formula and the use of glucocorticoids as growth promoters in beef cattle. Veterinary Research Communications 22, p. 457-465.

Brüggemann, O., Haupt. K., Ye, L., Yilmaz, E., and Mosbach, K., 2000. New configurations and applications of molecularly imprinted polymers. Journal of Chromatography A 889, p. 15-24.

Burgstaller, P., Jenne, A. and Blind, M., 2002. Aptamers and aptazymes: accelerating small molecule drug discovery. Current Opinion in Drug Discovery and Development 5, p. 690-700.

Cacciatore, G., Bergwerff, A.A. and Petz, M., 2004a. Development of screening assays for veterinary drug residues utilising surface plasmon resonance-based biosensors. In: Proceedings of EuroResidue V, edited by R. Stephany and A. Bergwerff, National Institute for Public Health and the Environment & Utrecht University, Utrecht, the Netherlands, p. 143-150.

Cacciatore, G., Petz, M., Rachid, S., H. R. and Bergwerff, A.A., 2004b. Development of an optical biosensor assay for detection of ß-lactam antibiotics in milk using the penicillin-binding protein 2x. Analytica Chimica Acta 520, p. 105-115.

Crooks, S.R., Baxter, G.A., O'Connor, M.C. and Elliot, C.T., 1998. Immunobiosensor—an alternative to enzyme immunoassay screening for residues of two sulfonamides in pigs. Analyst 123, p. 2755-2757.

Danyi, S., Degand, G., Granier, B., Elliott, C.T., Maghuin-Rogister, G. and Scippo, M.-L., 2004. Development of non-radioactive multi-analyte methods, based on the use of a recombinant human β_2-adrenergic receptor, for the detection of β-agonists and antagonists residues in food-producing animals. In: Proceedings of EuroResidue V, edited by R. Stephany and A. Bergwerff, National Institute for Public Health and the Environment & Utrecht University, Utrecht, the Netherlands, p. 159-164.

Dijkmans, R., Kormelink, D., Sultan, K., Jongerius, B., Blankvoort, B., Schilt, R., Witkamp, R., Haagsman, H. and Bergwerff, A., 2002. Protein fingerprints reporting abuse of veterinary drugs. Poster presentation at the 4th International Symposium on Hormone and Veterinary Drug Residue Analysis, Antwerp, Belgium.

Draisci, R., Marchiafava, C., Palleschi, L., Purificato, I. and Ferretti, E., 2000. Accelerated solvent extraction and liquid chromatography-tandem mass spectrometry quantitation of corticoid residues in bovine liver. In: Proceedings of EuroResidue IV, edited by L. Van Ginkel and A. Ruiter, National Institute for Public Health and the Environment, Laboratory for Residue Analysis, Bilthoven, the Netherlands, p. 385-391.

Elliott, C.T., Baxter, G.A., Hewitt, S.A., Arts, C.J., van Baak, M., Hellenas, K.E. and Johannson, A., 1998. Use of biosensors for rapid drug residue analysis without sample deconjugation or clean-up: a possible way forward. Analyst 123: 2469-2473.

Elliott, C.T., G.A. Baxter, S.R.H. Crooks and W.J. McCaughey, 1999. The development of a rapid immunobiosensor screening method for the detection of residues of sulphadiazine. Food and Agricultural Immunology 11, p. 19-27.

Gaudin, V. and Maris, P., 2001. Development of a biosensor-based immunoassay for screening of chloramphenicol residues in milk. Food and Agricultural Immunology 13, p. 77-86.

Godfrey, M.A.J., 1998. Immunoaffinity extraction in veterinary residue analysis - a regulatory viewpoint. Analyst 123, p. 2501-2506.

Groot, M., Schilt, R., Haasnoot, W. and Berende, P.L.M., 1996. Veal calves treated with a low dosage of prednisolone - histological and analytical data. In: Proceedings of EuroResidue III, edited by N. Haagsma and A. Ruiter, Utrecht University, the Netherlands, p. 440-445.

Groot, M.J. and Biolatti, B., 2004. Histopathological effects of boldenone in cattle. In: Proceedings of EuroResidue V, edited by R. Stephany and A. Bergwerff, National Institute for Public Health and the Environment & Utrecht University, Utrecht, the Netherlands, p. 512-515.

Gustavsson, E. and Sternesjo, A., 2004. Biosensor analysis of beta-lactams in milk: Comparison with microbiological, immunological, and receptor-based screening methods. Journal of AOAC INTERNATIONAL 87, p. 614-620.

Haasnoot, W., Streppel, L., Cazemier, G., Salden, M., Stouten, P., Essers, M. and vanWichen, P., 1996. Development of a tube enzyme immunoassay for 'on-site' screening of urine samples in the presence of beta-agonists. Analyst 121, p. 1111-1114.

Haasnoot, W., Kemmers-Voncken, A., Van Rhijn, H. and Schilt, R., 2000. Immunofiltration as alternative for immunoaffinity chromatography, In: In: Proceedings of EuroResidue IV, edited by L. Van Ginkel and A. Ruiter, National Institute for Public Health and the Environment, Bilthoven, the Netherlands, p. 496-500.

Haasnoot, W. and Schilt, R., 2000. Immunochemical and receptor technologies. In: Residue analysis in food - principles and applications, edited by M. O'Keeffe, Harwood Academic Publishers, Amsterdam, the Netherlands, p. 107-144.

Haasnoot, W. and Verheijen, R., 2001. A direct (non-competitive) immunoassay for gentamicin residues with an optical biosensor. Food and Agricultural Immunology 13, p. 131-134.

Haasnoot, W., Loomans, E.E.M.G., Cazemier, G., Dietrich, R., Verheijen, R., Bergwerff, A.A. and Stephany, R.W., 2002. Direct versus competitive biosensor immunoassays for the detection of (dihydro)streptomycin residues in milk. Food and Agricultural Immunology 14, p. 15-27.

Haginaka, J., 2004. Molecularly imprinted polymers for solid-phase extraction. Analytical and Bioanalytical Chemistry 379, p. 332-334.

Hagrén, V., Crooks, S.R., Elliott, C.T., Lövgren, T. and Tuomola, M.. 2004. An all-in-one dry chemistry immunoassay for the screening of coccidiostat nicarbazin in poultry eggs and liver. Journal of Agricultural and Food Chemistry 52, p. 2429-2433.

Haughey, S.A., Baxter, G.A., Elliott, C.T., Persson, B., Jonson, C. and Bjurling, P., 2001. Determination of clenbuterol residues in bovine urine by optical immunobiosensor assay. Journal of AOAC INTERNATIONAL 84, p. 1025-1030.

Hogendoom, E.A., Huls, R., Dijkman, E. and Hoogerbrugge, R., 2001. Microwave assisted solvent extraction and coupled-column reversed-phase liquid chromatography with UV detection use of an analytical restricted-access-medium column for the efficient multi-residue analysis of acidic pesticides in soils. Journal of Chromatography A 938, p. 23-33.

Hoogenboom, L.A., Hamers, A.R. and Bovee, T.F., 1999. Bioassays for the detection of growth-promoting agents, veterinary drugs and environmental contaminants in food. Analyst 124, p. 79-85.

Hoogenboom, L.A.P., de Haan, L., Hooijerink, D., Bor, G., Murk, A.J. and Brouwer, A., 2001. Estrogenic activity of estradiol and its metabolites in the ER-CALUX assay with human T47d breast cells. APMIS 109, p. 101-107.

Janosi, A., Maghuin-Rogister, G. and Degelaen, J.. 2000. A sensitive screening method for the multi-analyte determination of β-lactam residues in swine tissue. In: Proceedings of EuroResidue IV, edited by L. Van Ginkel and A. Ruiter, National Institute for Public Health and the Environment, Bilthoven, the Netherlands, p. 639-644.

Kim, Y.M., Oh, S.W., Jeong, S.Y., Pyo, D.J. and Choi, E.Y., 2003. Development of an ultrarapid one-step fluorescence immunochromatographic assay system for the quantification of microcystins. Environmental Science and Technology 37, p. 1899-1904.

Kootstra, P.R., Kuijpers, C.J.P.F., Wubs, K.L., Van Doorn, D., Stephany, R.W. and Sterk, S.S., 2004. The analysis of beta-agonists in bovine muscle using molecular imprinted polymers (MIPS) with ion trap LCMS detection. In: Proceedings of EuroResidue V, edited by R. Stephany and A. Bergwerff, National Institute for Public Health and the Environment & Utrecht University, Utrecht, the Netherlands, p. 624-631.

Korpimaki, T., Hagrén, V., Brockmann, E.C. and Tuomola, M., 2004. Generic lanthanide fluoroimmunoassay for the simultaneous screening of 18 sulfonamides using an engineered antibody. Analytical Chemistry 76, p. 3091-3098.

Korsrud, G.O., Papich, M.G., Fesser, A.C.E., Salisbury, C.D.C. and MacNeil, J.D., 1996. Laboratory testing of the Charm test II receptor assays and the Charm Farm test with tissues and fluids from hogs fed sulfamethazine, chlortetracycline and penicillin G. Journal of Food Protection 59, p. 161-166.

Korsrud, G.O., Boison, J.O., Nouws, J.F. and MacNeil, J.D., 1998. Bacterial inhibition tests used to screen for antimicrobial veterinary drug residues in slaughtered animals. Journal of AOAC INTERNATIONAL 81, p. 21-24.

Kroll, S., Usleber, E., Zaadhof, K.-J., Schneider, E. and Märtlbauer, E., 2000. Evaluation of commercial rapid tests for ß-lactam antibiotics in raw milk. In: Proceedings of EuroResidue IV, edited by L. Van Ginkel and A. Ruiter, National Institute for Public Health and the Environment, Bilthoven, the Netherlands, p. 693-697.

Legler, J., van den Brink, C.E., Brouwer, A., Murk, A.J., van der Saag, P.T., Vethaak, A.D. and van der Burg, B., 1999. Development of a stably transfected estrogens receptor-mediated luciferase reporter gene assay in the human T47d breast cancer cell line. Toxicological Sciences 48, p. 55-66.

Macrì, A., Abete, M.C., Agrimi, U., Angeletti, R., Biolatti, B., Brambilla, G., Calvarese, S., Citti, G., Corrao, A., Cozzani, R., Fedrizzi, G., Feliciangeli, P.G., Moca, S., Montana, M., Nachtmann, C., Oliviero, G., Palermo, D., Peruccio, C. and Testa, C., 1993. Strategy in the screening and confirmation of ß2 agonist drugs in animal productions: experience of veterinary public health labs network. In: Proceedings of EuroResidue III, edited by N. Haagsma and A. Ruiter, Utrecht University, Utrecht, the Netherlands, p. 474-478.

McConnel, R.I., Lamont, J.V., FitzGerald, S.P., Farry, L.T. and Mills, J.A., 2000. The development of biochip technology for the detection of drug residues. In: Proceedings of EuroResidue IV, edited by L. Van Ginkel and A. Ruiter, National Institute for Public Health and the Environment, Bilthoven, the Netherlands, p. 746-752.

Mellgren, C. and Sternesjo, A., 1998. Optical immunobiosensor assay for determining enrofloxacin and ciprofloxacin in bovine milk. Journal of AOAC INTERNATIONAL 81, p. 394-397.

Murk, A.J., Legler, J., Denison, M.S., Giesy, J.P., van de Guchte, C. and Brouwer, A., 1996. Chemical-activated luciferase gene expression (CALUX): a novel in vitro bioassay for Ah receptor active compounds in sediments and pore water. Fundamental and Applied Toxicology 33, p. 149-160.

Myllyniemi, A.L., Sipila, H., Nuotio, L., Niemi, A. and Honkanen-Buzalski, T., 2002. An indirect conductimetric screening method for the detection of antibiotic residues in bovine kidneys. Analyst 127, p. 1247-1251.

Naegeli, H., Briand-Schumacher, S. and Buterin, T., 2004. Reporter gene-based bioassays. In: Proceedings of EuroResidue V, edited by R. Stephany and A. Bergwerff, National Institute for Public Health and the Environment & Utrecht University, Utrecht, the Netherlands, p. 87-92.

Nielen, M.W., Elliott, C.T., Boyd, S.A., Courtheyn, D., Essers, M.L., Hooijerink, H.H., van Bennekom, E.O. and Fuchs, R.E., 2003. Identification of an unknown beta-agonist in feed by liquid chromatography/bioassay/quadrupole time-of-flight tandem mass spectrometry with accurate mass measurement. Rapid Communications in Mass Spectrometry 17, p. 1633-1641.

Noot, D., Thompson, T., Keenliside, J., Henderson, J. and Kendall, J., 2004. A multi-residue method for veterinary antimicrobials in pork kidney using LC-MS/MS. In: Proceedings of EuroResidue V, edited by R. Stephany and A. Bergwerff, National Institute for Public Health and the Environment & Utrecht University, Utrecht, the Netherlands, p. 706-711.

O'Keeffe, M., 1996. Strategies for the detection of veterinary drug residues. In: Residues of veterinary drugs and mycotoxins in animal products - new methods for risk assessment and quality control, G. Enne, H.A. Kuiper and A. Valentini, Wageningen Pers, Wageningen, the Netherlands, p. 31-40.

O'Keeffe, M., 2000. Introduction. In: Residue analysis in food - principles and applications, edited by M. O'Keeffe, Harwood Academic Publishers, Amsterdam, the Netherlands, p. 1-15.

O'Keeffe, M.J., Foran, S., Kane, M. and O'Keeffe, M., 2000. A rapid method for the on-site screening of pork samples for sulphamethazine. In: Proceedings of EuroResidue IV, edited by L. Van Ginkel and A. Ruiter, National Institute for Public Health and the Environment, Bilthoven, the Netherlands, p. 797-801.

Okerman, L., Croubels, S., Cherlet, M., De Wasch, K., De Backer, P. and Van Hoof, J., 2004. Evaluation and establishing the performance of different screening tests for tetracycline residues in animal tissues. Food Additives and Contaminants 21, p. 145-153.

Pallaroni, L. and von Holst, C., 2003. Determination of zearalenone from wheat and corn by pressurised liquid extraction and liquid chromatography-electrospray mass spectrometry. Journal of Chromatography A 993, p. 39-45.

Peippo, P., Hagrén, V., Lovgren, T. and Tuomola, M., 2004. Rapid time-resolved fluoroimmunoassay for the screening of narasin and salinomycin residues in poultry and eggs. Journal of Agricultural and Food Chemistry 52, p. 1824-1828.

Schumacher, S.B., Van den Hauwe, O.,. Van Peteghem, C.H and Naegeli, H., 2003. Development of a dual luciferase reporter screening assay for the detection of synthetic glucocorticoids in animal tissues. Analyst 128, p. 1406-1412.

Shelver, W.L. and Smith, D.J., 2003. Determination of ractopamine in cattle and sheep urine samples using an optical biosensor analysis: comparative study with HPLC and ELISA. Journal of Agricultural and Food Chemistry 51, p. 3715-3721.

Situ, C., Crooks, S.R.H., Baxter, A.G., Ferguson, J. and Elliott, C.T., 2002. On-line detection of sulfamethazine and sulfadiazine in porcine bile using a multi-channel high-throughput SPR biosensor. Analytica Chimica Acta 473, p. 143-149.

Stead, S., Sharman, M., Tarbin, J.A., Gibson, E., Richmond, S., Stark, J. and Geijp, E., 2004a. Meeting maximum residue limits: an improved screening technique for the rapid detection of antimicrobial residues in animal food products. Food Additives and Contaminants 21, p. 216-221.

Stead, S., Tarbin, J.A., Richmond, S., Stark, J., Geijp, E. and Sharman, M., 2004b. Premi®Test coupled to scanner technology - a new approach for detection of antimicrobials in food. In: Proceedings of EuroResidue V, edited by R. Stephany and A. Bergwerff, National Institute for Public Health and the Environment & Utrecht University, Utrecht, the Netherlands, p. 219-224.

Stolker, A.A.M. 2000. Current trends and developments in sample preparation. In: Proceedings of EuroResidue IV, edited by L. Van Ginkel, and A. Ruiter, National Institute for Public Health and the Environment, Bilthoven, the Netherlands, p. 148-158.

Verheijen, R., Stouten, P., Cazemier, G. and Haasnoot, W., 1998. Development of a one step strip test for the detection of sulfadimidine residues. Analyst 123, p. 2437-2441.

Verheijen, R. and Haasnoot, W., 2000. Development of one step strip tests for residue analysis. In: Proceedings of EuroResidue IV, edited by L. Van Ginkel and A. Ruiter, National Institute for Public Health and the Environment, Bilthoven, the Netherlands, p. 1103-1107.

Verheijen, R., Osswald, I.K., Dietrich, R. and Haasnoot, W., 2000. Development of a one step strip test for the detection of (dihydro)streptomycin residues in raw milk. Food and Agricultural Immunology 12, p. 31-40.

Watanabe, H., Satake, A., Kido, Y. and Tsuji, A., 2001. Monoclonal-based enzyme-linked immunosorbent assay and immunochromatographic rapid assay for salinomycin. Analytica Chimica Acta 437, p. 31-38.

Watanabe, H., Satake, A., Kido, Y. and Tsuji, A., 2002. Monoclonal-based enzyme-linked immunosorbent assay and immunochromatographic assay for enrofloxacin in biological matrices. Analyst 127, p. 98-103.

Weijman, J., Ramaekers, F.C.S., Elsinghorst, T.A.M., Van Wichen, P.J. and Zwart, P., 1990. Immunohistochemistry of cytokeratins in prostate glandular epithelium as a sensitive method to detect (illegal) administration of oestrogens in cattle. In: Proceedings of Euroresidue I, edited by N. Haagsma, A. Ruiter and P.B. Czedik-Eysenberg, Utrecht University, Utrecht, the Netherlands, p. 196-201.

Zacharewski, T., 1997. In vitro bioassays for assessing estrogenic substances. Environmental Science and Technology 31, p. 613-623.

Zoontjes, P.W., Stolker, A.A.M. and Van Ginkel, L.A., 2000. The use of automated SFE in routine analysis of steroids in animal tissues. In: Proceedings of EuroResidue IV, edited by L. Van Ginkel and A. Ruiter, National Institute for Public Health and the Environment, Bilthoven, the Netherlands, p. 1152-1157.

Rapid immunoassays for the detection of pesticides

M.F. Wilson and R.J. Fussell
Central Science Laboratory, Sand Hutton, York, YO41 1LZ, United Kingdom

Abstract

Contaminants and natural toxic substances can enter the food chain at many points in the production process from the field, through storage and to the processing or manufacture site (Scanlan, 2004). Testing raw materials and finished products is a key part of assuring consumer safety. However, the need for comprehensive testing regimes needs to be balanced with the costs of the testing process such fresh, wholesome foods are available to consumers at a cost that they can afford. Rapid testing methods based on sensors and often real-time offering monitoring of product quality, can provide a solution.

Keywords: Rapid immunoassays; pesticides; lateral flow device; generic antibody-based assay

1. Introduction

The use of immunoassays as the basis for rapid testing for pesticide residues in foods offers the laboratory potential benefits in terms of speed, portability and low cost analyses. Set against this is the relatively long lead-in time for method development in comparison with adaptable chromatographic or other instrumental techniques and the prerequisite for a known target, or targets, for a method to be well defined. Thus, it is not generally possible for a laboratory to develop a novel immunoassay-based test immediately in response to, for example, a specific food scare. But if tests - or at least the appropriate antibodies and a suitable platform - are available then validation can be comparable with instrumental techniques and their application can generate large amounts of data as a screening tool and as an invaluable contribution to any risk assessment.

Lower cost is often cited as a major factor in favour of immunoassays. Again, a strict comparison would be not between the immunoassay test, often costing very little on a per test basis, and a chromatography based test costing between €100 - €500 depending on the matrix, levels of determination and turnaround time but between the costs of individual data obtained from each method. For example, an immunoassay for up to five organophosphorus pesticides residues may cost €10 per test. A general purpose, multi-residue GC-MS screening method may cost €200 but can detect at least 100 pesticides. The unit cost per data point is €2 in both cases and MS data is quantitative.

Time taken for the test can vary from a few minutes, typically 30 minutes, for an immunoassay test to many hours for a conventional chromatographic analysis. However, development in sample preparation techniques and rapid MS/MS techniques can reduce these times considerably, also to a matter of minutes.

Immunoassay methodologies have the great advantages of portability and a reduced requirement for very costly equipment and the expertise needed to use it. This makes these methods highly suitable for field studies and pre-screening routines where it is necessary to test large numbers of samples for a limited range of analytes. However, even in this context, the rate of positives expected and the need for quantification/confirmation must be taken into account. For example, if a screening test for the fungicide imazalil was used on oranges it might show that 89% of samples were positive (Annual report of the WPPR, 1999). Should these data need confirmation by mass spectrometry, and official surveillance data generally do, the tests would have been best carried out by MS initially since most samples would have to be re-tested in any event. Conversely, if the same orange samples were investigated using a screening test for organophosphate insecticides where the number of positives is between 11% (chlorpyriphos) and about 1.5% (diazinon) (Annual report of the WPPR, 1999), then the low positive rate would make the use of an immunoassay based pre-screen a cost-effective option.

Overall, the advantages of immunoassay technology can be summarised as the possibilities (Shan *et al.*, 2002) for:
- low detection limits;
- high analyte selectivity and targeting;
- high throughput of samples and cost effectiveness;
- reduced sample preparation and handling allowing possible use in the field.

and taking into account the limitations:
- matrix interferences;
- cross reactivity to structural analogues and other chemicals;
- current unsuitability for multi-analyte methods;
- long method development lead in time necessitating a known 'problem' and a (relatively) large number of samples.

It is not within the scope of a short presentation to describe the details of immunoassay method development. Key stages (Shan *et al.*, 2002) are:
- design and synthesis of the haptan;
- conjugation of the haptan to antigenic macromolecular carriers;
- immunisation of a host animal and subsequent generation of antibodies;
- characterisation of antibodies;
- assay optimisation;
- assay validation.

Immunoassays fall into two types, direct and indirect competitive formats. The means by which these are applied can also vary between an immobilised antigen ELISA format, and immobilised antibody ELISA or a 'sandwich' immunoassay. Such methods have been successfully applied to the determination over 50 pesticides including selective, targeted methods for widely used (Pesticide Usage Statistics for Great Britain, available from the Central Science Laboratory, UK) herbicides such as isoproturon and the triazines (e.g. atrazine), fungicides such as imazalil and procymidone and insecticides. In the case of the latter much effort has been put into generic immunoassays for groups of insecticides such as the organophosphates and the synthetic pyrethroids.

2. Case study 1: lateral flow device (LFD) for tebuconazole

The lateral flow device or LFD is a convenient format for portable, field testing kit which does not require a great deal of scientific skill or equipment to use. They have found extensive use in medical diagnostics and in the on-site detection of agricultural plant pathogens (Danks and Barker, 2000). Where the need for the measurement of a pesticide is (a) predictable, that is, there is a long-term requirement for the measurement of a specific molecule, and (b) such measurements are best done *in situ* and in large numbers, test kits such as those based on LFDs, come into their own.

For example, the fungicide tebuconazole has been extensively used on cereals (Figure 1). As with any pesticide usage, the ability to measure residues during or immediately following use is important in providing farmers with information on the levels of residues in the crop. These can be used as part of strategies aimed at minimising pesticide use by linking observed pest control to real residue levels. An immunoassay based method for the determination of tebuconazole was developed (Danks *et al.*, 2001). The assay characteristics are shown in Figures 2, 3 and 4. Both the standard curve and a comparison with a gas chromatography-mass spectrometry (GC-MS) demonstrate that the assay performs well. As with many immunoassay methods, results obtained are generally slightly higher than those obtained with comparable chromatographic techniques, presumably due to cross-reactivity of the immunoassay with other chemicals present in the samples.

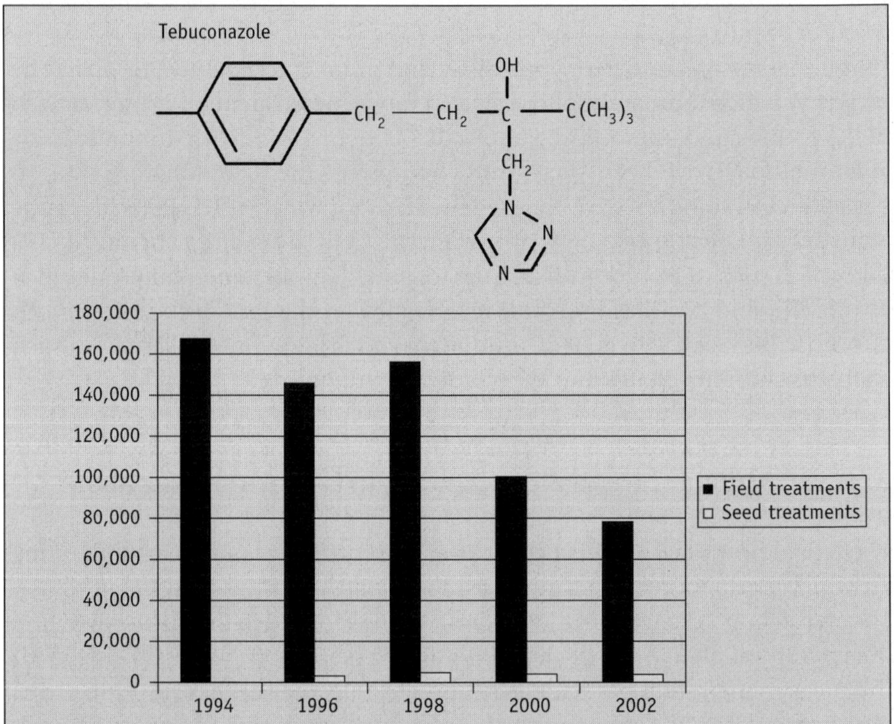

Figure 1. Tebuconazole - breakdown of estimated annual usage since 1990 for all cereal crops in Great Britain (kg ai applied).

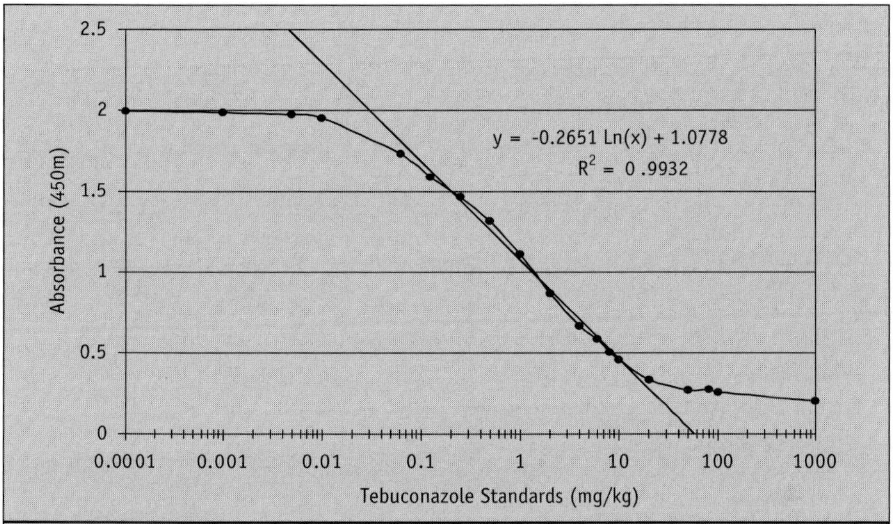

Figure 2. Standard curve for immunoassay response with tebuconazole standards, linear over the range 0.1-10 mg/kg.

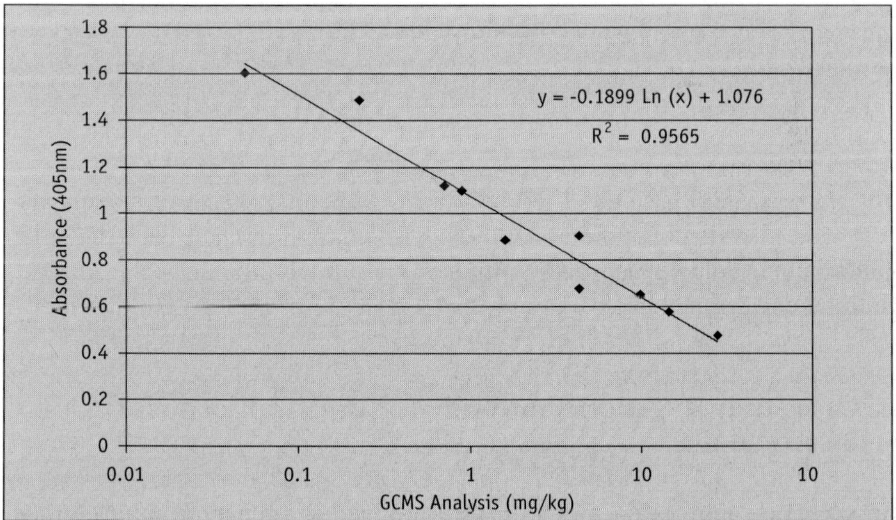

Figure 3. Comparative response for immunoassay for tebuconazole and GC-MS analysis.

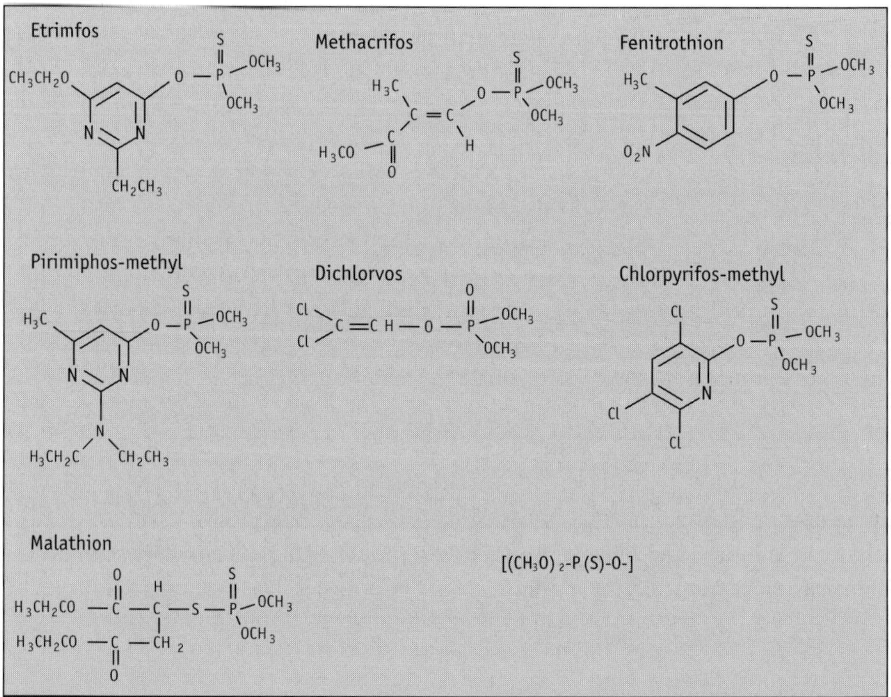

Figure 4. Common organophosphorus insecticides determined by a generic antibody.

The immunoassay gave good reactivity with tebuconazole, another conazole fungicide, triadimenol but importantly showed no cross-reactivity with the other conazoles including penconazole, propiconazole, myclobutanil, epoxiconazole, flutiafol and hexaconazole.

3. Case study 2: generic antibody for organophosphate insecticides

An alternative approach would be the development of a generic or multi-residue antibody. However, this has only been achieved for a limited range of pesticides, for example for a sub-set of organophosphorus insecticides (Banks *et al.*, 1998) or more recently with selected type I pyrethroids (Watanabe *et al.*, 2001). Such methods are based on the careful selection of the hapten. That is,

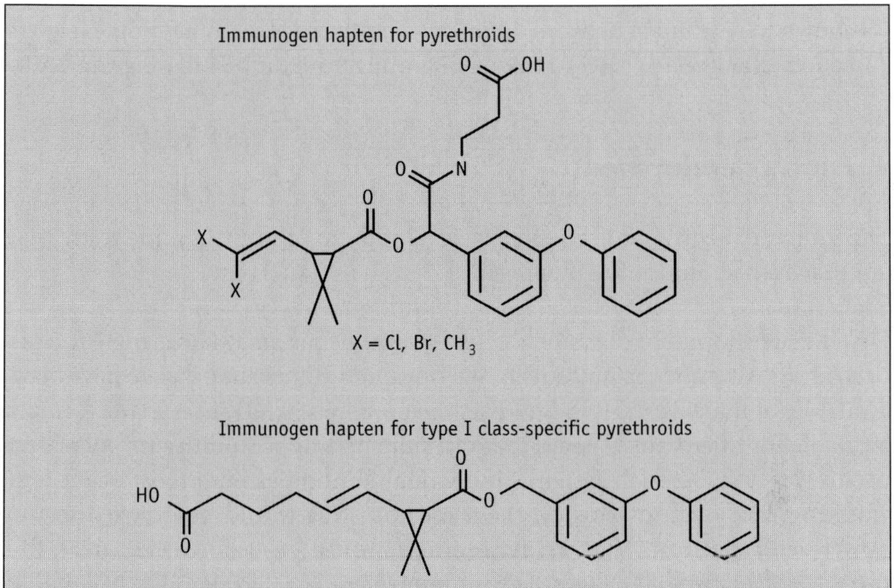

Figure 5. Hapten design for pyrethroid insecticides.

the creation of a molecule which carefully retains those common part of the chemical series which characterise it interactions.

However, generic methods have found only limited application in the determination of pesticide residues in real samples. This is possibly because of a combination of several factors Wilson *et al.*, 2002):

- Pesticides are a very heterogeneous group of molecules with a range of chemical structures and properties and, likewise, a considerable range of biological effects. There is no common measure of "pesticidal" activity.
- By its very nature, a generic antibody-based assay is more prone to interactions with both other pesticides and, more importantly, with other components found in complex biological matrices. This also applies to other technologies including amperometric and potentiometric sensors using ranges of enzymes as the basis for sensors. Many successful sensors, which have worked well on standards, fail when exposed to real samples.
- Differential reactivity or the difficulty in obtaining a semi-quantitative result due to the different responses of a system to the same concentration of a range of analytes. For example, organophosphorus insecticides have been

shown to respond at between 20%-90% of the response of the simple hapten spacer (Banks *et al.* 1998), making quantification for the full range difficult.

4. Future development

The future for sensors must lie in the development of multi-residue tests. Two approaches will undoubtedly prevail.

First, the development of systems capable of multiple assays in a single format or in arrays. Commercial test-kits based on a 96-well microtitre plate have been available for many years. It is, however, the more recent advances in the formats available for micro-array assays that offer the best opportunities for increased use in this area. In such systems, individual antibodies with fluorescent tags would replace nucleic acids on the array, the array would be exposed to the sample and electronic pattern recognition would be used to recognise the presence of a range of analytes. The limiting factor is the availability of suitable antibodies.

Surface plasma resonance (SPR) based systems such as the Biacore® instrument, employ a method of monitoring the progress of bio-molecular interactions in real time. The development of multi-channel SPR instruments offers some possibility for multi-residue analysis but still some way off from the 200-300 residues capabilities of GC-MS.

Gyrolab® is a system based on the CD-ROM format. Within a "Gyrolab micro-laboratory" (see http://www.gyros.com/technology/tech_natural_platform_3.html), applications such as immunoassays and derivatisations may in the future be miniaturised and integrated into single, streamlined procedures. This, seemingly elegant technique offers the capability of performing a combination of chemical and biological analyses on a single disc in a hand-held system.

Second, the future of sensors may not lie exclusively with the direct measurement of chemicals but with the measurement of effect. The inhibition of acetylcholine esterase has long been used to determine the presence of organophosphate residues. The ability to test a sample for, for example, estrogenic, anti-androgenic or some other relatively broad, biological effect and then to correlate this to a general but unequivocal chemical characterisation such as mass spectrometry must represent a step-change in food and environmental testing.

References

Annual report of the Working Party on Pesticide Residues (WPPR) 1999. Supplement to the Pesticides Monitor 2000, p.130-135.

Banks, J.N., Chaudhry, M.Q., Matthews, W.A., Haverly, M., Watkins, T. and Northway, B.J., 1998. Production and characterisation of some polyclonal antibodies to the common moiety of some organophosphorus pesticides and development of a generic type ELISA. Food and Agric. Immun. 10, p. 349-361.

Danks, C. and Barker, I., 2000. On-site detection of plant pathogens using lateral-flow devices. Bulletin OEPP/EPPO Bulletin 30, p. 421-426.

Danks, C., Chaudhry, M. Q., Parker, L. and Banks, J. N., 2001. Development and validation of an immunoassay for the determination tebuconazole residues in cereal crops. Food and Agricultural Immunology 43, p. 151-159.

Scanlan, F.P., 2004. Why rapid testing? In: A. van Amerongen, D. Barug and M. Lauwaars, eds., Rapid Methods for Food and Feed, p. ...-... Wageningen Academic Publishers, Wageningen, the Netherlands.

Shan, G., Lipton, C., Gee, S.J. and Hammock, B.D., 2002. In: Handbook of residue analytical methods for agrochemicals, P.W. Lee, ed. Immunoassay, biosensors and other nonchromatographaic methods, p. 623-679. John Wiley & Sons, Chichester.

Watanabe, T., Shan, G., Stoutamire, D.W., Gee, S.J. and Hammock, B.D., 2001. Development of a class-specific immunoassay for the type I pyrethroid insecticides. Anal. Chim. Acta 444, p. 119-129.

Wilson, M.F., Reynolds, S.L. and Fussell, R.J, 2002. Advances in methods for pesticide residue in food In: Handbook of residue analytical methods for agrochemicals, P.W. Lee (ed.), p. 727-752. John Wiley & Sons, Chichester.

Rapid methods for dioxins in food and feed

P.A. Behnisch
Eurofins GfA ,Otto-Hahn Str. 22, 48161 Münster-Roxel, Germany

Abstract

In the past few years several dioxins and PCBs crises in food and feed have occurred and brought about a sudden drop in consumer confidence in the food/feed safety. Now it's the time to establish and evaluate faster and cheaper methods for analysing dioxins and PCBs. Therefore currently several bio-analytical and chemical analytical methods are available on the market as pre-screening method for green/red light decisions and to reduce the numbers of the high costly confirmatory method by chemical analysis.

Keywords: dioxins, feed, food, bioassays, rapid methods

1. Introduction

At the end of the year the European Union will expand the definition of dioxins to include 29 compounds that produce similar biochemical effects in test animals, 7 polychlorinated dibenzodioxins (dioxins, or PCDDs), 10 polychlorinated dibenzofurans (furans, or PCDFs), and 12 polychlorinated biphenyls (PCBs). These compounds are collectively referred to as dioxin-like compounds. At the same time most of European countries agreed to accept the Stockholm Convention of Persistent Organic Pollutants (POPs), therefore also reducing dioxins, PCBs and other POPs for both consumer and environment protection. This will be one of the major goals of this international agreement in the next few years.

Including the fact that several new member states of the European Union have so far only few experiences with dioxin-like compounds it seems to be necessary to establish cheaper, easier and faster methods for these compound classes to ensure the safety of food and feed for human health.

1.1 Basic facts about dioxin-like compounds

Dioxin contamination occurrences are not always clear. However, there are some basic facts about dioxins that should be understood to examine the ways to protect consumer products and consumers from incidents by feed/food (see e.g., van den Berg *et al.*, 1998). Dioxin is the generic name for a group of 210 organic compounds that have two linked benzene rings. These two benzene rings are linked with either two oxygen atoms (dibenzodioxins), or one oxygen atom (dibenzofurans). Although there are 210 individual chlorinated dioxin and furan molecules, only 17 compounds in this class - 7 dioxins and 10 furans - are considered toxic.

Dioxin risk assessment involves complex classes of dioxin-like compounds: dioxin, furan, and some polychlorinated biphenyls. The most toxic individual compounds are 2,3,7,8-tetrachlorodibenzodioxin and 1,2,3,7,8-pentachlorodibenzodioxin. To estimate the total toxic activity of dioxins, furans, and PCBs, especially when found in mixtures, a weighted factor called the toxic equivalence quotient (TEQ) is used. The toxicities of the 17 most dangerous compounds have been estimated. From these estimates, weighting factors - called toxic equivalence factors (TEFs) - have been calculated. To calculate the TEQ of a sample by chemical analysis, concentrations of each compound must be measured and multiplied by each TEF, while in case of bioassays the calculation is usually done by comparing the results to 2,3,7,8-TCDD and extrapolating the results to this marker congener.

1.2. Biological mechanism of dioxin toxicity

The current model of the mechanism of dioxin toxicity is based on the binding to the Ah receptor, which exists as a multi-protein complex.

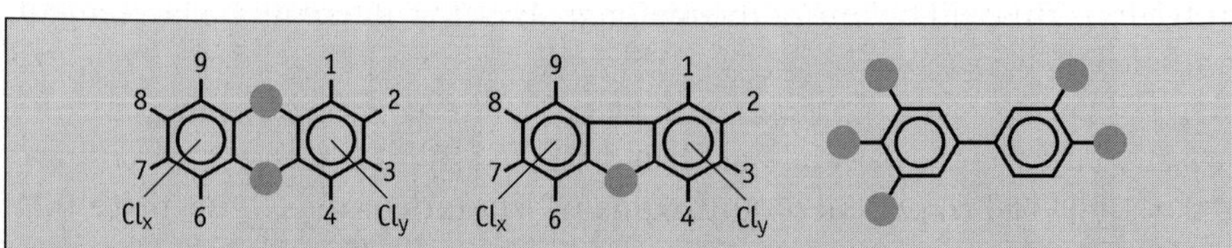

Figure 1. Molecular structures of polychlorinated dibenzodioxins, polychlorinated dibenzofurans and polychlorinated biphenyls (e.g. dioxin-like PCB No. 126).

Release of the liganded AhR (aryl-hydrocarbon receptor) from this protein complex and its subsequent dimerisation with a related nuclear protein called Arnt (Ah receptor nuclear translocator) converts the AhR into is high affinity DNA binding form. Binding of the heteromeric ligand:AhR:Arnt complex to its specific DNA recognition site, the dioxin responsive element (DRE), upstream of a responsive gene (such as cytochrome P4501A1 (CYP1A1)) stimulates its transcription and results in an increase in accumulation of its gene products (i.e. cytochrome P4501A1) that can be readily detected in cells (see Figure 2; for a review see e.g., Behnisch *et al.*, 2001a-c; Hoogenboom *et al.*, 1999; Hoogenboom, 2002; Whyte, 2004).

1.3. Sources and pathways of dioxin-like compounds

In the past, uncontrolled incinerators, open fires, wood pulp bleaching operations, and some chemical syntheses were the major sources of dioxin. Most of human exposures come now from legacy or reservoir sources such as sediments and contaminated soil. With dioxin disappearing, PCB, a liquid once used as the dielectric cooling fluid in transformers, will become a new pollution source to study and regulate more strictly. Food is the primary route of human exposure, so trace analysis of complex food samples will be needed to measure the extremely low levels of dioxin and PCB found in some foods.

1.4. Dioxin crisis situations

In the last few years the number of dioxin crisis situations in feed/food has steadily increased. The reasons for dioxins crisis situations has shown a wide variety from sources in different countries, such as: ball clay in USA/1996; illegal disposal of capacitor fluid in Belgium/1999; improper drying of feed leads to contamination via dioxin-containing fumes, e.g., citrus pellets from Brazil/1999, green feed from two feed producers in Germany/1999 or eggs and rabbits meat from Luxemburg/2001; large fire accidents involving higher amounts of chlorine-containing materials (e.g., PVC) caused local dioxin contaminations; sewage sludge in feed in France/1999, zinc oxide, copper oxide in feed premixes (several countries 2000-2002), PCP contaminated saw dust/choline chloride case in Spain/2000, Carbosan in France and USA 2002, farm-raised fishes/fish feed in several countries (2003/4) or eggs from free-ranging chicken in the Netherlands and France (2003/4).

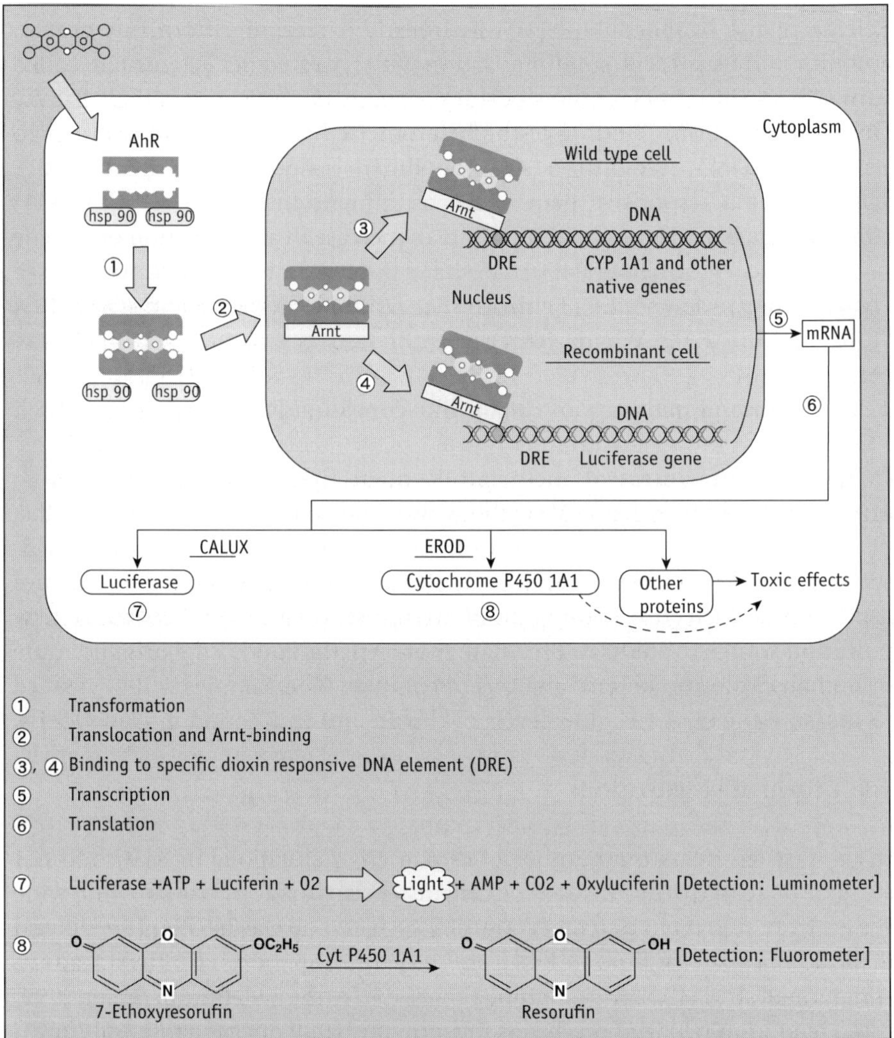

Figure 2. Biological mechanism of the dioxin toxicology and description of different bioassays such as CALUX and EROD.

1.5. European Guidelines for dioxins in feed and food

Therefore the European Union developed a set of maximum allowed limits for feed, food and ingredients and made it effective July 1 2002. At the moment only Korea also adopted maximum levels for dioxins in food/feed. It is planned to revise these levels in 2004 to integrate some dioxin-like PCBs. Furthermore, it is foreseen that by the end of 2006 the maximum levels will be revised, aiming for a significant reduction. It is planned to reduce the human exposure to dioxins by at least 25 % until end of year 2006.

1.6. Confirmatory method for dioxin analysis

In most countries the dioxin levels have decreased by 80% since they were first monitored in the 1980s. Today, the levels people are concerned with are lower, parts-per-trillion (picograms per gram) in most samples. Therefore a high sensitivity, low limits of detection, high selectivity, high specificity and high accuracy/precision are needed. The determination should provide a valid estimate of the true concentration in a sample. Furthermore, an acceptable degree of agreement between measurements must be achieved.

The classic way to measure dioxin at these low levels is by using selective extraction and then capillary gas chromatography with high-resolution mass spectrometry (HRGC/HRMS). This is a time-consuming procedure that requires several days and can cost 500 € or more per test. There is clearly the need for faster, simpler methods to measure dioxin-like species. For the analysis of dioxins in food and feed ultra trace concentrations in the sub pg/g range have to be safely quantified (pg = 10^{-12} g). Separate laboratories are needed for sample matrices with different dioxin or PCB levels to avoid cross contamination (a range of fg/m^3 in outdoor air up to µg/g in chemical waste, e.g. separation of ultra trace, trace, medium/high laboratories or laboratory parts; fg = 10^{-15} g). The traditional method is very sensitive using expensive equipment and instruments (e.g. high resolution gas chromatography/mass spectrometry, HRGC/HRMS, sensitivity below 100 fg TCDD absolute). With this technology high maintenance costs are connected with the HRMS instruments and $^{13}C_{12}$-labelled internal standards are used. High efforts in the clean-up of the sample extracts to separate the matrix and interfering compounds are necessary (3-4 steps). Therefore highly experienced and qualified personnel is needed and high demand for internal and external quality assurance.

1.7 Modern extraction and clean-up procedures

In the last few years faster extraction and clean-up procedures have been established and evaluated to speed up the sample pre-treatment. In chemical analysis the extraction and clean-up needs about 50-75% of the time of the complete analysis procedure. Sample extraction times can be reduced from about 12-24 hours by soxhlet to about 20-30 min by using Accelerated Solvent Extraction (ASE) and the traditional sample clean-up can be speed up from about 6 hours to about 1.5 hours by using an automated clean-up system called Power Prep.

2. Strong needs for rapid methods for dioxin analysis

Monitoring activities in the industry and of governmental bodies concerning the levels of dioxins and PCBs in food/feed started now and revealed a strong need to establish also rapid detection methods for screening purpose. There is a need for such technologies for safer decision-making with regard to complex food and feed crisis situations. Considering the relatively high price and long delivery time of the results of traditional dioxin analysis as well as the low laboratory capacities in many of the countries in the European Union, the development and validation of rapid screening methods for dioxins in food and feed was given a high priority by the EU and WHO commissions.

Table 1. Rough estimate of possible fastest turnaround time (TAT). Assumption: only machine steps and cell culture. (* runtime doubled when including dl-PCBs).

Sample set	ASE	Evaporation	PowerPrep	Turbovap	Measurement [hrs]	TAT [hours h, days d]
1	20 min	30 min	90 min	15 min	HRMS: 1	HRMS: 4 h
					CALUX: 24	CALUX: 30 h
					ELISA: 4-5	ELISA: 7-8 h
6	120 min	60 min	90 min	45 min	HRMS: 6*	HRMS: 11-12 h
					CALUX: 24	CALUX: 30 h
					ELISA: 4-5	ELISA: 7-8 h
100	4 days	2,5 days	6 days	1.5 days	HRMS: 144*	HRMS: 7-14 d
					CALUX: 24	CALUX: 3-6 d
					ELISA: 8-10	ELISA: 4-6 d

2.1. Quality and performance criteria from the EU Commission Directives 2002/69 and 70/EC from 26 July 2002 for testing of foodstuffs and feed

Aim is to perform a strategy involving a screening method in order to select those samples with levels of dioxins and dioxin-like PCBs that are less than 30-40% below or exceed the level of interest. Strict requirements for the confirmatory method have been described in detail and only minimum requirements for the screening method. False positive results should be low enough to make use of a screening tool advantageously. Positive results always have to be confirmed by HRGC/HRMS. In addition samples from a wide TEQ-range should be confirmed by HRGC/HRMS (2-10%). These methods are specially designed to avoid false negatives (below 1%). The coefficient of variation of the method should be below 30%. In case of the confirmation method it is required that the control laboratory shall analyse the sample in duplicate analysis in case the first analysis is less than 20% below or above the maximum level, and calculate the mean of the results. Continuous participation in interlaboratory studies for dioxins and dioxin-like PCBs in the relevant feed/food matrices is mandatory (Matrix dependant once/year) (see also Malisch *et al.*, 2001; Behnisch *et al.*, 2001c).

2.2 Rapid methods for dioxin-like compounds based on molecular biology

Many of these new technologies are based on well-known mechanisms of action and utilise well-defined end points (see Table 2). Several of these methods are used since many years for the evaluation of toxicological effects of dioxin-like compounds (Table 2).

These new methods are based on several approaches such as: a) bioassays which measure a cellular response produced by protein or enzyme activation; b) the binding of specific antibodies followed by indirect measurement of bound material; c) DNA binding measured by PCR technology; or d) other proteomic endpoints.

a. Bioassays

At the moment the only method which demonstrated already that it fulfils the quality criteria set by the EU for bioanalytical methods measuring dioxins in food/feed are reporter gene assays such as the so-called reporter gene systems, e.g. CALUX or P450. CALUX stands for Chemically Activated Luciferase

Table 2. Milestones of the development of rapid methods for dioxin analysis.	
1968:	AHH-Bioassay
1975:	EROD-Bioassay
1979:	US-FDA uses EROD screening
1987:	Dioxin-Antibodies
1993:	CALUX-Bioassay
1995:	Accelerated solvent extraction (ASE)
1996:	Ah-Immunoassay
2002:	AhRC-PCR; PowerPrep with SFE/C18
Future:	Proteomics testing

Expression bioassay (see e.g., Aarts *et al.*, 1993; Archer *et al.*, 2003; Denison *et al.*, 1993; Aarts *et al.*, 1996; Hoogenboom *et al.*, 2001; Cooke *et al.*, 2000). It uses liver cells from mouse/rat as a sensor, a firefly gene as the reporter and the bioassay is based on the mechanism of action at the mammalian cellular level. The dioxin-like compounds bind on the Ah receptor and the complex binds on the firefly gene added in the dioxin responsible elements on the DNA level in the nucleus of the cells. These leads to luciferase activity and after adding luciferin to a light emission which is measurable with a luminometer on a 96-well plate.

The second commercial system is P450 RGS and is based on a human cell line (Anderson *et al.*, 2000). The sensitivity of the CALUX system is higher than the P450 system, because in general human cell lines are less sensitive than mammalian cell lines such as rat or mouse used by the commercial available CALUX systems.

The advantage for using a reporter gene assay for dioxin analysis is the possibility for using it for high-through-putout screening as red-green light decision. Cheaper and faster yes/no decisions are possible. It covers all possible dioxin-like compounds (also brominated dioxins). Lower equipment costs and less training is necessary. A reported disadvantage for using a reporter gene assay for dioxin analysis is that several non-dioxin-like effects (some natural compounds such as spices/vegetables, see e.g., Denison *et al.*, 1998, 1999) also lead to false positive results or usual higher values than the confirmation method. Several studies reported already antagonistic effects of several WHO-PCB congeners (e.g. PCB-118). Additional recovery correction is required (Brown *et*

Table 3. EU Guidelines for specific matrices, typical LOQ for this matrix, comparison of WHO-TEQ and CALUX-TEQ (literature review).

Matrix	EU-Limit	CALUX LOQ	CALUX-TEQ/ WHO-TEQ	Correlation CALUX to HRMS	Reference
Fish	4.0	0.1	1.7	0.89	Yabushita *et al.*, 2002
Pork	3.0	0.7	1.5	0.85	Jeong *et al.*, 2001
Food	0.75-4.0	0.1-0.7	max. 3-fold	0.92	Cederberg *et al.*, 2002
Feed	0.75	0.1	2	0.75	Jeong *et al.*, 2001
Cow milk	3.0	1.1	1.6		Overmeire *et al.*, 2003
Mother milk			max. 2 fold		Laier *et al.*, 2001

al., 2002). So far, there is no successful international round robin study confirming interlaboratory results for feed/food.

In the past few decades the induction of 7-ethoxyresorufin-o-deethylase (EROD) activity in rat hepatoma cells H4IIE has been by far the most extensively used cell bioassay system for the detection of TCDD and other dioxin-like compounds in unknown sample extracts (reviewed in e.g., Behnisch *et al.*, 2002; Whyte *et al.*, 2004). In this bioassay, H4IIE cells are grown in 96-well microplates and exposed to chemicals or chemical extracts and incubated for 24-72 h after which cells are washed and EROD activity determined in liver or lysed cells. Since CYP1A1-dependent EROD activity results in the deethylation of its substrate 7-ethoxyresorufin to a single easily measured and sensitive fluorescent product, resorufin, the level of EROD activity in cells following exposure to a chemical extract is proportional to the level of induction of CYP1A1. This is proportional to the concentration of TCDD or related HAH in the original sample extract. The assay is rapid, quantitative, reproducible, accurate, precise and sensitive as the reporter gene assays, the cells and reagents are readily available and no known legal restrictions or limitations to free use exist for these cells and the bioassay system. Limitations of the H4IIE cell bioassay are that it requires trained individuals to perform the bioassay, long chemical exposure time, cell culture equipment, chemical and sterile hoods, and a fluorometric plate reader, all of which can be costly. It has been observed that many of the EROD inducing chemicals are substrates for

CYP4501A1 and can inhibit EROD activity, which could result in underestimation of EROD activity and ultimately chemical concentrations. In addition to H4IIE cells, induction of EROD activity has also been reported in a variety of other cell lines and many of these cells have also been used for dioxin bioassay analysis.

b. ELISA technology

Kit-based bioassays commonly used in the feed/food testing are currently established and tested in several laboratories for dioxin measurement in feed/food. They offer the advantages of a cell-free system, which reflects TEQ-values; it is an ELISA-type kit which is simple to use, the effects are measured by a 405 nm plate reader in around 5 hours as assay time. At the moment the main problems to be solved seem to be the extraction, clean-up and finally sensitivity of the antibody for dioxin-like compounds. Kit-based bioassays available are basically antibody-based or using Ah receptor parts for their detection of dioxin-like compounds.

In case of the immunoassays the sample has to be extracted, the solvent exchanged and the extract is allowed to bind with an immobilised dioxin antibody. After an incubation period, excess sample extract is washed away. A dioxin-enzyme conjugate is added to bind to antibody sites not occupied by dioxin or furan. After a second rinse to remove remaining competitor conjugate, a colour-forming enzyme substrate is added that generates a colour proportional to the amount of bound enzyme. The colour formed is inversely related to the original sample concentration.

The advantages of these kit-based bioassays are the possibility of a cheaper and faster method than CALUX, because of shorter incubation time and no time consuming cell culture. ELISA technologies are widespread and standardised. So far, they are only successfully applied on highly contaminated matrices with sufficient results (false negatives for soils = 1%; false positives = 3%). First results of replicate analyses for soils, fly ashes or sediments are below 15%. Standard curves between 1-64 pg TCDD-TEQ/well ($R^2 = 0.997$) have been reported. The drawbacks of these methods are the lack of sensitivity for feed/food and so far not enough data are available for the different feed and food matrices. Extraction/clean-up methods for all kinds of feed/food are currently under investigation. In most studies a consistent overestimation (~10-fold) depending on sample type is reported.

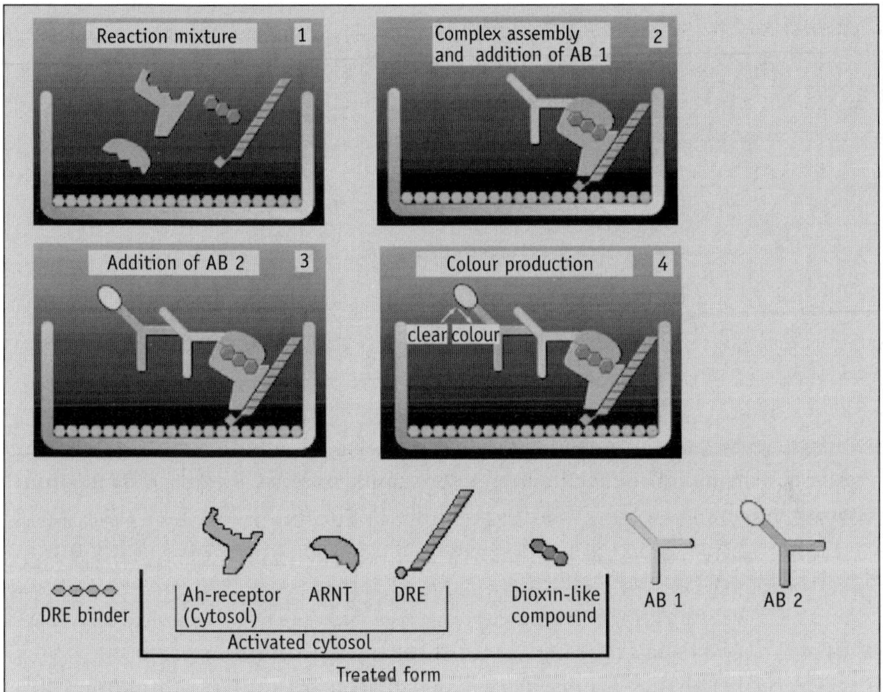

Figure 3. Scheme for sample analysis by the Ah-Immunoassay (from Biosense, Norway).

c. PCR technology

Since a few years also ongoing research is conducted with the Polymerase Chain Reaction (PCR) analysis, the so-called AhPCR kit. Dioxins bind to the Ah (aryl-hydrocarbon) receptor in a specific manner; the resulting DRE is amplified and detected by using real-time PCR. A brief description of the method: the human monocyte cell line U937 is used, centrifuged, washed with buffer and sonificated. Then the DRE Probe is added, incubated, transferred to a microwell of a capture strip, incubated and washed. The PCR kit is added on each well and measured on a real-time thermocycler (see Figure 4). The limit of detection of this method has been reported around 200 fg for TCDD (see www.hybrizyme.com).

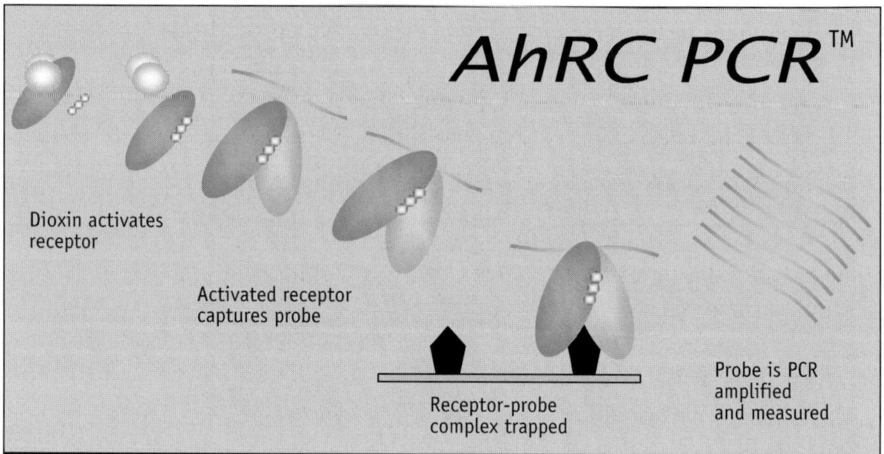

Figure 4. Scheme for the methodology for sample analysis by the AhRPCR (from Hybrizyyme, Inc).

With regard to kit-based bioassays, PCR technology or other proteomic tools, it was recognised that no evidence has yet been submitted of commercially available assays having sufficient sensitivity and reliability for screening for the presence of dioxins at the required levels in samples of foodstuffs.

d. Proteomic biomarker analysis

Proteomic biomarker research is currently trying to make correlations between concentrations of enzymes (e.g., superoxide dismutase), stress proteins (e.g., hsp60), receptors or cytosketal proteins (e.g., myosin).

2.3. Rapid methods for dioxin-like compounds based on chemical methods

Alternative chemical methods for dioxin analysis are the so-called comprehensive multidimensional gas chromatography (GC x GC or CMDGC) (Phillips *et al.*, 1999) and the use of other, less expensive MS-techniques such as the ion-trap MS/MS (Hayward *et al.*, 1999).

2.4. Differences between chemical and bio-analytical methods

An important difference between the chemical and bio-analytical developments is that information on isomer identity is usually maintained using the chemical methodologies, whereas it is a single value, typically a TEQ value or a total PCB value, that commonly results from application of the bioassay. Assuming that the chemical and bio-analytical screening assays have similar performance characteristics it is up to the user of the diagnostic information to decide which method is preferred.

2.5. New dioxin-like compounds

These new biotechnologies offer also the possibility to screen for new dioxin-like compounds or to measure relative potencies of other known environmental pollutants. Therefore in the last few years several other potential toxic compound classes have been already tested by these technologies (Behnisch *et al.*, 2003). Other potent dioxin-like compounds are for example some mixed halogenated dioxins, PBDE-126 and some hexa-chlorinated polychlorinated naphtalines (e.g., Behnisch *et al.*, 2003).

3. International intercalibration studies

In the last couple of years several international round robin studies have been established to compare the results between individual methods, technologies and laboratories for HRMS analysis. Especially, the round robin studies of FAPAS, Folkehelsa and Umea University have demonstrated that most of the chemical analysis laboratories using the traditional method with HRGC/HRMS are able to measure all kinds of samples with a measurement uncertainty of around 20%. In case of international round robin studies for biotechnologies measuring dioxins, so far the finished round robin studies from Örebro University and from the EU did show significant higher measurement uncertainties.

Therefore at the moment it seems that these biotechnologies would be routinely mostly used for making yes or no decisions and the traditional chemical analysis is used as the confirmatory method of choice.

In another approach these technologies are currently compared to each other in an international project called Difference (see www.dioxins.nl). In a first round robin study measuring vegetable oils by HRMS, GCxMSxMS and CALUX the z-score has been below 2 for all three technologies demonstrating the possibilities of promising alternative methods.

Several publications compared already the results from traditional HRMS analysis with these new biotechnologies and in most case they respond to several classes of POPs and therefore the results are usually higher than the results from HRMS analysis (CALUX about double, kit-based bioassays are usually about one magnitude higher). Therefore such an approach gives a risk assessment not only for the regulated polychlorinated dioxins, but it is also possible to cover the whole range of dioxin-like compounds such as dioxin-like PCBs or polybrominated dioxins. Additionally it is possible to evaluate the persistency of compounds and therefore not persistent compounds such as PAHs can be also measured with the CALUX bioassay.

4. Crisis management using rapid methods

Several of these technologies did already help to manage crisis situations in several feed/food crisis in Europe in the last few years (see e.g., Hoogenboom, 2002). In case of the Belgian dioxin crisis in 1999 more than 2000 samples have been tested with CALUX and 87% has shown negative results. CALUX showed in 11% of the samples positive results from which 53% could be confirmed by HRMS and from 141 negative samples only one was actually positive (less than 1% false negatives).

In the case of the citrus pellets crisis, Brazil 1998, CALUX had to be used because of the low capacity of HRMS labs and the therefore too long delivery times. In this case most of the samples have been actually positive and therefore had to be confirmed by the HRMS method. Therefore the advantage of using CALUX has been limited.

5. Take-home messages and future outlook

Biological screening tests for dioxin-like compounds can be used according to new EU guidelines for feed/food, when the tested levels are higher or lower than 30-40% of the interesting concentration. The outcome of the ongoing research and studying will show us in the near future which new rapid methods will be able to fulfil the EU requirements and their results of international intercalibration studies for all kind of feed and food matrices.

Therefore its seems to be at the moment the best way to select the best technology and laboratory for dioxin testing concerning the quality and delivery time in crisis situation by continuously asking the suppliers about their current state of the art and performance; such as the lower limits of detection (which are set by the EU quality criteria to 1/5 of the level of interest) and the quality/number of the equipment to evaluate the turn around time in crisis situations.

As described above, these assays have numerous advantages over the more costly instrumental analysis for rapid screening application, but also have a variety of disadvantages. However, many of these methodologies are still in their early stages of development and given the rapid progress in biotechnology, it is certain that additional improvements in sensitivity, specificity and detection and chemical clean-up methods will occur in the very near future. In addition to technical improvements in the assays themselves, before these bioassays can be accepted for regulatory use, they must be subjected to full validation studies and must meet widely accepted performance criteria of national and international standards organisations and federal agencies (USEPA, EU). Approval of full validation studies by national (Sterlab, the Netherlands; DAR, Germany; Beltest, Belgium) and international agencies as well as the accreditation for laboratories according to ISO 17025, would also encourage their use. Without these criteria being met and standardisation in place, there will be resistance by regulatory agencies to acceptance of these new methodologies. There is certainly a strong need for rapid high-throughput screening analysis of large numbers of samples to both identify samples that are contaminated and to provide an estimate of the relative level of contamination. Therefore, rapid methods should be continuously reviewed by governmental bodies as well as the industry. Either as inexpensive pre-screening methods to complement and enhance instrumental analysis of unknown samples or to select the positive samples. Validation and standardisation of several rapid methods for measuring dioxin-like compounds are currently in progress; some have either received or are currently pending regulatory approval. To speed up dioxin analysis especially extraction and clean-up methods have to be modified. Several modern extraction (e.g., ASE) and clean-up procedures (e.g., PowerPrep) are already available to increase the speed of sample preparation.

The combination of a bioassay screening tool and mass spectrometry confirmation method has already been used successfully as crisis management tools in the food/feed field.

References

Aarts, J.M.M.J.G., Denison, M.S., de Haan, L.H.J., Schalk, J.A.C., Cox, M.A. and Brouwer, A., 1993. Ah receptor-mediated luciferase expression: a tool for monitoring dioxin-like toxicity. Organohalogen Compounds 13, p. 361-364.

Aarts, J.M.M.J.G, Cenijn, P.H., Blankvoort, B.M.G., Murk, A.J., Brouwer, A., Bovee, T.F.H., Traag, W.A., Hoogenboom, L.A.P., Patandin, S., Weisglas-Kuperus, N., Sauer, P.J.J. and Denison, M.S., 1996. Application of the chemical-activated luciferase expression (CALUX) bioassay for quantification of dioxin-like compounds in small samples of human milk and blood plasma. Organohalogen Compounds 27, p. 285-290.

Anderson, J.W., Jones, J.M. and McCoy, D.L., 2000. Cost-effective dioxin site characterization using a P450 human reporter gene system (HRGS; EPA 4425). Organohalogen Compounds 45, p. 208-211.

Archer, J.C., Mobley, R.J., Shojaee, S., Pence, L.M. and Earnheart, C., 2003. Alternative extraction and clean-up for CALUX analysis of feeds and feed components. Organohalogen Compounds 60, p. 207-210.

Behnisch, P.A., Hosoe, K. and Sakai, S., 2001a. Bioanalytical screening methods for dioxins and dioxin-like compounds a review of bioassay/biomarker technology. Environ. Int. 27, p. 413-439.

Behnisch, P.A., Hosoe, K. and Sakai, S., 2001b. Combinatorial bio/chemical analysis of dioxin and dioxin-like compounds in waste recycling, feed/food, humans/wildlife and the environment. Environ. Int. 27, p. 495-519.

Behnisch, P.A., Allen, R., Anderson, J., Brouwer, A., Brown, D.J., Campbell, T.C., Goeyens, L., Harrison, R.O., Hoogenboom, R., Van Overmeire, I., Traag, W. and Malisch, R., 2001c. Harmonized quality criteria for chemical and bioassay analyses of PCDD/PCDFs in feed and food: Part 2: General considerations, bioassay methods. Organohalogen Compounds. 50, p. 59-63.

Behnisch, P.A., Hosoe, K., Brouwer, A. and Sakai, S., 2002. Screening of dioxin-like toxicity equivalents for various matrices with wildtype and recombinant rat hepatoma H4IIE cells. Toxicol. Sci. 69, p. 125-130.

Behnisch, P.A., Hosoe, K. and Sakai, S.-I., 2003. Brominated dioxin-like compounds: in vitro assessment in comparison to classical dioxin-like compounds and other polyaromatic compounds. Environ. Int. 29, p. 861-77.

Brown, D.J., Nakamura, M., Chu, M.D., Denison, M.S., Murata, H. and Clark, G.C., 2002. Recovery determination for bioassay analysis, considerations and results. Organohalogen Compounds. 58, p. 357-360.

Cooke, M., Clark, G.C., Goeyens, L. and Baeyens, W. 2000. Environmental bioanalysis of dioxins. Today's Chemist at Work, Vol. 9, No. 7, 15, 16, 19.

Denison, M.S., El Fouly, M.H., Aarts, J.M.M.J.G., Brouwer, A., Richter, C. and Giesy, J.P., 1993. Production of novel recombinant cell line bioassay systems for detection of 2,3,7,8-tetrachlorodibenzo-p-dioxin-like chemicals. Organohalogen Compounds 13, p. 365-368.

Denison, M.S., Seidel, S.D., Rogers, W.J., Ziccardi, M., Winter, G.M. and Heath-Pagliuso, S., 1998. Natural and synthetic ligands for the Ah receptor. In: Molecular Biology Approaches to Toxicology (Puga, A. and Wallace, K.B., eds), pp. 393-410. Taylor & Francis, Philadelphia.

Denison, M.S., Seidel, S.D., Ziccardi, M., Rogers, W.J., Brown, D.J. and Clark, G., 1999. Ah receptor based bioassays: applications and limitations. Organohalogen Compounds 40, p. 27-30.

Hayward, D.G., Hooper, K. and Andrzejewski, D., 1999.Tandem-in-time mass spectrometry method for the sub-parts-per-trillion determination of 2,3,7,8-chlorine-substituted dibenzo-p-dioxins and -furans in high-fat foods. Anal. Chem. 71, p. 212-220.

Malisch, R., Baumann, B., Behnisch, P.A., Canady, R., Fraisse, D., Furst, P., Hayward, D., Hoogenboom, R., Hoggerbrugge, R., Liem, D., Papke, O., Traag, W. and Wiesmuller, T., 2001. Harmonized quality criteria for chemical and bioassay analyses of PCDD/PCDFs in feed and food: Part 1: General considerations, GC/MS methods. Organohalogen Compounds. 50, p. 53-58.

Hoogenboom, L.A.P., Hamers, A.R.M. and Bovee, T.F.H., 1999. Bioassays for the detection of growth-promoting agents, veterinary drugs and environmental contaminants in food. Analyst 124, p. 79-85.

Hoogenboom, L.A.P., Traag, W., Hoogerbrugge, R., Baumann, B. and de Vries, J., 2001. Control analysis of eel using CALUX and GC/MS. Organohalogen Compounds 54, p. 23-27.

Hoogenboom, R, 2002. The combined use of the CALUX bioassay and the HRGC/HRMS method for the detection of novel dioxin sources and new dioxin-like compounds. ESPR- Environ. Sci. and Pollut. Res. 9, p. 304-306.

Phillips, J.B., Gaines, R.B., Blomberg, J., van der Wielen, F.M., Dimandja, J.-M., Green, V., Granger, J., Patterson, D., Racovalis, L., de Geus, H.-J., de Boer, J., Haglund, P., Lipsky, J., Sinha, V. and Ledford, E.B., 1999. A robust thermal modulator for comprehensive two-dimensional gas chromatography. J. High Resolut. Chromatogr. 22, p. 3-10.

Van den Berg, M., Birnbaum, L.S., Bosveld, A.T.C., Brunström, B., Cook, Ph., Feeley, M., Giesy, J.P., Hanberg, A., Hasegawa, R., Kennedy, S.W., Kubiak, T., Larsen, J.C., van Leeuwen, F.X.R., Liem, A.K.D., Nolt, C., Peterson, R.E., Poellinger, L., Safe, S., Schrenk, D., Tillitt, D., Tysklind, M., Younes, M., Wærn, F., and Zacharewski, T., 1998. Toxic Equivalency Factors (TEFs) for PCBs, PCDDs, PCDFs for humans and wildlife. Environ. Health Perspect. 106, p. 775-792.

Whyte, J.J., Schmitt, C.J. and Tillitt, D.E., 2004 The H4IIE cell bioassay as an indicator of dioxin-like chemicals in wildlife and the environment. Crit. Rev. Toxicol. 34, p. 1-83.

Biosensor detection of sulfonamides: From specific to multi-sulfonamide assays

W. Haasnoot[1], M. Bienenmann-Ploum[1], T. Korpimäki[2], G. Cazemier[1], J. du Pré[1] and F. Kohen[3]
[1]RIKILT-Institute of Food Safety, Bornsesteeg 45, 6708 PD Wageningen, the Netherlands, [2]Department of Biotechnology, University of Turku, Tykistökatu 6A, FIN-20520 Turku, Finland, [3]Department of Biological Regulation, Weizmann Institute of Science, 76100 Rehovot, Israel

Abstract

One of the technologies for performing rapid analysis is based on commercially available automated multi-channel surface plasmon resonance (SPR) biosensors (Biacore) in which analysis can be performed within a few minutes. The application of these instruments in food control was evaluated for the detection of sulfonamides in chicken serum. Different antibodies (from specific to generic) were tested using biosensor chips with different coatings. The most robust assays were obtained using biosensor chips coated with low molecular weight ligands (sulfonamide-derivatives) and these chips could be used for hundreds of cycles. For the multi-sulfonamide detection in chicken serum three different assays were evaluated of which one, based on a mutant antibody, detected all tested sulfonamides (n=17) below the 100 ng/ml level.

Keywords: optical biosensor, immunoassay, sulfonamides, chicken serum

1. Introduction

The aim of a Dutch research project ("Development of monitoring and surveillance systems in the poultry meat chain") is to evaluate biosensor options for the on-line and simultaneous detection of pathogens and drug residues in slaughterhouses. The detection of antibodies against *Salmonella* and residues of sulfonamides in chicken serum was chosen as a first model and the research applied for sulfonamides is reviewed.

In the Netherlands, at least nine different sulfonamides are approved for veterinary application. For the medication of poultry five sulfonamides (sulfamethoxazole, sulfadiazine, sulfamethazine (sulfadimidine), sulfachloropyridazine and sulfaquinoxaline) are approved of which the first two are most frequently applied. Intensive use of these drugs in animal breeding can lead to unwanted residues in food. To establish safe limits for human consumption, the European Union established a maximum residue limit (MRL) of 100 $\mu g\ kg^{-1}$ for the total amount of sulfonamides in edible tissue (Commission Regulation, 1999).

For the detection of sulfonamides in slaughterhouses, body fluids (bile, urine and blood serum) can be employed as markers for residue presence in tissue (Haasnoot *et al.*, 1996; Fodey *et al.*, 1997; Crooks *et al.*, 1998 and Elliott *et al.*, 1999). These body fluids can normally be analysed directly without extraction. Biosensor applications were described for the determination of sulfamethazine and sulfadiazine in pig bile (Crooks *et al.*, 1998; Elliott *et al.*, 1999). However, due to the specificity of the antibodies used, these assays were specific for one of the sulfonamides and, to detect several sulfonamides, different assays had to be used. For multi-sulfonamide screening, several investigations were focused on the development of antibodies detecting the whole group of sulfonamides (Spinks *et al.*, 1999; Haasnoot *et al.*, 2000a,b). However, these approaches were partly successful because important sulfonamides (such as sulfamethazine) were not detected or at a high level only. This was the state of the art at the beginning of this research project in 2001 and our later experiences with the biosensor detection of different sulfonamides (Figure 1) are described in this overview.

2. Biosensor systems

One of the technologies for performing rapid analysis is based on commercially available automated multi-channel surface plasmon resonance (SPR) biosensors. SPR is a phenomenon that occurs in thin conducting films at an interface between media of different refractive index. In the Biacore systems used in this project, the media are the glass of the sensor chip and the sample solution, and the conducting film is the gold layer on the sensor chip surface. SPR causes a reduction in the intensity of reflected light at a specific angle of reflection. This angle varies with the refractive index close to the surface on the side opposite from the reflected light. Assays based on SPR detection rely in principle on the measurement of mass

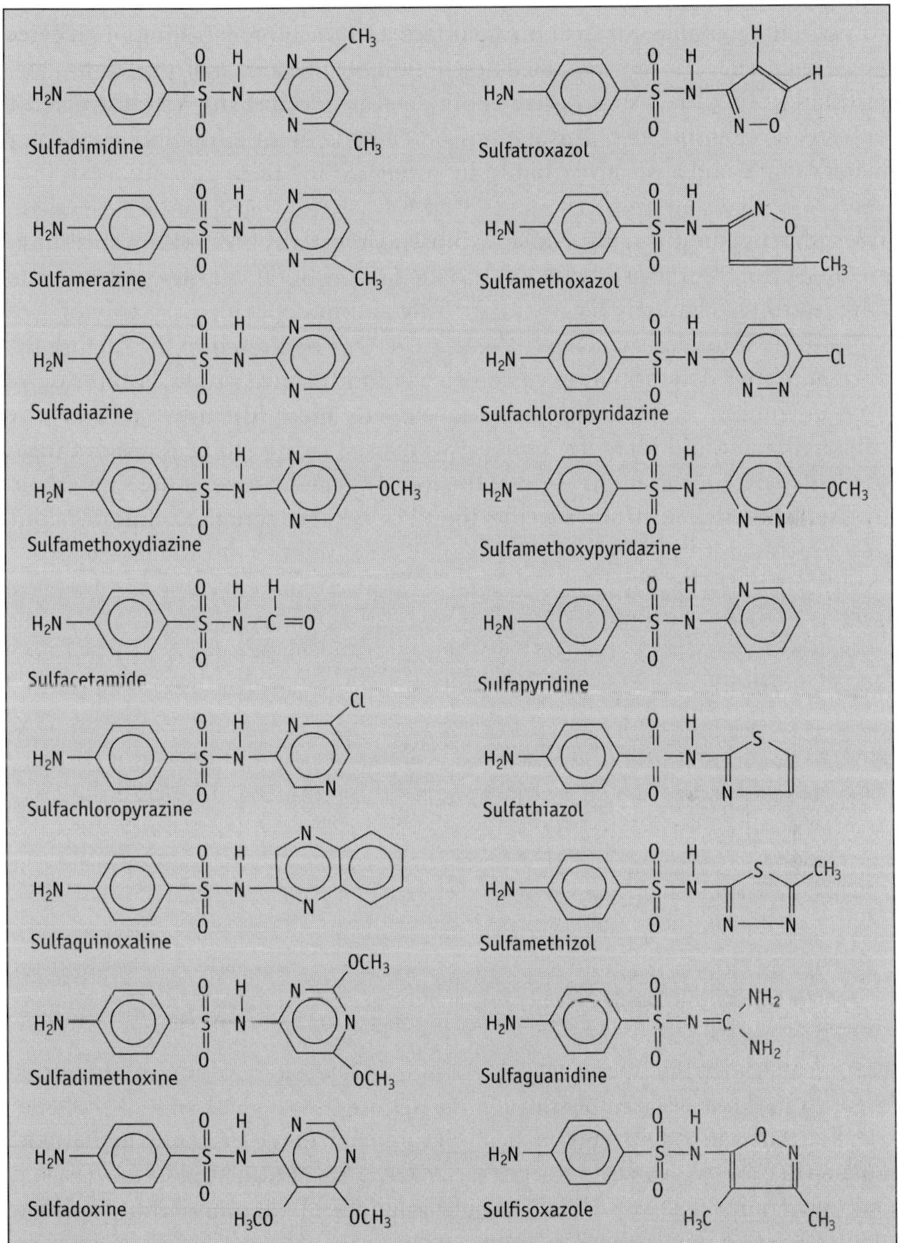

Figure 1. Structural formulas of the sulfonamides used throughout the research.

concentration changes at the sensor surface and therefore labelling of analytes is not required. For a more detailed description of SPR as an analytical technique, see Markey (2000). A sensorgram is obtained by plotting the response against time. As an example, two sensorgrams obtained with an inhibition assay for a blank sample and a positive sample are presented in Figure 2.

The different commercially available biosensor systems used within this study are the Biacore Q, the Biacore 2000 and the Biacore 3000. All these instruments have four flow channels (Fcs) and are fully automated with a capacity of 192 samples (two microtitre plates). The Biacore Q is dedicated to the qualitative or quantitative determination of analytes in food related products and can be used in combination with specially developed Qflex® Kits of which two (for sulfamethazine and for sulfonamides) were used in this study. A disadvantage of the Biacore Q is that only one of the four Fcs can be used at the same time. In the other two instruments the four Fcs can be serially connected and

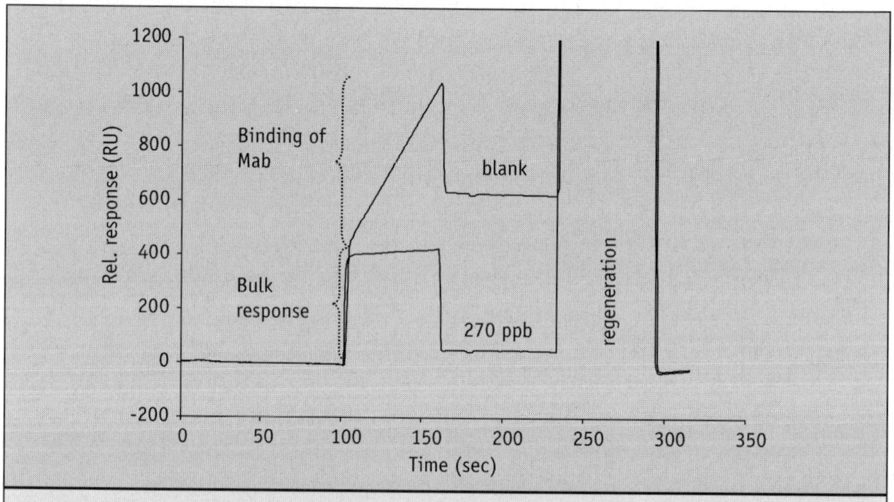

Figure 2. Two sensorgrams obtained in the Biacore Q using monoclonal antibody (Mab) 21C7 and a CM5 chip coated with a sulfonamide derivative (Qflex Kit Sulfamethazine) with a blank chicken serum (10 times diluted in extraction buffer) and that serum spiked with 270 ng sulfamethazine per ml. The diluted chicken serum was mixed with the antibody solution (1:1; v/v) and injected during 1 min at a flow rate of 40 μl/min. The regeneration solution (0.2 M NaOH + 20% acetonitril) was injected during 1 min. The relative responses obtained 10 sec prior to the injection of the regeneration solution were used for the calculations.

simultaneously detected. This gives the opportunity to run multi-assay analysis, as was proven with the simultaneous detection of five aminoglycosides using a mixture of four specific antibodies and four different aminoglycosides immobilised into the four flow channels (Haasnoot *et al.*, 2003). The Biacore 3000 is the most sensitive instrument and is also best suited for kinetic analysis and the direct detection of small molecules.

3. Biosensor chips

For drug residue detection, the Biacore CM5 biosensor chip is most frequently applied. This chip is coated with carboxymethylated dextran to which biomolecules or small ligands can be covalently linked. Sulfonamide-protein conjugates and sulfonamide-derivatives were immobilised using the routine amine coupling chemistry. With this method, the dextran matrix on the sensor chip surface is first activated with a mixture of 1-ethyl-3-(3-dimethylaminopropyl) carbodiimide (EDC) and N-hydroxysuccinimide (NHS) to give reactive succinimide esters. Ligand is then passed over the surface and the esters react spontaneously with uncharged amino groups or other nucleophilic groups to link the ligand covalently to the dextran. In the case of the immobilisation of carboxylated-sulfonamide derivatives (Figure 3), ethylene diamine was used during the coupling (Bienenmann-Ploum *et al.*, 2004).

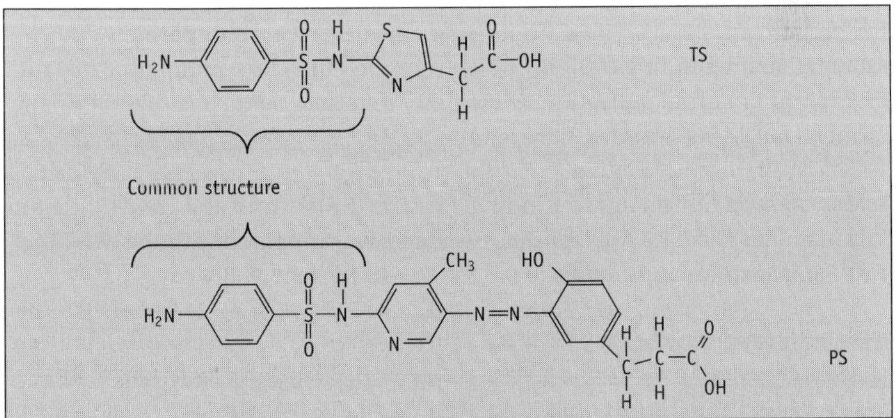

Figure 3. Structures of the haptens used for the development of multi-sulfonamide Mabs. TS = N^1-[4-(carboxymethyl)-2-thiazolyl] sulphanilamide, PS = N^1-[4-methyl-5-[2-(4-carboxyethyl-1-hydroxyphenyl)]-azo-2-pyridyl] sulphanilamide.

4. Antibodies

Monoclonal antibodies (Mabs) were developed against the common structure of sulfonamides by using two different derivatives (Figure 3) linked to carrier proteins via the carboxyl groups. Four different Mabs were obtained against TS of which the best two (Mabs 27G3 and 4E10) were used in this study. One of these Mabs (27G3) showed interactions with all 18 sulfonamides tested in an ELISA (Haasnoot *et al.*, 2000b). Two different Mabs (11H12 and 1D5) were obtained against PS (Haasnoot *et al.*, 2000a) and used in this study.

A mutant antibody (A.3.5) was made with protein engineering of the broad specificity sulfonamide binding Mab 27G3 (Korpimäki *et al.*, 2002; Korpimäki *et al.*, 2003). In a competitive time-resolved fluoroimmunoassay this mutant showed an improved sensitivity towards several sulfonamides compared with the wild-type Mab. However, the sensitivity towards important sulfonamides such as sulfamethazine, sulfadimethoxine and sulfaquinoxaline was still low. In order to improve the properties of this antibody, targeted saturation mutagenesis was used (Korpimäki *et al.*, 2004). In the time-resolved fluoroimmunassay, the best mutant isolated (M.3.4) showed an improved sensitivity for all thirteen sulfonamides tested and even sulfamethazine, sulfadimethoxine and sulfaquinoxaline could be measured below the 100 ng ml^{-1} level. The application of both mutants (A.3.5 and M.3.4) in the biosensor was evaluated.

A polyclonal antiserum raised against sulfamethazine (RIKILT 464b), previously applied in an ELISA (Haasnoot *et al.*, 1996), was compared with the sulfamethazine binding protein (SBP) from the Qflex Kit Sulfamethazine and with a monoclonal antibody (Mab 21C7) raised against sulfamethazine (Kohen *et al.*, 2000) in the Biacore Q.

A Mab named Sulfonamide Binding Protein (SBP) from the new Qflex Kit Sulfonamides (Biacore AB, Uppsala, Sweden) was compared with the two other multi-sulfonamide antibodies (Mab 21C7 and Mutant M.3.4).

5. Results

5.1 Biosensor chips coated with sulfonamide-protein conjugates

The first experiments were performed with a Biacore 2000 in which four Mabs (anti-TS Mabs 27G3 and 4E10 and anti-PS Mabs 11H12 and 1D5) raised against the generic structure of sulfonamides (Figure 3) were compared with Mab 21C7 raised against sulfamethazine (SMZ). At that time, the direct immobilisation of small molecules into the different flow channels (Fcs) was not recommended due to contamination problems in the integrated µ-fluidic cartridge (IFC). Therefore, protein-hapten conjugates (approx. 2000 RU of ovalbumin-TS, ovalbumin-PS and ovalbumin-SMZ) were immobilised in separate Fcs onto the CM5 chip surface. Solutions of the Mabs were diluted to obtain maximum responses of approx. 200 RU (using 40 µl injections at 20 µl/min). The antibodies were mixed with standard solutions of different sulfonamides (1:1; v/v) and the inhibitions of the maximum responses are described in Table 1.

The anti-TS Mabs were most sensitive towards sulfamethizole and high inhibitions (>50%) were observed with six sulfonamides. The anti-PS Mabs were most sensitive towards sulfadimethoxine and high inhibitions were seen with only two (Mab 1D5) or four (Mab 11H12) sulfonamides. Surprisingly, Mab 21C7 (raised against SMZ) showed high inhibitions with fourteen of the sulfonamides tested. Applying a mixture of three Mabs (21C7, 27G3 and 11H12) and connecting the three Fcs resulted in a combined assay in which most relevant sulfonamides could be measured in buffer at a level below 100 ng/ml (Haasnoot *et al.*, 2000c).

The advantage of sulfonamide-protein conjugates immobilised on the chip is that this can be performed in the Biacore with the option to immobilise different flow channels with different conjugates. A disadvantage is that, due to the regeneration procedures, the surfaces lose active groups in time resulting in decreasing responses. More important, with samples containing high protein levels (such as blood serum) unidentified binding compounds (probably protein-protein interactions) are observed having a negative effect on the assay performance in serum.

Nowadays, a device (Surface Prep unit) is available for the Biacore 3000 which makes it possible to immobilise small molecules on the chip surface in the different flow channels.

Table 1. Inhibitions (%) of the maximum responses obtained with two anti-TS Mabs, two anti-PS Mabs and with Mab 21C7 due to the presence of different sulfonamides (100 ng/ml buffer).

Sulfonamide	Anti-TS		Anti-PS		Anti-SMZ
	Mab 27G3	Mab 4E10	Mab 1D5	Mab 11H12	Mab 21C7
Sulfaquinoxaline	5	11	54	62	10
Sulfadoxine	61	6	26	35	0
Sulfachloropyrazine	4	3	1	2	74
Sulfamethoxazole	41	85	15	32	56
Sulfamethoxydiazine	50	76	19	20	61
Sulfamethazine	3	2	36	13	100
Sulfamerazine	15	41	19	17	100
Sulfaguanidine	12	0	0	0	0
Sulfatroxazole	2	11	16	62	95
Sulfapyridine	45	34	20	27	68
Sulfachloropyridazine	82	93	11	44	100
Sulfadimethoxine	6	41	87	90	79
Sulfadiazine	17	48	1	3	100
Sulfathiazole	77	63	0	0	93
Sulfisoxazole	10	48	0	6	100
Sulfamethoxypyridazine	72	82	31	70	77
Sulfamethizole	93	96	1	0	84
Dapsone	5	0	3	0	4

5.2 Biosensor chips coated with sulfonamide-derivatives

The multi-sulfonamide Mab 21C7 was applied in the Biacore Q in combination with the sulfamethazine derivative (supplied in the Qflex Kit Sulfamethazine) coated onto a CM5 biosensor chip. Compared with the previously used ovalbumin-SMZ coated chip, an eight times lower amount of the Mab injected resulted in approx. four times higher responses. This means that the direct coating of the SMZ derivative resulted in a much higher quantity of SMZ on the chip surface.

The performance of this Mab in the Biacore Q was compared with the Sulfamethazine Binding Protein (SBP) supplied in the kit and with an in-house developed polyclonal antiserum against SMZ (RIKILT 464b) using chicken serum as sample material (Haasnoot *et al.*, 2003). Using the SBP and the polyclonal antiserum the limits of detection (LOD) in chicken serum (analysed after a ten times dilution in buffer) were 30 ng/ml and using Mab 21C7 the LOD was 10 ng/ml. The assays based on SBP and the polyclonal antiserum were specific for SMZ, whereas the assay based on Mab 21C7 showed high cross-reactivities with sulfisoxazole (149%), sulfachloropyridazine (112%), sulfachloropyrazine (94%), sulfamerazine (87%), sulfadiazine (56%), sulfatroxazole (56%) and sulfathiazole (50%); low cross-reactivities (11-25%) were observed with six other sulfonamides.

The Mab based assay was found suitable for the detection of eight sulfonamides in chicken serum with LODs between 7 and 20 ng/ml and within a total run time of 7 min. Of the five sulfonamides approved for poultry in the Netherlands, this Mab recognised three of them (sulfamethazine, sulfadiazine and sulfachloropyridazine) with high cross-reactivity which made it a unique antibody at that time. However, the cross-reactivity with the other two sulfonamides approved for chicken was low (11% for sulfamethoxazole and <10% for sulfaquinoxaline). During these experiments the biosensor chip with the immobilised SMZ derivative was used for approximately 1100 cycles and no significant change in performance was observed which makes this assay, in combination with the high maximum responses (Figure 2), very robust.

Searching for an assay which could detect more of the relevant sulfonamides, the two genetically modifications of the wild-type Mab 27G3 (raised against the common structure of sulfonamides) were evaluated. In the Biacore 3000, the performances of the two mutants were compared with the wild-type Mab 27G3 using a CM5 sensor chip coated with TS (Figure 3) and chicken serum as sample material. To couple the carboxyl group of TS to the carboxylated dextran, ethylene diamine was used in combination with the amine coupling procedure. With this TS-coated chip, at least a ten times higher response was obtained compared with the previously used TS-ovalbumin coated chip injecting approx. the same amount of Mab 27G3. In this experiment low antibody dilutions resulting in high maximum responses (approx. 1500 RU) were applied to develop robust assays. The differences in inhibition due to the presence of different sulfonamides added to chicken serum at the 100 ng/ml level, are shown in Figure 4 for the three antibodies. With the wild-type Mab 27G3 six sulfonamides showed a significant inhibition of the response (>20%).

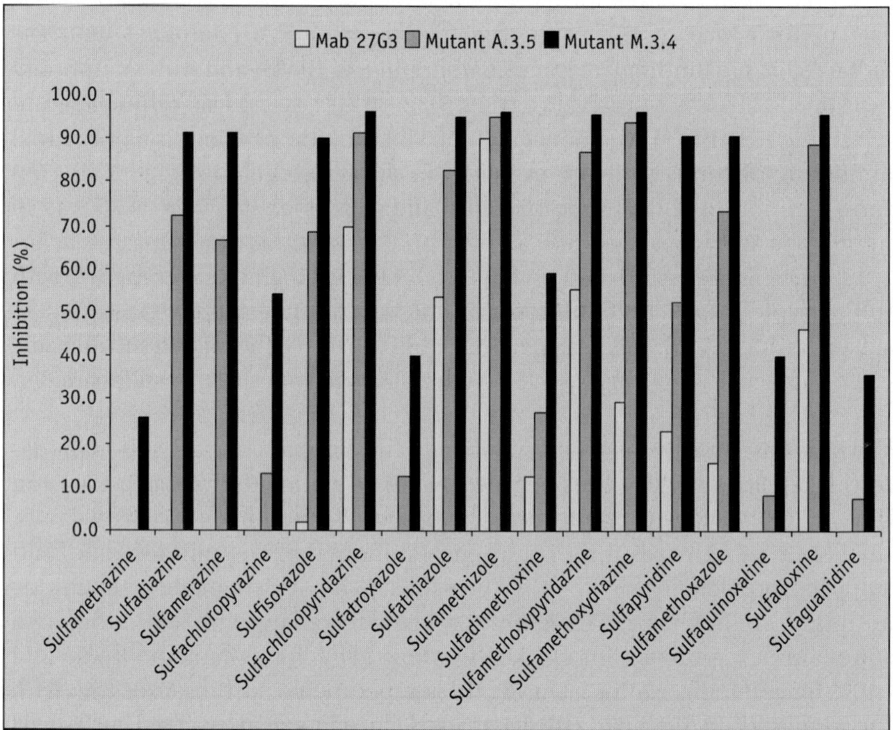

Figure 4. Inhibition (%) of the maximum responses due to the presence of different sulfonamides (100 ng/ml) added to a blank chicken serum using the wild-type Mab 27G3, mutant A.3.5 and mutant M.3.4. Chicken serum (20 µl) was diluted with 100 µl of the antibody-containing buffer of which 50 µl was injected at a flow rate of 20 µl/min.

With mutant A.3.5, twelve sulfonamides showed significant inhibitions and using mutant M.3.4, all seventeen sulfonamides tested showed significant inhibitions.

Mutant A.3.5 was used to analyse chicken serum and plasma samples obtained from an animal experiment in which broilers were treated with sulfamethoxazole and sulfadiazine. To quantify sulfonamide concentrations in these samples calibration curves were prepared for the two sulfonamides in a blank serum and plasma sample and comparable sensitivities for both sulfonamides were obtained (Figure 5).

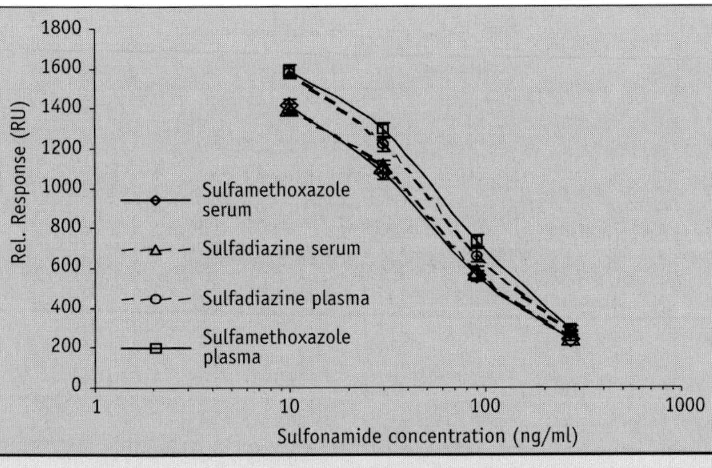

Figure 5. Calibration curves using four calibrants of sulfadiazine and sulfamethoxazole in chicken serum and plasma with the mutant A.3.5 based biosensor assay.

The responses and calculated concentrations in the serum and plasma samples obtained from the sulfamethoxazole animal experiment are presented in Figure 6. After analysing the 20 blank chicken sera and plasma samples the LOD for sulfamethoxazole was determined as 10 ng/ml and a good correlation (R^2 = 0.996) between the concentrations in serum and plasma was observed. The high variation in responses obtained in the samples taken after the treatment (days 6, 8 and 10 of the animal experiment) was due to the variation of the concentrations in the individual broilers (n=6) and not due to the assay performance (CV <5%).

The concentrations found in serum or plasma have to be correlated with the concentrations in edible tissue to establish predictive levels for the screening assay. The mutant M.3.4 based assay was more sensitive for sulfamethoxazole (Figure 4) and was used to analyse sulfamethoxazole in tissue samples obtained from the animal experiment using a calibration curve in tissue (Figure 7). Tissue samples (2 g) were extracted with buffer (8 ml) and centrifuged. Of the supernatant 20 µl was mixed with 50 µl of extraction buffer and 50 µl of a mutant solution in HBS-EP buffer of which 50 µl was injected at 20 µl/min. After analysing blank tissue samples (n=20) the LOD for sulfamethoxazole in leg meat was determined as 19 ng/g.

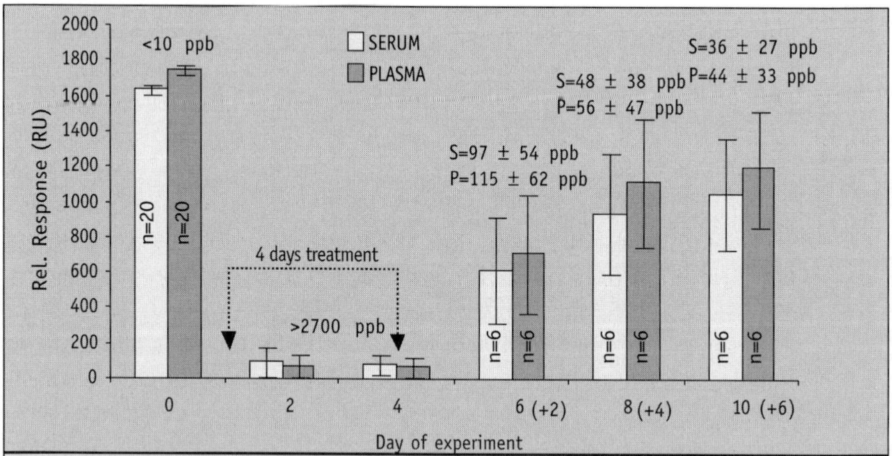

Figure 6. Responses and average concentrations of sulfamethoxazole found with the mutant A.3.5 based biosensor assay in serum (S) and plasma (P) samples of broilers treated with sulfamethoxazole. Sample were taken before treatment (day 0), during treatment (days 2 and 4) and 2, 4 and 6 days after treatment (days 6, 8 and 10 of the experiment).

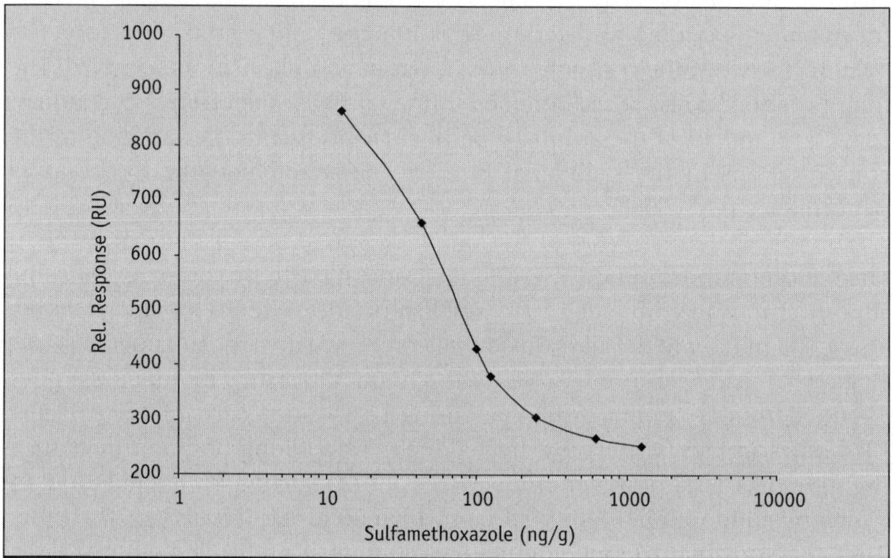

Figure 7. Calibration curve of sulfamethoxazole in meat using the mutant M.3.4 based assay.

In the leg meat samples from the animal experiment (after treatment) only four samples contained sulfamethoxazole concentrations higher than or comparable with the LOD (between 18-38 ng/g). The serum samples of these broilers contained 4.8 times higher concentrations of sulfamethoxazole (as was determined with the mutant A.3.5 based assay). All leg meat samples were analysed with LC-MS (LOD approx. 1 ng/g) and the correlation with concentrations found in chicken serum is shown in Figure 8. Also here, the average concentration found in chicken serum was 4.8 times higher compared with the average concentration found in leg meat. The leg meat samples taken during the treatment had to be diluted an extra ten times to fit into the range of the calibration curve and concentrations of sulfamethoxazole varied between 180 and 640 ng/g.

In October 2003 Biacore launched the Qflex Kit Sulfonamides. This kit is intended for screening analysis of sulfonamide residues in food products and at that time the kit handbook described the application in porcine muscle only. The kit was evaluated for the application in chicken serum (Bienenmann-Ploum *et al.*, 2004) and the results were compared with the two other multi-sulfonamide assays (based on Mab 21C7 and mutant M.3.4). For the 17 sulfonamides tested in the three assays, the LODs in chicken serum varied

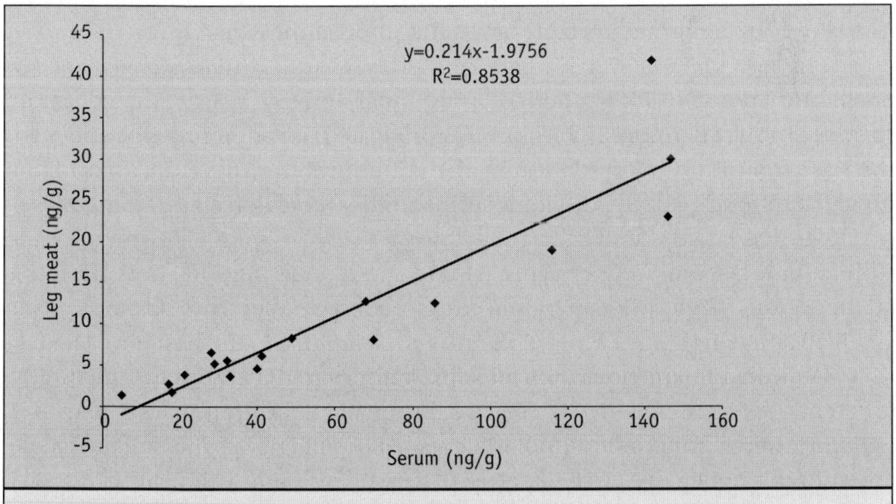

Figure 8. Correlation between concentrations of sulfamethoxazole found in serum (using the mutant A.3.5 based biosensor immunoassay) and in leg meat (analysed by LC-MS).

between 7 and >1000 ng/ml (Mab 21C7), 15 and 340 ng/ml (Qflex) and 4 and 82 ng/ml (Mutant M.3.4). The mutant antibody based assay was found to be the most sensitive towards most of the sulfonamides whereas the Qflex Kit Sulfonamides detected the five sulfonamides registered for application in poultry in the Netherlands within the narrowest measurements range (LODs between 19 and 62 ng/ml).

6. Discussion

Sulfonamides are a group of drugs for which there is considerable demand for a broad specificity assay detecting all members of the group. Coupling the sulfonamides via the common aromatic amino group (Figure 1) to carrier proteins resulted in specific antibodies (detecting one of the sulfonamides) only (Haasnoot *et al.*, 2000a). A surprising exception was Mab 21C7 raised against sulfamethazine coupled to carrier proteins after diazonium activation of the aromatic amino group (Kohen *et al.*, 2000). This Mab recognised a range of structurally different sulfonamides including their N^4-acetyl metabolites (Haasnoot *et al.*, 2003) and the central part of the sulfonamide structure must play an important role in the antibody binding. With this Mab based biosensor assay eight sulfonamides could be detected in chicken serum with LODs between 7 and 20 ng/ml. However, the cross-reactivities with two other important sulfonamides (sulfamethoxazole and sulfaquinoxaline) were low.

Sheth and Sporns (1991) produced polyclonal antisera against the common structure of sulfonamides (TS, Figure 3) conjugated to carrier proteins and after the selection of a sub-population of the antibodies they developed an ELISA in which nine structurally different sulfonamides were detected at a high level (in the μg/ml range). Assil *et al.* (1992) used a similar approach with another sulfonamide derivative (PS, Figure 3) and seven sulfonamides were detected at the μg/ml range. Instead of selecting a sub-population, Haasnoot *et al.* (2000ab) prepared Mabs against TS and PS. In an ELISA, the best anti-TS Mab (27G3) showed interactions with all sulfonamides tested and eight structurally different sulfonamides showed 50% inhibition at concentrations less than 100 ng/ml. However, important sulfonamides (such as sulfamethazine, sulfatroxazole and sulfachloropyrazine) were detected at a high level only. Cliquet *et al.* (2003) followed a comparable route and with the best anti-TS Mab (3B5B10E3), in a heterologous ELISA using PS-ovalbumin coated plates, most of the tested sulfonamides, except sulfamethazine and sulfamerazine, could be detected at the 100 ng/ml level in buffer and, compared with the Mab 27G3 based ELISA, an improved sensitivity towards important sulfonamides (such as sulfamethoxazole and sulfadiazine) was obtained.

Korpimäki *et al.* (2002, 2003 and 2004) made mutant antibodies with protein engineering of the wild-type Mab 27G3 and an improved sensitivity was observed for all sulfonamides. Using the best mutant (M.3.4) in the biosensor resulted in an assay in which all sulfonamides could be detected in chicken serum with LODs between 4 and 82 ng/ml. Just as with the wild-type Mab, these mutants showed low interactions with the N^4-acetyl metabolites of sulfonamides and the common aromatic amine structure must play an important role in the interaction.

The origin of the antibody used in the new Qflex Kit Sulfonamides (launched in October 2003) is unknown. However, as this antibody does not react with the N^4-acetyl metabolites, the aromatic amine structure must play an important role in the binding. With this kit all sulfonamides could be detected in chicken serum in which the LODs varied between 15 and 340 ng/ml.

7. Conclusions

Mabs have proven to be more successful in creating multi-sulfonamide assays than polyclonal antisera and, in combination with biosensor chips coated with sulfonamide-derivatives, robust biosensor assays were obtained. The application of antibody engineering was proven to be very successful by creating recombinant antibodies with improved sensitivities towards all the sulfonamides tested. The most sensitive assay was obtained with a mutant antibody (M.3.4) in which all sulfonamides could be tested in chicken serum with LODs varying from 4 to 82 ng/ml. Although less sensitive for the whole group of sulfonamides, the Mab in the Qflex Kit Sulfonamides detected the five sulfonamides registered for application in poultry in the Netherlands within the narrowest measurements range (LODs varied from 19 to 62 ng/ml).

Chicken serum seemed to be a good sample material for the detection of sulfonamides in a slaughterhouse. In the sera samples from an animal experiment with sulfamethoxazole about five times higher concentrations were found compared with the concentrations in meat. Additional experiments will be performed in which other tissue samples (liver and skin plus fat) will be involved.

Acknowledgements

The authors thank the Dutch Ministry of Agriculture, Nature and Food Quality for funding this research (performed within the LNV research program nos. 389 and 390).

References

Assil, H.I., Sheth, H. and Sporns, P., 1992. An ELISA for sulphonamide detection using affinity-purified polyclonal antibodies. Food Research International 25, p. 343-353.

Bienenmann-Ploum, M., Korpimäki, T., Haasnoot, W. and Kohen, F., 2004. Comparison of multi-sulfonamide biosensor immunoassays. in: Proceedings of the EuroResidue V conference, 10-12 May 2004, Noordwijkerhout, the Netherlands, in press.

Cliquet, P., Cox, E., Haasnoot, W., Schacht, E. and Goddeeris, B.M., 2003. Generation of group-specific antibodies against sulfonamides. J. Agric. Fd Chem. 51, p. 5835-5842.

Commission Regulation (EC) no. 508/1999, Official Journal of the European Communities, L60, p. 16-52.

Crooks, S.R.H., Baxter, G.A., O'Connor, M.C. and Elliott, C.T., 1998. Immunobiosensor - an alternative to enzyme immunoassay screening for residues of two sulfonamides in pigs. Analyst 123, p. 2755-2757.

Elliott, C.T., Baxter, G.A., Crooks, S.R.H. and McCaughey, W.J., 1999. The development of a rapid immunobiosensor screening method for the detection of residues of sulfadiazine. Food Agric. Immunol. 11, p. 19-27.

Fodey, T.L., Crooks, S.R.H., Elliott, C.T. and McCaughey, W.J., 1997. A comparison of porcine urine and bile as matrices to screen for the residues of two sulphonamides using a semi-automated enzyme immunoassay. Analyst 122, p. 165-168.

Haasnoot, W., Korsrud, G.O., Cazemier, G., Maneval, F., Keukens, H. and Nouws, J.F.M., 1996. Application of an enzyme immunoassay for the determination of sulphamethazine (sulphadimidine) residues in swine urine and plasma and their use as predictors of the level in edible tissue. Food Addit. Contam. 13, p. 811-822.

Haasnoot, W., Cazemier, G., du Pré, J., Kemmers-Voncken, A., Bienenmann-Ploum, M. and Verheijen, R., 2000a. Sulphonamide antibodies: from specific polyclonals to generic monoclonals. Food Agric. Immunol. 12, p. 15-30.

Haasnoot, W., du Pré, J., Cazemier, G., Kemmers-Voncken, A., Verheijen, R. and Janssen, B.J.M., 2000b. Monoclonal antibodies against a sulfathiazole derivative for the immunochemical detection of sulfonamides. Food Agric. Immunol. 12, p. 127-138.

Haasnoot, W., Kohen, F., du Pré, J., Cazemier, G., Kemmers-Voncken, A., Bienenmann-Ploum, M. and Verheijen, R., 2000c. Application of generic monoclonal antibodies against sulfonamides in optical biosensors. in: Proceedings of the EuroResidue IV conference, 8-10 May 2000, Veldhoven, the Netherlands, p. 501-505.

Haasnoot, W., Bienenmann-Ploum, M. and Kohen, F., 2003. Biosensor immunoassay for the detection of eight sulfonamides in chicken serum. Anal. Chim. Acta 483, p. 171-180.

Haasnoot, W., Cazemier, G., Koets, M. and van Amerongen, A., 2003. Single biosensor immunoassay for the detection of five aminoglycosides in reconstituted skimmed milk. Anal. Chim. Acta 488, p. 53-60.

Kohen, F., Gayer, B., Amir-Zaltsman, Y. and O'Keeffe, M., 2000. Generation of an anti-idiotypic antibody as a surrogate ligand for sulfamethazine in immunoassay procedures. Food Agric. Immunol. 12, p. 193-201.

Korpimäki, T., Rosenberg, J., Virtanen, P., Karskela, T., Lamminmäki, U., Tuomola, M., Vehniäinen, M. and Saviranta, P., 2002. Improving broad specificity hapten recognition with protein engineering. J. Agricultural and Food Chemistry 50, p. 4194-4201.

Korpimäki, T., Rosenberg, J., Virtanen, P., Lamminmäki, U., Tuomola, M. and Saviranta, P., 2003. Further improvement of broad specificity hapten recognition with protein engineering. Protein Engineering 16, p. 37-46.

Korpimäki, T., Brockmann, E-V., Kuronen, O., Sarate, M., Lamminmäki, U. and Tuomola, M., 2004. Engineering of a broad specificity antibody for simultaneous detection of 13 sulfonamides at the maximum residue level. J. Agricultural and Food Chemistry 52, p. 40-47.

Markey, F., 2000. Principles of Surface Plasmon Resonance. In: Real-time Analysis of Biomolecular Interactions. Applications of Biacore; Nagata, K., Handa, H., Eds.; Springer-Verlag: Tokyo, Japan, p. 13-22.

Sheth, H.B. and Sporns, P., 1991. Development of a single ELISA for the detection of sulfonamides. Journal of Agricultural Food Chemistry 39, p.1696-1700.

Spinks, C.A., Wyatt, G.M., Lee, H.A., Morgan, M.R.A., 1999. Molecular modelling of hapten structure and relevance to broad specificity immunoassay of sulfonamide antibiotics. Bioconjug. Chem. 10, p. 583-588.

Detecting trace levels of chlorophenols and chloroanisoles in wineries by immunodiagnostic techniques

B. Alfaro Redondo[1], B. Pérez-Villarreal[1], S. Setford[2], C. Nicholls[2], L. Bourdin[3], J. Wijdenes[3] and C. Vermot-Desroches[3]*
*[1]FUNDACIÓN AZTI, Txatxarramendi ugartea z/g 48395. Sukarrieta, Bizkaia. Spain, [2]Cranfield Centre for Analytical Science (CCAS), Cranfield University, Silsoe, Beds., United Kingdom, [3]DIACLONE, Besançon, France, *Current address: CEFAS Lowestoft lab, Pakefield Road, Lowestoft, Suffolk NR33 0HT, United Kingdom*

Abstract

The central aim of the FOODSAFE project has been the development and application of both conventional immunoassay procedures and also the use of a miniaturised liquid handling system compatible with immunosensor devices as tools for the rapid determination of chlorophenol fungicide residues, in liquid food samples, mainly within the wine sector. Appropriate sample methods have also been developed for analysing packaging materials, cardboard, paper, bottle washings and corks to allow the trace level detection of chlorophenols. Combination of an optimised sample preparation method with enzyme linked immunosorbent assay (ELISA) procedures has allowed determination of chlorophenols within the ng/L range. The measurement tools developments are low cost, simple to perform and do not require highly trained personnel and expensive equipment. Results from the comparative measurement of chlorophenols in liquid food samples by the combined sample preparation-ELISA method and GC/MS will be presented.

Keywords: pesticides, chlorophenols, ELISA, immunodiagnostic, winery materials

1. Introduction

Chlorophenols, particularly commercial preparations of pentachlorophenol (PCP) have been used as general biocides and wood preservatives for more than

50 years and are presently widespread in the environment. In addition to industrial production and usage, chlorophenols are produced from naturally occurring phenols as a result of chlorine bleaching of wood pulp in the paper industry and through the chlorination of domestic water supplies and swimming pools. Chloroanisoles are not used industrially, but it has been established that they are formed from the chorophenols due to fungal methylation. Many *Aspergillus* and *Penicillium* moulds are capable of methylating chlorophenols. Current methods of food analysis involve the measurement of crude physical parameters allied to low throughput and instrumentally complex analytical methodologies. Consequently, traceabiliy of hazardous materials within food production processes is both costly and time-consuming.

Food contamination by chlorophenols and chloroanisoles have been well reported and documented. These compounds have been found in a multitude of food products: wine, canned carbonated beverages, reconstituted fruit juice, poultry, bulk chocolate, modified flour, canned fruit, cocoa powder, packaged biscuits, chilled meat and so on. Contamination with chlorophenols and chloroanisoles at concentrations below 10^{-3} mg/kg is of toxicological concern and produces off-flavours in processed and packaged foods. Such low levels bring about the need for a careful selection of the isolation methods and for running the instrumental techniques near their limit of sensitivity.

The development of specific immunochemical methods that offer high sensitivity and high sample throughput whilst being low cost and simple to operate would aid traceability of toxic agents and hence reduce the levels of toxic components within processed food products. These rapid and cheap assay techniques would be ideal in testing raw materials prior to their integration into the food processing operation, and as such would prove ideal to trace materials of sub-standard quality and prevent their usage. The implementation of such technologies has the potential to provide the consumer with higher quality and healthier products.

FOODSAFE has centred on the creation of new and safer methods of food production through development of a new generation of rapid, low cost, sensitive and simple immunochemical-based measurement technologies to aid identification of the Hazardous Analysis Critical Control Points (HACCP) within liquid food processing operations.

2. Materials and methods

2.1. Monoclonal antibody generation

2.1.1 Preparation of the protein carrier conjugate

Two mg of the carrier protein Keyhole Limpets Hemocyanin (KLH) or Bovine Serum Albumin (BSA), (Pierce, Rockford, USA) was reconstituted with 200 µL of deionised water and 2 mg of the hapten PCP or PCA (Diaclone, Besançon, France) was dissolved in 0.5 ml of conjugation buffer (0.1 M MES, 0.9 M NaCl, 0.02% NaN_3, pH 4.7). Ten mg of EDC (1-ethyl-3-[3-dimethylamino-propyl]carbodiimide hydrochloride) was reconstituted in 1 ml of deionised water. For BSA conjugation, 1 ml of EDC, 500 µL of PCA and 200 µL carrier protein solutions were added. For KLH conjugation, 50 µL of the EDC solution, 500 µL of PCP and 200 µL carrier protein solutions were added. After stirring for 2 hours at 25 °C, the reaction mixture was loaded onto a PD-10 column (Pharmacia Biotech., Uppsala, Sweden) which was equilibrated with phosphate buffered saline (PBS). Monitoring of positive fraction was performed using UV detection at 280 nm and collected.

2.1.2 Immunisations of mice

The mice were immunised by intraperitoneal (I.P.) injection with 100 µL of the immunogen PCA-BSA or PCP-KLH (100 µg/ml) with an equal volume of complete Freund's adjuvant for the first I.P. At least, two other I.P. injections were done every two weeks with 100 µL of the immunogen (100 µg/ml) with an equal volume of incomplete Freund's adjuvant. Four days prior the spleen collection, the mice were boosted with 100 µL of the immunogen preparation (100 µg/ml) in the vein of the tail.

2.1.3 Production of monoclonal antibody

Splenocytes from the immunised mice were mixed with murine myeloma cells (X63-AG8) at a ratio of 5:1 and fused by the procedure described by Köhler and Milstein (1975) with some modifications following the Diaclone methodology. The fused cells were cultured in hypoxanthine aminopterin thymidine medium (HAT medium). Following the immunogen used, the culture medium of each hybridoma was screened for anti-PCA or -PCP antibody by ELISA. Hybridoma cell populations secreting anti-PCA or -PCP antibody were cloned at least three times by the limiting dilution method. Isotyping of antibody was performed with a mouse mAb isotyping kit (Diaclone, France). The cloned hybridoma cell line was injected into Freund-primed mice. The anti-PCA or -PCP mAbs were purified from the ascitic supernatants by affinity chromatography on a protein A-agarose gel (Pharmacia Biotech., Uppsala, Sweden).

2.1.4 Hybridoma selection by indirect enzyme immunoassays

The indirect enzyme immunoassay (ELISA) was used to identify culture wells containing hybridoma secreting antibodies against PCP-KLH or PCA-BSA following fusion and cloning. Each well of a microtiter plate was coated with 100 µL of PCP-KLH or PCA-BSA (500 ng/ml) in phosphate buffer saline overnight at 4 °C. The plate was then washed four times with distilled water containing 0.05% Tween 20 (buffer T). The unbound active sites in each well were blocked with PBS containing 5% BSA (250 µL per well) for 2h at room temperature. After washing four times with the buffer T, 100 µL of the optimum dilution of hybridoma supernatant was added to each well, and the plates were incubated for 1 h at room temperature. Unbound antibodies were removed by washing four times with the buffer T and 100 µL of the sheep anti mouse IgG1, IgG2a, IgG2b, IgG3 and IgM - peroxidase conjugate solution at the optimum dilution (The Binding Site, Birmingham, UK) was added to each well. The plates were incubated for 30 min at room temperature and then washed four times with the buffer T. Then, 100 µL of 3,3′,5,5′-TetraMethylBenzidine (TMB) (Moss, Pasadena, USA) solution was added to each well. After incubation for 10 min at room temperature, the reaction was stopped with 100 µL per well of 1N H_2SO_4 (Merck, Damstadt, Germany). The optical density in each well was measured with a microplate reader MRX Revelation (Dynex) at nm.

2.1.5 Indirect competitive enzyme immunoassay

The indirect competitive ELISA was used to verify the specificity of antibodies against free PCP or PCA. Briefly, each well of the microtiter plates were coated with PCP-KLH or PCA-BSA as described for the indirect ELISA procedure. Next 100 µL of the standard solution of free PCP, PCA or BSA at different concentrations diluted in PBS or blank buffer (PBS) was added. Then 100 µL of the optimum dilution of hybridoma supernatant or purified antibody was added to each well. After immediate mixing, the plate was incubated for 1 h at room temperature. The assay was then completed as described for the indirect ELISA procedure.

2.2. Determination of chlorophenols and chloroanisoles in wine samples by gas chromatography detection

Reagents: Petroleum ether (PE) and hexane were pesticide residues analysis grade. Potassium carbonate, anhydrous sodium sulphate, sulphuric acid and acetic anhydride were analytical reagent grade. Chlorophenol and chloroanisole reference standards were purchased from Dr. Ehrenstorfer GmbH and from Supelco. Standard solutions were prepared in hexane and stored at 4°C.

Sample preparation procedure: The primary extraction was specific for each substrate. For wine: introduce 200 ml of samples into a separatory funnel. Add 3 ml 0.05 mol.l^{-1} H_2SO_4 and extract with 180 ml of petroleum ether (PE). Re-extract the aqueous layer with 50 ml of PE. Collect the two organic layers. After pre-preparation, transfer in aqueous carbonated solution by extracting PE layer with 50 ml 0,1 mol.l^{-1} K_2CO_3 aqueous solution.
Organic phase: contains chloroanisoles. Filter with anhydrous Na_2SO_3 and concentrate by rotavapor under N_2. Afterwards, filter by silica gel column and evaporate (until a few millilitres are obtained). Analyse by GC.
Aqueous phase: contains chlorophenols, it is necessary to perform a derivatisation of these compounds. Derivatisation: Mix the aqueous K_2CO_3 layer with 0,5 ml acetic anhydride. Extract the PCP acetate twice with hexane (10 ml). Dry hexane layer on anhydrous Na_2SO_3 before concentrating under N_2 (until a few millilitres are obtained). Analyse by GC.

Instrumental analyses: Extracts were analysed with a Hewlett-Packard HP-5890 series II coupled to an electron capture detector. Confirmation was carried out by mass selective detector (GC/MS).

2.3. Samples preparation procedures before enzyme linked immunosorbent assay (ELISA)

Water (distilled or deionised), solvents and chemicals were all of pesticide grade. Two sample blanks were included with each analytical batch to confirm that chemicals are free of contaminants.

2.3.1 Extraction of chlorophenols from wine using solvent extraction (Procedure 1)

Apparatus: Wrist-action flask shaker (e.g. Stuart Scientific); Ultrasonic cleaning bath; Measuring cylinder (500ml); Pyrex volumetric flasks (500 ml); Glass vial or tube for back extraction (20ml) fitted with conical base and screw cap with PTFE seal; Tubes for sample storage (2ml) were polypropylene or similar.

Procedure:
- Wash all Glassware to a standard appropriate for pesticide residue analysis (USEPA 40CFR 136 or similar). Procedures will normally involve soaking glassware overnight in dilute alkaline-detergent and acid solutions.
- Transfer 200 ml of the wine sample into a clean 500 ml volumetric flask.
- Add 20 ml pentane (Merck Suprasolv), shake for 1 h using the flask shaker.
- Allow flasks to stand for 20 min to ensure separation of the solvent layers.
- Add 1 ml 2-propanol (Merck Suprasolv) to each flask to remove any surface foam.
- Dilute the flask contents with pesticide grade water (pour in with care to avoid mixing) until the upper solvent layer enters the neck of the flask. In some cases emulsions may form at the interface of the aqueous and organic phases. These may be reduced by placing the flask in an ultrasonic cleaning bath for a few minutes.
- Allow flask to stand for 2 h or overnight to ensure that phase separation is complete.
- Transfer 10 ml of the upper pentane layer into a suitable vial or tube.
- Add 1 ml 0.1 M sodium carbonate solution cap tightly and shake vigorously for 10 minutes using the flask shaker.
- Allow to stand for 20 minutes to ensure phase separation.
- Pipette 0.5 ml of the lower aqueous phase into clean polypropylene vial and add 0.5 ml 0.1 M sodium dihydrogen orthophosphate solution for pH control (pH 7 approx.).
- Sample extracts should be stored at 4°C and tested within 48 h of extraction.

2.3.2. Extraction of chlorophenols and chloroanisoles from wine using steam distillation and solid phase extraction (Procedure 2)

Chemicals: Sulphuric acid and glacial acetic acid (both AR grade), acetone, methanol and propanol, dichloromethane (Merck Suprasolv), phosphate buffered saline (PBS) buffer tablets (Sigma P-4417), Tween20 (Polyxyethylenesorbitan monolaurate) (Sigma P-7949).

Reagents: 2 M Sulphuric acid, 50% v/v methanol (mix equal volumes of methanol and water in a suitable glass container). Elution solvent - 1% v/v glacial acetic acid in 50% v/v acetone/ dichloromethane. 0.1% Tween20 PBS/sample buffer solution: Dissolve 0.2 g Tween20 and one PBS tablet to 200 ml.

Apparatus: Dedicated steam distillation apparatus (alternatively Quickfit™ glassware may be used). An essential feature of the apparatus must include the ability to externally heat the sample vessel. Solid phase extraction (SPE) apparatus comprising SPE cartridges (IST Env+, 100 mg), ultrasonic cleaning bath, measuring cylinder (100 ml), Pyrex volumetric flasks (25 ml), glass vial or tube for SPE sample collection and concentration (20 ml), fitted with conical base and screw cap with PTFE seal.

Steam distillation procedure:

- Wash all glassware to a standard appropriate for pesticide residue analysis.
- Set-up the steam distillation apparatus and allow to generate steam for 20 minutes prior to analysis.
- Transfer 50 ml of blank or wine sample to a clean 250 ml steam distillation sample flask.
- Add 4 ml 2 M sulphuric acid and mix.
- Transfer 10 ml 0.1 M sodium carbonate solution to a 50 ml collection flask and attach to the steam distillation apparatus.
- Pass steam through the sample flask and heat externally to boiling (e.g. Bunsen burner).
- Distil sample until 50ml of distillate has been collected.

Solid phase extraction (SPE) procedure:

- Activate the SPE cartridge by introducing 10ml of methanol.
- Allow to drain, but liquid level should not fall below the top of the resin bed.
- Add 10 ml water and allow to drain.
- Fit a suitable reservoir to the top of the SPE tube.
- Add 4 ml 2 M sulphuric acid to each sample extract.
- Apply sample distillate to the SPE cartridge at a flow rate of approximately one drop/sec.
- Allow to drain but liquid level should not fall below the top of the resin bed.
- Wash the SPE tube using 10 ml 50% v/v methanol/water.
- Allow to drain and evacuate using low vacuum.

- Place the SPE tube in a small oven at approximately 40°C and dry by passing N_2 gas through the tube for 20 minutes.
- Elute the retained material using 2 ml of elution solvent mixture, collecting the eluent in a suitable glass tube or vial with a conical base.
- Evaporate the contents of the sample tube until no liquid can be seen in the tube (care must be taken not to leave tube unattended as losses may occur if evaporation of dried residues is continued).
- Add 0.5 ml 0.1% Tween20/PBS reagent and sonicate to re-dissolve the sample extract.
- Sample extracts should be stored at 4°C and tested within 48 h of extraction.

2.4. Determination of pentachlorophenol (PCP) by enzyme linked immunosorbent assay (ELISA)

Introduction: This ELISA assay is based on competitive antibody binding between the antigen (PCP) and a synthetic antigen-enzyme conjugate produced by chemically bonding horseradish peroxidase (HRP) to a tetrachlorohydroxyphenyl moiety (HRP-PCP conjugate). The retained HRP conjugate is determined using the fluorescent substrate QuantaBlu™ (Pierce, USA) with fluorescence detection. Preparation of the PCP-HRP conjugate is described elsewhere (see Final report to European Commission).

Reagents:
- Phosphate buffered saline solution (PBS) - Dissolve one tablet PBS (Sigma P-4417) in 200 ml of water to obtain 0.01 M phosphate buffer pH7 in saline solution.
- Monoclonal antibodies anti-PCP (Diaclone BL36 -1 mg/ml approx., in storage buffer). Dilute 10 µl of the antibody solution with 10 ml PBS using aseptic techniques, mix.
- PCP-HRP conjugate - Add 1 ml stock solution (containing 0.01 mg/ml approx. PCP-HRP conjugate in PBS) to 10 ml PBS buffer using aseptic techniques, mix.
- PCP stock standard solution (approximately 1000 µg/ml) - Weigh accurately approximately 100 mg PCP solid and dilute in 100 ml pesticide grade methanol.
- PCP standard solution (approximately 10 µg/ml) - Dilute 1 ml 1000 µg/ml PCP stock standard solution to 100 ml in PBS.
- PCP working standard solution (approximately 100 ng/ml) - Dilute 1 ml, 10 µg/ml PCP standard solution to 100 ml in PBS.

- PCP standard solutions - Prepare eight standard reference solutions in PBS by a serial dilution of the 100 ng/ml working standard solution at the following approximate concentrations: 64, 32, 16, 8, 4, 2, 1 ppb. Prepare an identical blank solution vial containing PBS.
- Wash buffer - 0.1% (w/v) Tween 20 in PBS buffer.
- Blocking buffer - 1% (v/v) mouse serum (Dako - X0910) in PBS.
- HRP substrate - QuantaBlu™ Fluorogenic Peroxidase substrate, stable peroxide and stop solution (Pierce - 15169).

Apparatus:
- ELISA micro-plates (Pierce Reacti-Bind™ goat anti-mouse coated 96 well plates).
- Automatic plate washer (e.g. Denley Well-wash 4).
- ELISA fluorescent plate reader (e.g. Perkin Elmer LS50B with plate reader accessory).
- Small polypropylene test tubes, approximately 1 ml volume.
- Tube vortex mixer.
- Micro-plate mixer (e.g. Flow Laboratories Titretek DSG).

Preparation of microtitre plates:
- Wash micro-plates three times using 300 µl wash buffer.
- Remove any residual buffer by inverting the plate and blotting against clean paper towelling.
- Block plate by adding 200 µl blocking buffer per well. Incubate for 2 h at room temperature using the microplate mixer.
- Wash micro-plate three times using 300 µl wash buffer, blot.

Assay procedure:
- Prepare standards and sample extracts for analysis and check pH. This should be approximately pH 7, adjust pH if required using 0.1 M NaH_2PO_4 or 0.1 M Na_2CO_3.
- Pipette 200 µl of each standard and sample solution into a small tube.
- Add 200 µl diluted HRP-conjugate solution and mix using the vortex mixer.
- Prepare blanks by adding 200 µl PBS buffer and sample to a separate tube.
- Pipette 100 µl of each standard, blank or sample solutions into each well. All measurements must be performed in triplicate.
- Cover with adhesive strip or plastic cover and incubate for 1 h using micro-plate mixer.
- Wash micro-plate three times using 300 µl wash buffer, blot.

- Add 1 ml of the QuantaBlu^{TM} stable peroxide to the QuantaBlu^{TM} substrate solution and mix using the vortex mixer.
- Pipette 100 µl of the mixture into each well, cover and incubate for 1 h using the micro-plate mixer.
- Add 100 µl QuantaBlu^{TM} stop solution and mix for 1 min using the micro-plate mixer.
- Measure the emitted light from each well using the Perkin Elmer LS50B fluorescent plate reader or similar using the following instrumental conditions: Excitation wavelength 325 nm; Excitation slit width 8.0; Emission wavelength 420 nm; Emission slit width 8.0; Emission cut off filter 390 nm.
- Inspect results and average all triplicate data values.
- Plot calibration graph using semi-log graph paper or a suitable computer based data analysis program. Inspect graph and determine concentrations of blank subtracted samples from the graph. Correct for any dilution factors etc. (Figure 1a).
- Check sample blanks to identify any sample extracts with natural fluorescence. (Samples with positive blanks may contain residues of fluorescent contaminants such as polycyclic aromatic hydrocarbons. Confirm results using an alternative analytical method).

2.5. Determination of pentachloroanisole (PCA) by enzyme linked immunosorbent assay (ELISA)

Procedure: The basis of the assay is similar to that employed for PCP, described above, except for the fact that PCP related components are substituted by corresponding PCA components. Anti-PCA antibody was produced by Diaclone SAS (Diaclone B-R28 -1mg/ml approx., in storage buffer). Preparation of the PCA-HRP conjugate followed a six-stage synthesis from 2,3,5,6 tetrachlorohydroquinone and is described elsewhere (Köhler and Milstein, 1975).

The sample to be tested and the PCA-HRP conjugate are initially mixed and applied to the wells of a microtitre plate coated with anti-PCA antibodies. The sample PCA and PCA-HRP conjugate are incubated for 30 minutes and compete for the available antibody sites. If the sample contains no PCA then the antibody sites will be fully occupied by the PCA-HRP conjugate. If some PCA is present in the sample proportionally less of the conjugate is bound. Unbound materials are then removed by washing and the substrate for the

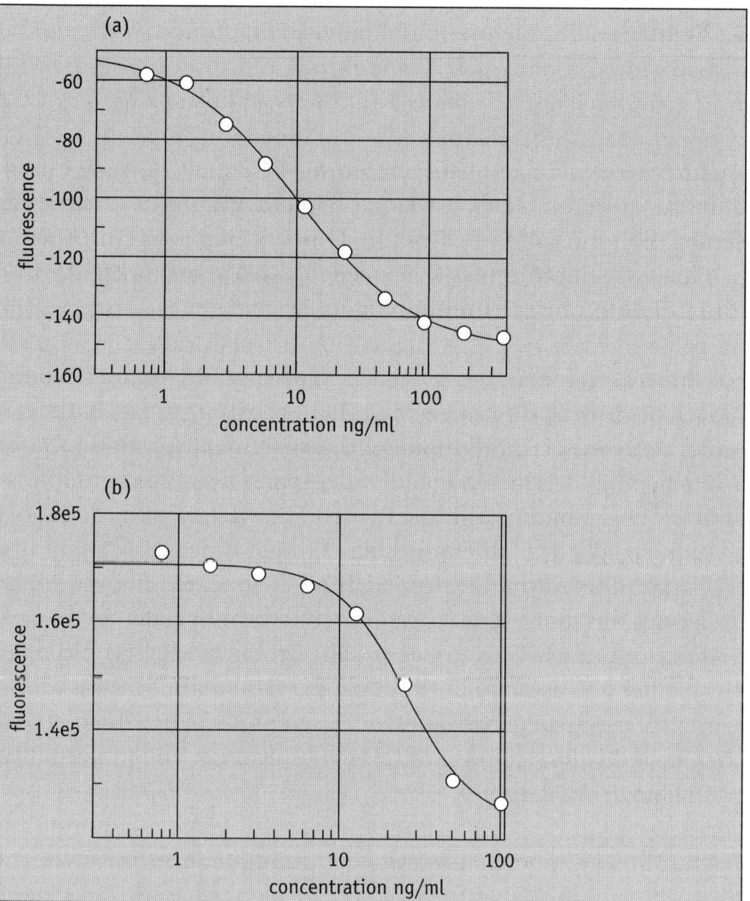

Figure 1. Standard ELISA calibration graphs for (a) PCP and (b) PCA.

reporter conjugate label is added to each well. In this way an inverse concentration dependent relationship is established that may be used to determine PCA concentrations in sample extracts (Figure 1b).

3. Results

3.1.1 Generation of the immunogens and mouse immunisation

Due to the small size of the molecules PCP or PCA, they cannot initiate an immune response. For these small haptens, common carriers such as Bovine Serum Albumin (BSA) or Keyhole Limpets Hemocyanin (KLH) and a spacer of 4 to 5 carbon molecules were needed to make the molecules more accessible for recognition by the immunological system.

For this reason the antigens needed to be specially synthesised for this project (data not shown). It was decided that the chloride molecule at C-4 for both molecules was sacrificed to allow the attachment of the spacer to C-4 on the aromatic ring. On the other end of the spacer a carboxyl group was synthesised to allow cross-linking with the carrier. The starting point for the synthesis was a commercially available molecule with chloride at the positions 2,3,5,6 and NH_2 at position 1 on the aromatic ring. It appeared that the synthesis of both molecules was much more complicated than anticipated in the beginning. Four different approaches gave intermediate products which could not be processed any further downstream until finally the right approach was found. The whole synthesis requires actually twelve different steps of which step twelve is the change of the OCH_3 group to OH on position 1. The molecules were ready to conjugate to the carriers.

The coupling agent 1-ethyl-3-(3-dimethylaminopropyl) carbodiimide hydrochloride (EDC) was used to conjugate the hapten (PCP or PCA) with the carrier (KLH or BSA). The carbodiimide initially reacted with available carboxyl groups on the hapten to form an active O-acylurea intermediate. This unstable intermediate reacted with a primary amine to form an amide bond with the release of a soluble urea derivative. The optimum format for immunogen preparation was KLH conjugated pentachlorophenol (PCP-KLH) and BSA conjugated pentachloroanisole (PCA-BSA).

Five mice were immunised for each target compounds: PCP-KLH or PCA-BSA. Spleens from immunised mice were collected.

3.1.2 Monoclonal antibody production

The fusion between splenocytes and the myeloma cell line X6.3 AG8 has been done. All the spleens have been used. Five fusions were done for PCP (F452, F453, F455, F456, F457) and three for PCA (F460, F461, F462).

The screening was done in an ELISA format with antigens coated onto the plate. In the first selection, the hybridoma cells secreting Abs to PCP-KLH (immunisation PCP) or to PCA-BSA (immunisation PCA) were selected. In the second selection, using the irrelevant conjugated antigens such as IL-7-KLH, or MTBE-BSA, the hybridoma cells secreting Abs which only react with PCP-KLH and/or with PCA-BSA were selected.

Following the PCP immunisation, two hybridomas among 8478 hybridomas were selected and cloned. These mAbs of IgG1 isotype were called B-L36 and B-P31. They recognised the conjugated antigen PCP-KLH. A moderate reactivity was observed with PCA-BSA, whereas no reactivity was observed with the irrelevant antigen MTBE-BSA (Figure 2). Similar results were obtained with B-P31 (data not shown).

Three hybridomas among 6431 hybridomas tested were selected following the PCA immunisation. After cloning, the mAbs B-A46, B-G40 and B-R28 were available. Whereas no reactivity was obtained with PCP-KLH, these Abs only recognise the conjugated antigen PCA-BSA (Figure 3).

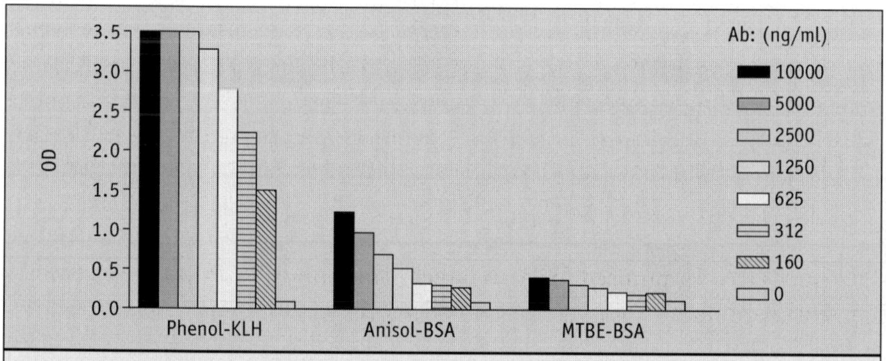

Figure 2. Detection of PCP-KLH with B-L36 mAb. The reactivity of the mAb B-L36 was tested with the antigens PCP-KLH, PCA-BSA or MTBE-BSA immobilised on the plastic surface by ELISA.

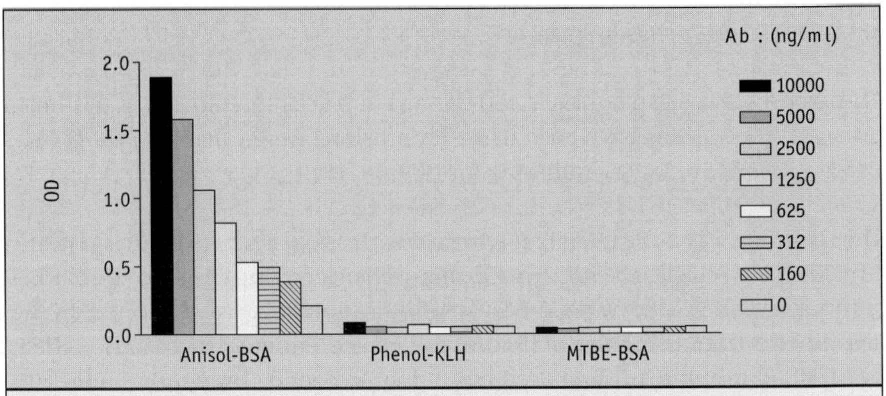

Figure 3. Detection of PCA-BSA with B-G40 mAb. The reactivity of the mAb B-G40 was tested with the antigens PCA-BSA, PCP-KLH or MTBE-BSA immobilised on the plastic surface by ELISA.

As shown in Figure 4, the binding of B-L36 to the coated PCP-KLH was strongly inhibited in the presence of free PCP. A weak inhibition was obtained with the highest concentration of PCA. Inversely, the binding of the mAb B-G40 strongly decreased in the presence of PCA.

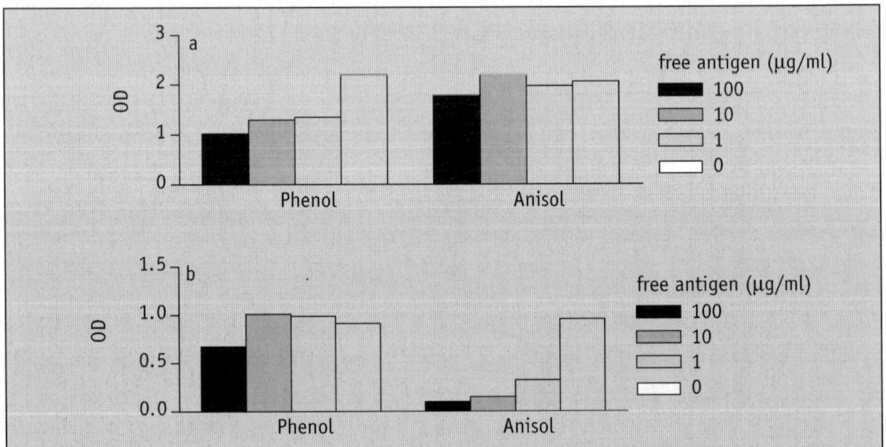

Figure 4. Detection of the free PCP and PCA. The reactivity of the mAbs B-L36 (Figure 4a) and B-G40 (Figure 4b) to free PCP and PCA were analysed by indirect competitive enzyme immunoassay.

3.2.1 Determination of chlorophenols in commercial wines by gas chromatography

Wine samples (65 bottles of commercial wines) were analysed. After preparation of wine samples (as is described in materials and methods section), extracts were analysed with a Hewlett-Packard HP-5890 series II gas chromatography with a electron capture detector and mass selective detector.

Results obtained indicate that all wine samples had trace levels of any chlorophenol or chloroanisole. As indicated in Figures 5-6 most of the wine samples contained pentachlorophenol at a concentration from 16.7 ng/L to 474 ng/L. Moreover, samples presented 2,4,6-TCP at levels from 5 to 712 ng/L.

3.2.2 Determination of chloroanisoles in commercial wines by gas chromatography

Results obtained for chloroanisoles (Figures 7 and 8) indicate that the principal chloroanisoles presented in wine samples are 2,4,6-trichloroanisole, 2,3,4,6-tretrachloroanisole and pentachloroanisole. Most of them contained pentachloroanisole at a concentration from 5.3 to 256 ng/L and 2,4,6-trichloroanisole at levels from 5.3 ng/L to 14 ng/L.

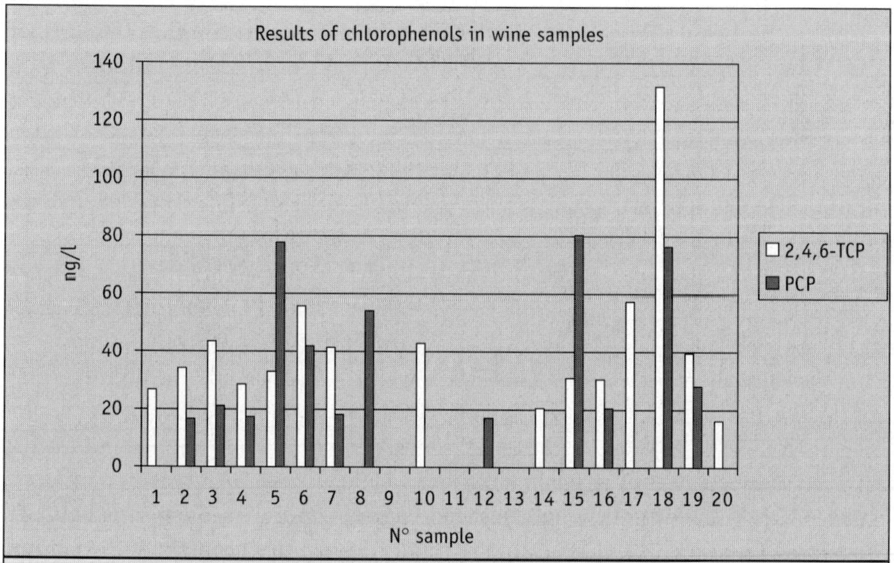

Figure 5. Concentrations of endogenous chlorophenols found in commercial bottled wines (ng/L) analysed by GC/MS (samples n°1-20).

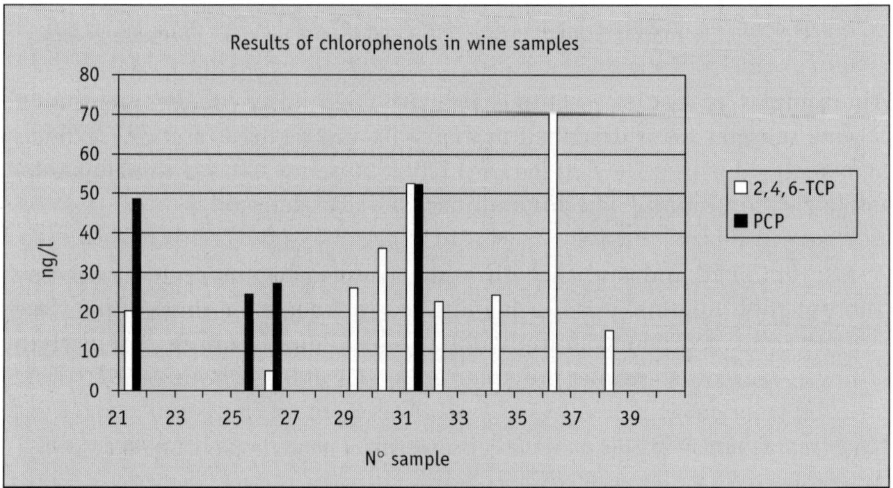

Figure 6. Concentrations of endogenous chlorophenols found in commercial bottled wines (ng/L) analysed by GC/MS (samples n°21-40).

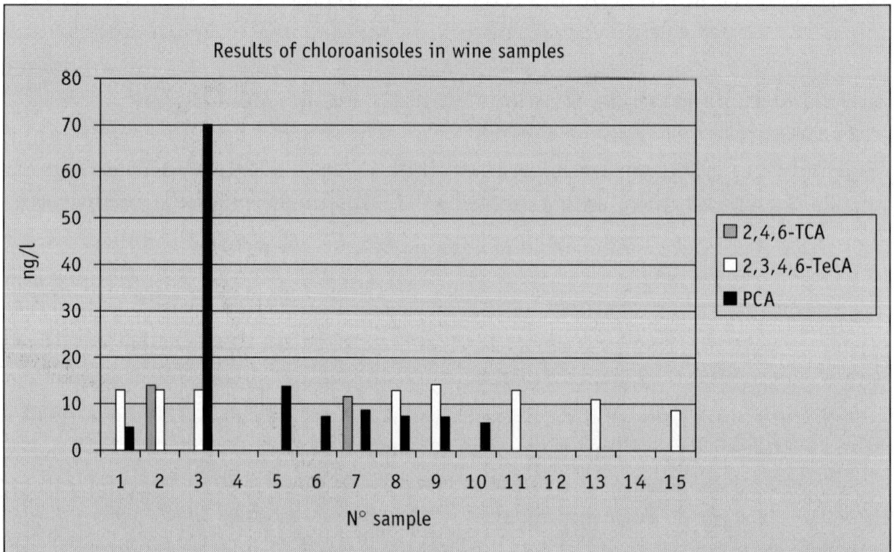

Figure 7. Concentrations of endogenous chloroanisoles found in commercial bottled wines (ng/L) analysed by GC/MS (samples n°1-15). Samples 02 contained 256 pg/ml of pentachlorophenol and it is not represented in the Figure.

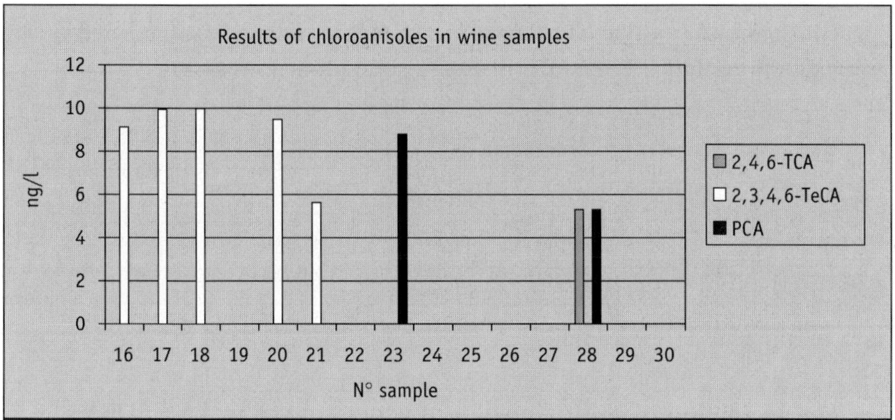

Figure 8. Concentrations of endogenous chloroanisoles found in commercial bottled wines (ng/L) analysed by GC/MS (samples n°16-30).

Samples number 02 and 03 were seen as tainted by consumers, these wines contained high levels of pentachloroanisole, 256 and 70.13 ng/L, respectively. Furthermore, sample 02 contained also 14 ng/L of 2,4,6-trichloroanisole.

3.3. Wine analysis: PCP ELISA versus gas chromatography

Samples were extracted using method P1 or P2. Extracts were analysed by the standard GC procedure and by the PCP ELISA method. Results are shown in Tables 1 and 2 (P1 and P2 extraction methods, respectively).

Table 1. Comparative GC and ELISA analyses of PCP in wine samples prepared using extraction procedure 1 (pentane extraction, sodium carbonate back extraction).

Wine	Type	GC (ng/ml)	ELISA (ng/ml)	ELISA +Spike (0.14 ng/ml)
Calvet	Bordeaux	0.14	0.12	0.20
La Foret Hilaire	Bordeaux	0.20	0.20	0.24
Melini	Chianti	0.19	0.14	0.20
Marques de Grinon (A)	Rioja	0.14	0.16	0.20

Table 2. Comparative GC and ELISA analyses of PCP in wine samples prepared using extraction procedure 2 (steam distillation + solid phase extraction).

Wine	Type	GC (ng/ml)	ELISA PCP (ng/ml)	ELISA +Spike (0.10 ng/ml)
Campo Viejo	Rioja	<0.02	<0.04	>0.1
Marques de Grignon (B)	Rioja	<0.02	<0.04	>0.1

3.4. Determination of chloroanisoles in wine samples by PCA ELISA

Preliminary examinations indicate that ELISA performance is similar to that obtained by the PCP assay but slightly less responsive over the range 1 to 10 ng/ml PCA. The effective working range of this assay appears to be from 5.0 to 50 ng/ml, approx. five-fold less sensitive than the PCP assay.

4. Discussion

4.1 PCP ELISA

The monoclonal antibody anti-PCP used in this assay was B-L36, developed during the FOODSAFE project by partner Diaclone SAS, whilst the novel PCP-HRP conjugate was synthesised by Cranfield University. Use of the HRP enzyme label allowed high sensitivity detection using the fluorescent HRP substrate QuantaBlu. The assay format was that of an indirect competitive ELISA performed within microtitre plates. The effective working range of the PCP-ELISA was found to be 1.0 to 10 ng/ml.

Whilst the literature widely supports the concept of performing ELISAs in minimally pre-treated samples, it was realised at an early stage in the programme that the complexity of the wine matrix, coupled to the low analyte detection limits demanded by the end application, that some form of sample preparation was required. The extraction methods developed served the dual purpose of both pre-concentrating and purifying the samples and were designed within the remit that the methods should be within the scope of a competent chemical technician operating within a simple winery laboratory using a readily available low-cost kit.

A number of extraction methods were developed during the programme, specifically aimed at wine samples, cork materials, other packaging materials (cardboard, wooden crates, wooden pallets, paper filters, pine wood and wine bottle washings). The scope of the extraction methods was purposely broad, supporting the general aims of the FOODSAFE programme in developing rapid at-line methods for traceability studies with respect to the chlorophenol and chloroanisole contamination of wine.

Table 1 indicates good agreement between the GC-MS and ELISA methods for total chlorophenol content of the wine extracts, indicating the efficacy of the P1 extraction method coupled to the simple ELISA test for chlorophenols determination. Spiking of the P1 extracts and re-analysing by ELISA indicates that there are matrix effects within the wine extracts that affect ELISA performance. The PCP ELISA data indicated that PCP levels in extracts spiked to an additional 0.14 ng/ml PCP had been elevated by only 0.04-0.08 ng/nl. These lower than expected recovery values indicate that the assay is best used for the semi-quantitative determination of trace level chlorophenol contamination of wine, or material used in the wine making process. Given that most small wineries do not have routine access to expensive state of the art analytical facilities, a method that is able to identify an appreciable chlorophenol contamination (0.14 ng/ml in this example) does have value. Current research efforts are focused on improving assay reliability, primarily by reduction in matrix interference effects through further development of the sample preparation processes. It is not intended for this assay to replace standard laboratory analytical method; rather it is to place technologies in the hands of end-users where no current viable alternative exists.

Wine extracts prepared by the steam distillation method (Table 2) and containing very low levels of chlorophenols yielded negative values by both the GC and ELISA methods. The 2 wine extracts were then spiked to an additional 0.1 ng/ml PCP and re-assayed. In both cases, the ELISA method was able to identify this increase in contamination level. The spiked wine samples gave optical densities slightly in excess of the 10 ng/ml standard used, hence the reported values of >10 ng/ml.

4.2 PCA ELISA

The monoclonal antibody anti-PCA used in this assay was B-R28, produced within the FOODSAFE project by Diaclone SAS, whilst an entirely novel PCA-HRP conjugate was also synthesised within this project by Cranfield University. The ELISA principle and format was identical to that described for the PCP ELISA described above (with the obvious addendum that PCA specific components replaced the PCP-specific counterparts). The working range of the PCA ELISA was from 5.0 to 50 ng/ml, equivalent to a five-fold reduction in detection limit compared to the PCP ELISA.

Preliminary work on the performance of the PCA ELISA method for wine, cork and wood based packaging materials indicates that matrix interference, as with the PCP ELISA, is also a major issue to be addressed. To enable the PCA ELISA to be utilised with 'real-world' samples, alternative sample extraction methods have been evaluated for wine, corks and packaging materials, again with the proviso of applicability within a simple winery laboratory.

The extraction of wine samples for PCA analysis using steam distillation followed by solid phase extraction or steam distillation with pentane back extraction was also used in conjunction with the PCA ELISA test method. Results obtained by testing wine directly or using a single SPE of liquid/liquid solvent extraction were considered to provide falsely elevated results. Further work is required to investigate the robustness of the procedure for different sample types.

5. Conclusion

The contamination of wine by chlorophenols and chloroanisoles can have several origins. Application of good manufacturing practices and standard operating procedures should allow accurate control of wine processing at different levels: the local environment and facilities, production issues (raw materials, equipment, winemaking, packaging operations), storage and distribution.

The implementation of immunodiagnostic tools detailed here will reduce the risk of chemical contamination by chlorophenols or/and chloroanisoles. It has been shown that, with appropriate low-cost and simple sample preparation methods, it is possible to use instrumentally simple ELISA procedures to identify chlorophenols in materials associated with the wine production industry. The approaches described are not intended as a replacement for standard gas chromatographic procedures, but rather are intended to allow smaller wineries to use low-cost simple methods for identifying possible contamination problems during the wine production process.

The findings presented here are within the context of the wider FOODSAFE project, which was aimed at addressing traceability within food processing operations. By analysing the whole food (in this case wine) production process, the ingress of contaminants into the food production process can be identified. Measurement of contaminants in the final product has benefits, but the product is unusable. The rationale behind FOODSAFE has been to identify contamination in the raw materials, which in many cases may be excessively high. The methods described here should be within the capabilities of small concerns and competent technicians, show considerable promise for acting as the first line of defence in assessing raw material quality and thus protect both the producer and consumer alike with respect to food quality and safety from 'the farm to the fork'.

Acknowledgements

This project is a shared cost RTD project, supported by the European Commission under the Framework 5 programme. (Contract number: QLK1-2000-00518). We would also like to express our gratitude to our Scientific Officer, Mrs. Alkmini Katsada.

References

Final report to European Commission, 'Rapid Assessment of Food Safety via Novel At-line Measurements', November 2003. QLK1-2000-00518.

Köhler, G. and C. Milstein, 1975. Continuous cultures of fused cells secreting antibody of predefined specificity. Nature 256, p. 495-497.

New and future at- and on-line sensors in food production: EU research results

F. Holm
FoodGroup Denmark, Sciencepark Aarhus, Gustav Wieds Vej 10, DK-8000 Aarhus, Denmark

Abstract

Quality control is essential in the food industry and efficient quality assurance is becoming increasingly important. European food regulations, directives and standards are expanding fast, requiring full control of quality and safety before release of the products. This implies that on-line / real-time technologies are essential in the future food factory. This challenge has been adopted within the EC Framework Programmes, which has supported about 20 research projects aimed to develop new sensor technologies for food quality and safety. Among the most promising principles are electromagnetic sensors, several biosensor types, sensors based on "artificial antibodies" (MIT), "lab on a chip" sensors, and multi-sensor systems using more orthogonal sensor types, often non-specific sensors.

Keywords: on-line sensors, food quality, EU Framework Programmes

1. Introduction

Quality and safety control is essential in the food industry and efficient quality assurance is becoming increasingly important. Consumers expect adequate quality at a given price, good shelf-life and high safety while food inspections require good manufacturing practices, safety, labelling and compliance with the regulations. Further, food producers are increasingly asking for efficient control methods, in particular through on-line / real-time or at-line quality sensors, firstly to satisfy the consumer and regulatory requirements and secondly to improve the production feasibility, quality sorting, automation and reduction of production cost and production time (increased throughputs).

Thus, all three drivers of on-line quality control, consumers, authorities and food producers, have great interest in the development of new sensing systems which are beyond the traditional on-line technologies, like control of weight, volume, temperature, pH, viscosity, colour and appearance.

Additionally, many new food safety concepts and key quality parameters have arisen during the last decade: Hazard Analysis Critical Control Points (HACCP), Total Quality Management (TQM), ISO 9000 certifications, traceability, and authentication all require improved control methods. Besides, pathogenic microorganisms, BSE, GMO and many pollutants are specific safety problems, which require intensive control, data logging and data treatments. And they all call for in-time and on-line sensors for control, new data systems, warning systems, tight feedback loops for automation of the production, etc..

The great challenge is indeed to develop new and cheap real-time sensors and data systems surveying processes and products, controlling the automated process and the raw material stream, but also sensing the final product quality and typing the product labels, nutritional and health information, and much more.

Today, the most important quality parameters and concepts in food production control are:

- sensory quality (appearance, flavour, taste, texture, stability, etc.);
- nutritional quality, including health implications, such as "high in fibre", "lower in cholesterol", "GMO free", food allergens;
- composition and labelling, including additives, quality claims, ethical claims (like ecological information);
- pollutants, including environmental pollutants, veterinary drugs, agricultural chemicals, BSE-prions and mycotoxins;
- foreign body detection, such as stones, glass or metal;
- microbial safety, in particular *Listeria, Salmonella, Campylobacter*, *E. coli* and *Yersinia;*
- shelf-life (microbial, sensory, chemical, sterility testing, F_0-values);
- production hygiene, cleaning, contamination;
- HACCP, traceability and authentication (from 2005);
- process parameters: machine settings, temperature, pressure, flow, aseptic conditions and many physical parameters;
- package control: integrity, pin holes, gas permeability, migration control.

The great challenge for the future food factory is a stringent real-time control system surveying, reporting and controlling the raw materials, the process and the final product quality and safety. This challenge has been adopted by the EC Framework Programmes, offering opportunities for new scientific progress in sensor-technologies, e.g. biosensors, electromagnetic sensors and sensors based on new nanotechnologies. Thus, the EC has supported during the last years more than 20 research projects within the area, often resulting in new prototypes tested successfully in industry, and new initiatives are expected within the Framework Programme 6.

2. Food sensor types and the sensor market

In sensor sciences, the ideal sensors are often the very elegant nature's sensors, e.g. sensing the odour, where chemicals are bound to the surfaces of sensing cells, implying an electrical signal (the bio-transducer) which is recognised and interpreted (learning by doing) by special cells (e.g. in the brain). But in artificial sensors we often need much more details and recognition which is not delivered by the nature's sensors, e.g. we need specific molecular information on pollutants, identification of microbial species, nutritional information and forecasts of quality changes.

Artificial food quality sensors may be classified in different ways according to their operation principles, e.g. biological, chemical and optical sensors, but recently a European Concerted Action called *ASTEQ* suggested the following classification:

- Biosensors, incorporating a **biological** material, like enzymes, cell components or antibodies.
- Sensors based on an **electric** signal, for example Potentiometric Chemical Sensors, Metal Oxide Semiconductors (MOS), Field Effect Transistors (FET) or Conducting Polymer Sensors (CPS).
- Sensors based on interaction by **electromagnetic** waves, in particular sensors using visible, ultraviolet and infrared (NIR, NIT, FTIR, Thermography) waves, microwaves, radiowaves, X-rays and high frequency waves (Nuclear or Electronic Magnetic Resonance).
- Sensors based on interaction with **ultrasound** waves (100kHz-1MHz).
- Sensors based on variations in **frequency**, for instance Quartz Crystal Microbalance (QCM) and Surface Acoustic Wave (SAW).
- Sensors involving **selective agents**, such as molecular films or complexing films.

However, because of the large market potential and fast market growth, many new sensor types and concepts are under development, which may need an extended classification. In particular, the new opportunities within nanotechnologies are very promising, but also other sensor types are a part of the future, e.g. the "Lab On A Chip" concept which may include many operations, e.g. concentration as well as measuring, and the multiple sensing systems using several orthogonal sensors in combination with chemometric methods and artificial intelligence.

The global market for sensors has been estimated by Sensor Technology Centre, Copenhagen (Lading, 2002) to 50-200 Billion € / year, depending on the definition of a sensor. Further, the yearly growth rate is estimated to about 15%, but the food and agricultural area only constitute a minor part of the market. The same author has also estimated the market foresight for major market sectors (Figure 1), showing a strong future growth rate for the food sensors.

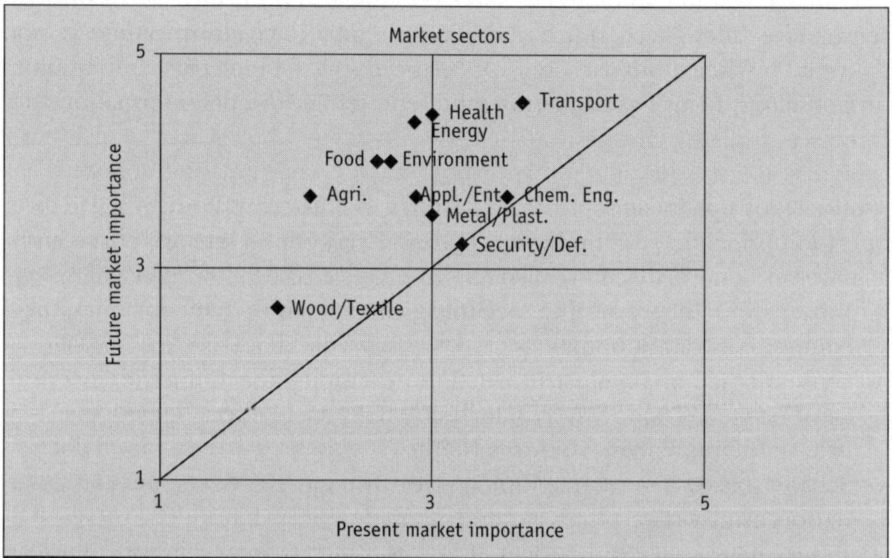

Figure 1. Sensor foresight, market (Sensor Technology Centre, Copenhagen).

3. EU food research - new findings

About 20 research projects funded by the European Commission within the 4th and 5th Framework Programmes deal with the development of new, efficient and real-time sensors, measuring important quality characteristics of foods. They all belong to the classes mentioned above. However, the majority focus on sensors based on the interaction of foods with electromagnetic waves (11 projects) or on electronic noses / tongues, which are arrays of non-specific sensors, often MOS, FET or CPS sensors (4 projects). In the following section, the EU research projects are briefly introduced. References to the scientist co-ordinating the work as well as to the websites of the projects (where existing) are also listed.

3.1 Electromagnetic sensors (EMW sensors)

Electromagnetic sensors may be classified on the basis of the wavelength of the electromagnetic waves applied. Sensors exist, which use: visible (400-700 nm), ultraviolet (10nm-400 nm), infrared (700nm-30.000 nm: NIR, NIT, FTIR, MIR) waves, microwaves (1-10 cm), radio waves (1-10 m), X-rays (100 pm-1 nm) and high frequency waves (Magnetic Resonance, MHz-GHz).

Each class may be further sub-divided according to the molecular information obtained through the interaction. Thus, infrared waves may be subdivided into near-infrared (700-2500 nm), mid-infrared (2500-30000 nm), far-infrared (up to 1,000,000 nm), and thermography (1-15 μm), which all extract different information from the molecules (sample) interacting with the light. We may also classify the sensors according to their precise type of interaction: absorbance, transmittance or reflectance.

Sensors based on an interaction with electromagnetic radiation have been on the market for many years, in particular for laboratory purposes. On-line examples are also numerous: X-rays for foreign body detection, visible light sensors for colour sensing or machine vision inspections, near-infrared sensors for quality inspection and temperature measurement, or microwave sensors for measurement of water content.

During the last 10-20 years, the scientific and technological challenge has been to develop simple, cheap, stable and on-line EMW sensors, in particular optical sensors. Another goal has been to develop advanced mathematical models, statistical (chemometric) methods, and efficient algorithms in order to extract and utilise the food quality information included in the transmitted or scattered signals from the food in question. A third important trend has been the construction of chips and advanced biosensors including biological components such as enzymes, antibodies or cell components, often using EMW detection or electrochemical detection. Most EU-sponsored research work has focused on the NIR/VIS sensors, SPR sensors (Surface Plasmon Resonance) and NMR sensors (pulsed and low resolution), but work has also been done on fluorescence sensors, MIR and Raman sensors, Fourier Transform NIR sensors, thermography based sensors and sensors based on combinations of two or more sensor principles.

Within the completed Concerted Action ASTEQ, scientists focused on several sensor principles measuring fruit quality, in particular internal and external quality and flavours. The aims were to develop, extend and enhance the value of instrumental methods in order to determine consumer-perceived macroscopic quality factors. The selected instrumental techniques were 1) vision techniques in the ultraviolet, visible and infrared domains; 2) imaging techniques such as NMR, ultrasonic, or microwave tomography; 3) various artificial noses. Among the results were new and improved software and methodologies, and improved information extraction by fusion of two or more technologies. They also studied the efficiency of:

- NMR pulses for measurement of water, oil, water activity and ripeness of apples;
- NIR/VIS sensors for measurement of mealiness, bruises (3-5 minutes), ripeness, glassiness and respiration rate of apples (transmittance, reflection and Fourier-transform spectra);
- VIS-fluorescence sensors for mealiness, freshness, and ripeness of apples;
- thermography (about 10 μm range) measuring mechanical damage, maturity and microbial infestation of several fruits;
- acoustic impulse sensors for measurement of ripeness, mealiness and for fruit grading, using the attenuation of acoustic impulses;
- electronic noses for measuring fruit quality (discussed below).

The NIR/VIS region has been used in several other projects. Thus, in the GLOVE project, scientists designed a glove-shaped apparatus equipped with various miniaturised sensors providing information on fruit quality parameters, namely, sugar content, maturity, mechanical properties (firmness, stiffness) and internal colour. The sugar content and internal colour are measured by a miniaturised spectrometer (NIR/VIS) coupled with optical fibres. A sound sensor measures the mechanical properties and the size is measured by a potentiometer placed at the hand aperture. These sensors are coupled to a microcomputer that delivers processed information about the fruit quality, based on previous learning of the variety and quality classes. The weight of the glove prototype is 400 g and the electronics are held in a rucksack weighing 1000 g. The glove may be used preharvest to control the growth and the harvest date, at harvest to select specific qualities or post-harvest to control and measure quality.

In the OPTIVEG project, scientists worked on methods using chlorophyll fluorescence and reflectance in the visible and near infrared spectrum. The quality factors dealt with were the mechanical properties of green beans, broccoli and carrots. They also used the sensors for measuring the effect of pasteurisation of the vegetables, by measurement of the remaining enzymatic activity, in particular lipoxygenase and peroxidase activities.

In the DIFFRUIT project, Time-resolved Diffuse Reflectance Spectroscopy (TDRS) in the NIR/VIS range was used to measure internal quality of fruit and vegetables. The project group developed statistical models for the relationship between the TDRS signals and the firmness, sugar, and acid content of kiwifruit, tomato, apple, peach, nectarine and melon. They also developed classification models to sort apples, peaches, kiwifruits and tomatoes into 3 quality classes. Using a pulsed laser diode (70-200 pico-seconds/pulse), the measurement time was 100 milliseconds. The absorption coefficient was related to the tissue constituents, while the scattering coefficient was related to firmness and fibre content. The equipment price was estimated at around US$ 25,000.

Finally, in a new project called NIQAT, real-time sensors in the NIR/VIS range are used to measure quality traits such as internal diseases and defects, maturity or flavour, and to sort potatoes, apples and peaches on the basis of their quality.

Surface Plasmon Resonance (SPR) is another optical phenomenon used in a number of new sensors, often in biosensors involving antibodies or enzymes (Rand *et al.*, 2002). The optical range used is most often in the visible part of the spectrum, but may also be in the NIR range. Traditionally, SPR devices detect minute changes in the refractive index of the sensing surface and its immediate vicinity. They may be constructed by a diffraction grating or prism on a glass slide or an optical waveguide carrying a thin metal layer, for instance gold. The metal layer carries a sensitising layer, like immobilised antibodies or other layers binding the analyte specifically, which is in contact with the sample. In the device, a collective excitement of electrons in the metal film occurs which leads at a specific wavelength to a total absorption of light at a particular angle of incidence. This angle is dependent on the refractive indices on either side of the metal film. Molecules binding to the sensitising layer change the refractive index and this changed angle of total absorption is measured and correlated to the concentration of the analyte (Figure 2).

The SPR technique has been used in FOODSENSE. In this project a methodology for on-line/at-line detection of veterinary drug residues (hormones and antibiotics) in dairies (milk stream) and slaughterhouses (bile or urine, including a commercial handling robot) has been developed and demonstrated. Clenbuterol and ethinyl-estradiol in bovine urine, sulfamethazine (SMT) and sulfadiazine (SDZ) in porcine bile, and SMT, SDZ and enrofloxacin in milk have been detected by the technique. The biosensor operates in real time and it may detect up to 8 different veterinary drugs simultaneously with a throughput of up to 600 samples per day. The project participants have established a new company to produce and further develop the sensor systems, and several new and elegant designs of SPR sensors are under development by other groups.

The SPR sensor principle has also been used in the MICROSENSOR project. The sensor developed has been applied to the quantification of mycotoxins, *Listeria* and markers for growth hormones (recombinant bovine somatotrophin, rBST). The sensor technology developed uses a miniature electrophoretic system (mFFE) that separates and concentrates the analyte in question, using a number of electrophoretic methods: preparative zone, interval zone,

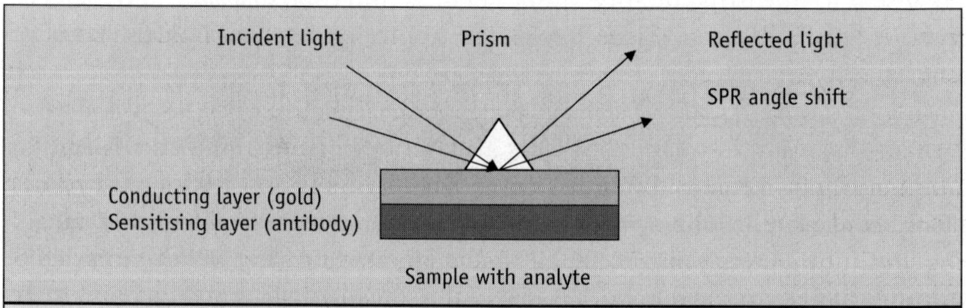

Figure 2. Principle of Surface Plasmon Resonance (SPR).

isotachophoresis, isoelectric focusing. The design of the mFFE system may be a plain glass substrate of 1.5 mm in thickness and a cross-linked polydimethylsiloxane (PDMS) layer with micromachined sample channels. The central separation chamber (12 x 4 x 0.15 mm) is connected to 34 inlet channels for sample injection and 36 outlet channels for sample collection. A syringe pumps the samples perpendicularly to the electric field (Figure 3). This example is an excellent example of a Lab On A Chip sensor type, which is an important step forward. The detector unit may be based on several principles. In the case of *Listeria*, the detector unit may be a well-known ATP luminescence detector. For other analytes, the SPR detector system may be used, immobilising a biospecific layer, for example an antibody (Patel, 2000).

Because of the cost and complexity of production of many sensitising layers in SPR technology and other biosensors, material science has developed new and much cheaper sensitising materials based on plastic polymers. Molecular Imprinted Polymers (MIP) are polymers produced by imprinting recognition sites of predetermined specificity in cross-linked synthetic polymers. The sites are tailor-made *in situ* by co-polymerisation of functional monomers with cross-linkers in the presence of print molecules, often called template molecules. After the polymerisation, the print molecules are extracted from the polymer, leaving recognition sites complementary to the print molecule in the positioning of the functional groups and in shape. The polymer is subsequently able to selectively re-bind the print molecule (Mosbach and Ramström, 1996). These kinds of sensor materials have been called "artificial antibodies" (Figure 4). The MIP technology has particular strengths for smaller molecular analytes up to about 400 Dalton.

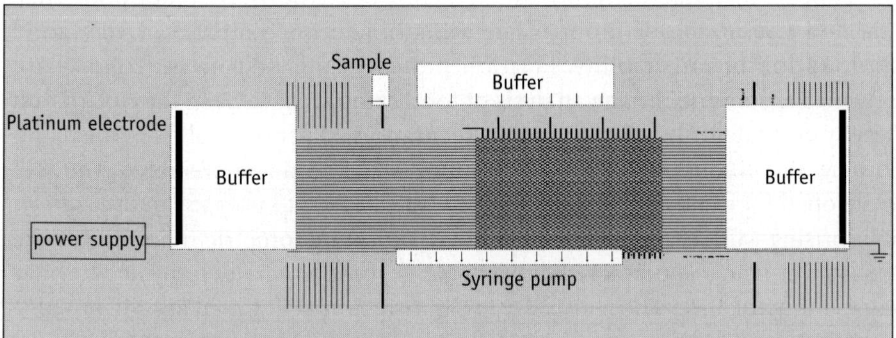

Figure 3. mFFE layout (MICROSENSOR project).

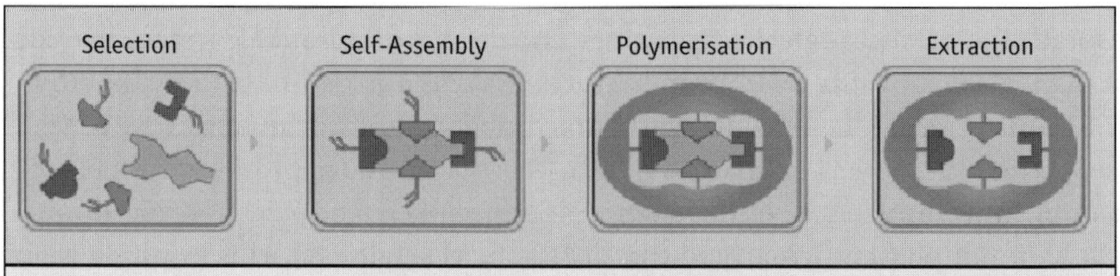

Figure 4. Molecular imprinting technology. Principle: The analyte and suitable plastic monomers are mixed and polymerised followed by extraction of the analyte. The fixed analyte sites are then ready for binding analyte in food samples.

This technology has been used in the CREAM project, the objective of which is to develop and optimise a plug-in detection cartridge supporting a molecularly imprinted polymer assay. The technology may be used to bind and detect many chemicals polluting our foods, like pesticides and veterinary drugs in meat and dairy products. In the project, the focus is on at-site detection of the different β-lactam antibiotics in milk. The sensor consists of a microfabricated column accommodating an optical detection window. Molecular imprinted polymers in the form of beads are used as packing materials and recognition elements. Analyte binding is detected by fluorescence.

The same MIP technology was used in another project, the overall objective of which was to develop novel and robust MIP-based techniques, in combination with sensor technology, for real-time measurement of food contaminants, in particular β-lactam antibiotics and clenbuterol. The results of the study indicate that MIPs can be used to prepare both selective and general recognition matrices for either individual analytes (like penicillin) or groups of compounds (like the β-lactam antibiotic group). The MIPs developed have shown very good reproducibility and stability. Practical protocols for their use as solid phase selective absorbents for the analysis of food samples have been developed and briefly evaluated. Specific MIP-based separation of clenbuterol from meat and kidney, and of β-lactam antibiotics from milk has been demonstrated. The MIP developed for clenbuterol has been successfully used to prepare a novel sensor comprising MIP as the selective element and amperometric detection as the transducer. The responses from several such sensors were determined to have a variability of 10%. The feasibility for an oxacillin MIP-based sensor was also demonstrated.

Among the optical sensor systems developed and demonstrated in industrial environments are machine or artificial vision. In the CRAFT project NIPCO, the main achievement was a system for olive sorting, using a traditional vision camera and three CCD colour sensors, sensing shape, size and colour. A new algorithm allowed the olives to be sorted into 4 classes with a speed of 132 olives/sec and 6 images/sec.

In the STAS project, scientists used Mid-Infrared (MIR) spectroscopy as well as Raman scattering on-line for quality assessment in bakeries, breweries, dairies and the fruit industry, but also investigated the market needs.

Finally, NMR technology is used in the project QTEPACK. The objective of this ongoing project is to develop and test innovative, low cost nuclear magnetic resonance techniques and fruit handling mechanisms for testing the quality level of horti-fruit products in the grading and packaging line. The participants focus on early detection of decay due to fungal rots, for example by *Penicillium digitatum*, in an orange packaging line, measuring the frequency displacement of the NMR proton line.

3.2 Electric and frequency sensors

Electronic noses and electronic tongues are the common names of sensors responding to flavour/odour (volatiles) or taste (solubles), using an array of simple and non-specific sensors and a pattern recognition software system. Historically, the sensors used were advanced mass spectrometers or gas/liquid chromatographs producing a fingerprint of the analyte for recognition. Nowadays, these sensors have been substituted by an array of up to 25 simple electrical and/or frequency sensors or sensors measuring changes in voltage or frequency as a response to the food contact. Essentially, each odour or taste leaves a characteristic pattern or fingerprint on the sensor array, and an artificial neural network is trained to distinguish and recognise the odours or the tastes. Pattern recognition is gained by giving the network known flavour mixtures and a library of flavours is built. Thus, the e-noses and tongues are sensor types intended to simulate a sensory response, a specific flavour, sourness, sweetness, saltiness, bitterness, etc. (Barlett *et al.*, 1997; Tan *et al.*, 2001).

The sensor types most often used are Potentiometric Chemical Sensors, such as ion selective sensors, and:

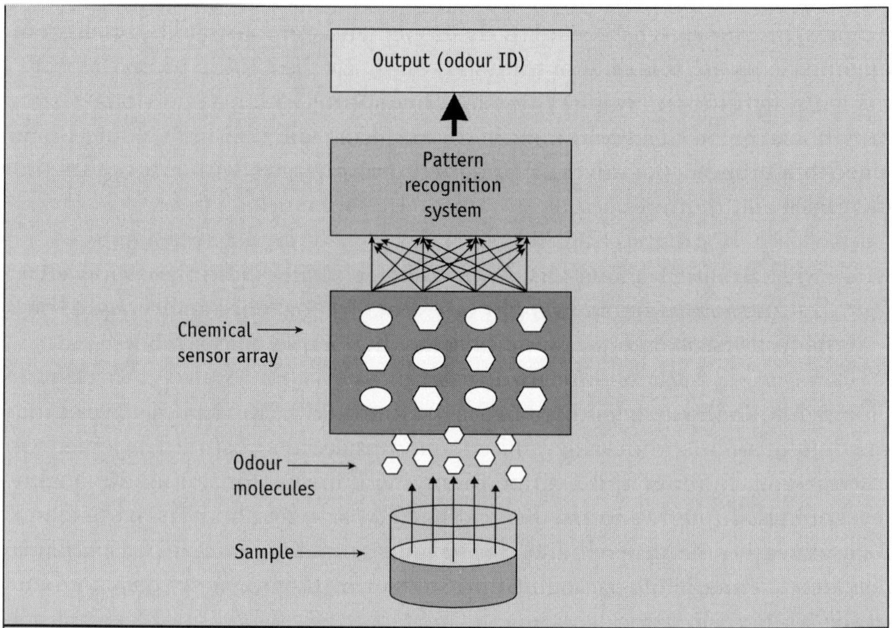

Figure 5. Principles of the electronic nose (Ellis and Goodacre, 2001).

- Metal Oxide Semiconductors (MOS), which are sensors made of a ceramic former heated by a heating wire and coated with a semiconducting film. Many gases are bound to the sensor surface and a change in the electric signal is monitored.
- Conducting Polymer Sensors, sensors obtained by electropolymerisation of a film of polymer between gold-plated electrodes. Many chemicals are bound to the surface, changing the conductivity of the polymer, and the change in voltage across the conductive polymer is monitored.
- Quartz Crystal Microbalance Resonators, sensors based on the piezoelectric effect. When molecules are absorbed on the coated layer of the quartz crystal surface, the frequency of oscillation changes in proportion to the amount of mass absorbed and consequently, the change in current is measured.

Electronic noses and tongues are produced commercially by many companies. The sensors are used in the food production and quality control of different products, typically for laboratory tests or at-line control, but may be further developed for in-line operation in the future. Testing times are often in the range of a few minutes and their largest drawback is the lack of sensor stability. Even

when they have been successful in some applications, others have been more difficult. Examples of claimed successful applications today are:
- discrimination between single volatile compounds;
- tracking of aroma evolution of ice-stored fish or meat;
- tracking of the evolution of cheese aroma during ageing;
- classification of wines;
- determination of boar odour (androsterone) in pork fat;
- classification of peaches and other fruits;
- differentiation of spices from different areas;
- general raw material control;
- testing coffee, soft drinks and whisky;
- control of beer quality and faults.

The following EU research projects have focused on different aspects of electronic noses and tongues.

ENOSEFOODMICRODETECT: The objective of this project is to develop electronic noses for early detection of bacteria, yeasts, filamentous fungi and off-odours in dairy and bakery products. Among the expected results are:
- Rapid discrimination between activity of different spoilage microorganisms, and toxigenic and non-toxigenic species in milk, cheese and bakery products *in vitro* and *in situ,* and comparison with bioassay systems
- Relationship between legislative requirements for microbial levels and sensitivity of e-nose systems in the above mentioned products
- Correlation for the first time between existing colony count methods and enzymatic assays with e-nose systems
- Rapid methods for the early detection of individual and mixed populations of spoilage microorganisms in food matrices
- Potential usefulness for screening the efficacy of preservatives for improved shelf-life
- Evaluation of the use of e-nose systems for accurate quality control systems *in situ* in processing plants in relation to the production of milk, cheese and bakery products

According to the first reports from the scientists, it has been possible in initial studies to discriminate between various microorganisms and concentrations of microorganisms in dairy and bakery products, grains and flour. Preliminary results with blue cheeses indicate that it is possible to differentiate between different cheese types, different dairies, different maturation stages and different microbial contaminants. The first results also indicate that it is possible to detect microbial contamination already at day 4 in the maturation of blue cheeses. The same positive results are coming from the groups working with bread products.

Considerable interest has been generated in the development of cheap, portable, electronic noses to discriminate and to detect, on-line or at-line, the sensory quality of many foods. For olive oil producers, there would be a tremendous advantage in detecting oil quality, shelf life and consumer preferences, and in classifying the oil in quality classes (e.g. extra virgin olive oil). This was the objective of a project in which SMEs and scientists from olive producing countries developed and tested electronic noses in olive production plants with great success.

In other EU research projects, e-noses have been under development for a variety of purposes. Thus, PROFILE-QD project focused on development of a new sensor-type, Discotic Liquid Crystal Sensor, to improve the sensor response in detecting microorganisms in the water supply or measuring the freshness of fish and fish products.

The ASTEQ Concerted Action worked with several sensor principles, in particular optical sensors and electronic noses. In their work on electronic noses, fruits and fruit juices were the main foods analysed. The ASTEQ project used Metal Oxide Semiconductors (MOS), Polymer Semiconductors, Quartz Microbalance Resonators and a new Ion Mobility Sensor (IMS). These were applied to pears, apples, peaches and fruit juices. The objective was to measure different quality factors, like mealiness, maturity, bruises and damage of the fruits or the flavour from the juices. The groups found in particular new MOS sensors of commercial interest for different fruits and IMS sensors in combination with new mathematical algorithms of interest for fruit juice characterisation.

3.3 Immunosensors

Other sensor principles to detect or measure different quality parameters have been studied in the EU Framework Programmes for research, in particular immunosensors not mentioned previously in this review. The FOODSAFE project is focusing on the development of new, at-line immunological sensors using amperometric detection of the resulting antibody-antigen complexes formed. The target quality factor in this project is toxic chlorophenolic fungicides and their chloroanisole breakdown products at the part per billion levels in potable water, wine and fruit juices. The electrochemical immunosensor uses monoclonal antibody preparations. By screening 14,000 hybridoma cell lines, two antibody clones to pentachlorophenol and two to pentachloroanisole have been identified, characterised and propagated. The investigations into the effect of the liquid food matrices on electrochemical transduction processes indicate that horseradish peroxidase is a suitable label for interrogation of the analyte-antibody immune complex using amperometry and in-house fabricated screen-printed electrodes.

Finally, another project, called RADAR, focuses on immunosensors detecting hormonal substances for growth promotion. The sensor is to be used prior to slaughter, measuring or detecting testosterone, methyltestosterone, 19-nortestosterone, stanozolol and trenbolone in biological fluids such as blood. Analysis time is expected to be 30 minutes, compared to 24-36 hours used in laboratories today.

4. Conclusions

During the last 8 years, EU research programmes have supported a large number of projects, focused on on-line or at-line detection of a large number of food quality factors. In particular, the emphasis has been on sensors using interactions between electromagnetic radiation and food, and subsequent spectroscopic analysis of the transmitted or reflected radiation. Electronic noses or tongues, using cheap non-specific solid-state sensor arrays have also been the target for further development and testing, as well as many other sensor principles, in particular immunological sensors and sensors based on Molecular Imprinted Polymer technology (artificial antibodies). Further, new trends have been seen in the concept "Lab On A Chip", e.g. including a concentration step as well as a measuring step, both in a miniature chip design. The trend also includes new sensor types based on polymers and new progress in nanoscience.

Many of the EU projects have been highly successful, producing new prototypes, which have been tested in industry, and protecting their innovations through licences or establishment of new sensor companies. These may implement the findings in the food industry with consequent improvements in food safety and food quality in general.

Acknowledgements

The author is grateful to Dr. Douglas Neil Rutledge, Dr. Gerry Downey and Ms. Marta Vidal for revising the manuscript and to the European Commission for funding the preparation of this document through the Flair-Flow Europe project (No QLK1-2000-00040) within the 5th Framework Programme under the Quality of Life and Management of Living Resources, Key Action 1.

Projects

Assessment of olfactory characteristics of olive oil by electronic sensing. FAIR-98-9540. Benavent Chafer, Oleicola El tejar Nuestra Senora de Araceli Sci, Carretra Cordoba-Malaga, 14915 Cordoba, Andalucia, Spain. Tel: +34 95 7530163

ASTEQ, Artificial sensing techniques for the evaluation of quality. FAIR-97-3516. Dr. Douglas Neil Rutledge, Institute National Agronomique, Paris-Grignon, 16 Rue Claude Bernard, Fr-75005 Paris, France. Tel/e-mail: +33 144081648 / rutledge@inapg.inra.fr. URL: http://www.inapg.inra.fr/ens_rech/siab/asteq/

CREAM, Cartridges with molecularly imprinted recognition elements for antibiotic residue monitoring in milk. QLK1-1999-00902. Dr. Maria Kempe, Lunds Universitet, Biomedical Center, B 12, Getingevagen 60, 22184-Lund, Sweden. Tel/e-mail: +46 46 2220857 / +46 462221410 / maria.kempe@medkem.lu.se. URL:http://www.samlab.unine.ch/Activities/Activity.htm

Development of novel and robust molecular imprint-based technology for the real-time analysis of food contaminants and components. FAIR-96-1219. Dr. Pradip Patel, Leatherhead Food International, Randalls Road, KT22 7RY Leatherhead, Surrey, United Kingdom. Tel/e-mail: +44 1372 376761 / ppatel@ifra.co.uk

DIFFRUIT, Diffuse reflectance spectroscopy for the evaluation of fruit and vegetable internal quality. FAIR-96-1060. Dr. Claudio Zarotti, Vela Srl, Viale teodorico 2, 20149 Milano, Italy. Tel /e-mail: +39 0232 4879 / velasrl@tin.it.

ENOSEFOODMICRODETECT, Rapid detection of microbial contaminants in food products using electronic nose technology. QLK1-2000-01763. Prof. Naresh Magan, Cranfield University, MK43 OAL, Silsoe, Bedford, UK. Tel/e-mail: +44 1234 754339 / n.magan@cranfield.ac.uk. URL: http://www.e-nose.net/

FOODSAFE, Rapid assessment of food safety via novel at-line measurements. QLK1-2000-00518. Dr. Steven Setford, Cranfield Centre for Analytical Science, Cranfield University, MK45 4DT, Silsoe, UK. Tel/e-mail: +441525863549 / s.j.setford@cranfield.ac.uk. URL: http://www.cranfield.ac.uk/ibst/ccas/foodsafe/

FOODSENSE, The use of optical biosensors for the detection of veterinary drug residues in foods - a demonstration project. FAIR-98-3630. Dr. Karl-Erik Hellnaes, National Food Administration, Chemistry Division III, 751 26 Uppsala, Sweden. Tel/e-mail: +46 18 175708 / kahe@slv.se. URL: http://www.slv.se/foodsense2/

GLOVE, Development of an instrumented glove to assist the quality control and the picker in selecting fruits according to their maturity. FAIR-97-3399. Michel Crochon, Cemagref-Itap, BP 5095, 361 rue JF Breton, F-34033 Montpellier, cedex 1, France. Tel/e-mail: +33 4 67046300 / michel.crochon@cemagref.fr. URL: http://www.montpellier.cemagref.fr/teap/projets/caporal/projetglove.htm

MICROSENSOR, Construction of miniaturised free flow electrophoresis (mFFE) incorporating dedicated sensors for real-time analysis of food contaminants. QLK1-1999-00343. Dr. Pradip Patel, Leatherhead Food International, Randalls Road, KT22 7RY Leatherhead, Surrey, United Kingdom. Tel/e-mail: +44 1372 376761 / ppatel@ifra.co.uk. URL: http://www.leatherheadfood.com/lfi/submenu.asp?item=2782&subsection=75

NIPCO, New Image processing for characterisation of olives and other fruits. FAIR-97-9505. Ricardo Diaz Pujol, Instituto Tecnologico Agroalimentario, Parque Tecnologico de Valencia, Benjamin Franklin 5-11, 46980-Paterna (Valencia), Spain. Tel/e-mail: +34 961 366 090 / rdiaz@ainia.es. URL: http://pomelo.ivia.es/index.html

NIQAT, Non-destructive NIR technology for fruit and vegetable internal quality assessment, eliminating the skin disturbing effect. QLK1-2000-00455. Dr. Raina Chalucova, Institute for Horticulture and Canned Foods, Vassil Aprilov Blvd. 154, 4000 Plovdiv, Bulgaria. Tel/e-mail: +359 32 952109 / cic@evro.net.

OPTIVEG, On-line optical measurements as a basis for control during mild heat processing of vegetables. FAIR-98-3155. Prof. Josse De Baerdemaeker, Katholieke Universiteit Leuven, Department Agrotechniek en Economie, Kard. Mercierlaan 92, 3001 Heverlee, Leuven, Belgium. Tel/e-mail: +32 16321445 / josse.debaerdemaeker@agr.kuleuven.ac.be. URL: http://www.agr.kuleuven.ac.be/aee/amc/research/process/optiveg/optiveg.htm

PROFILE-QD, Electronic profiling of food and beverage odour for quality determinations. FAIR-98-3436. Dr Tim Gibson, University of Leeds, 175 Woodhouse Lane, LS2 3AR Leeds, Yorkshire, UK. Tel/e-mail: +44 113 2332599 / T.D.Gibson@leeds.ac.uk

QTEPACK, Advanced electromagnetic solution for quality testing of packaging for horti-fruit products. QLK1-2000-00936. Dr. Jose Maria Martinez-Iglesias, Talleres Daumar SA, Wifredo 794-796, 08918 Badalona, Spain. Tel/e-mail: +34 934601593 / mziglesias@daumar.es. URL: http://qtepack.rmee.upc.es/

RADAR, Development of single and multi-analyte affinity sensors for rapid detection of androgen residues in live and post-mortem animals. QLK1-2001-01670. Dr. George Guilbault, University College Cork, Western Road, Cork, Ireland. Tel/e-mail: +353 214902403 / g.guilbault@ucc.ie. URL: http://intel.ucc.ie/sensors/androgen/

STAS, The development of spectroscopic techniques as advanced sensors for the optimisation and control of food and paper processing. FAIR-96-1169. Pierre Grenier, Cemagref, BP 5095, 34033 Montpellier cedex 1, France. E-mail: grenier@dir.montpellier.cemagref.fr. URL: http://www.montpellier.cemagref.fr/teap/projets/caporal/projetstas.htm

References

Barlett, P.N., Elliott, J.M., Gardner, J.W., 1997. Electronic noses and their application in the food industry; Food Technology, 51(12), 44-48

Ellis, D.I., Goodacre, R., 2001. Rapid and quantitative detection of the microbial spoilage of muscle foods: current status and future trends; Trends in Food Science and Technology, 12, p. 414-424

Garth Rand, A., Ye, J., Brown, C.W., Letcher, S.V., 2002. Optical biosensors for food pathogen detection; Food Technology, 56(3), p. 32-39

Lading, L., 2002. Sensor Technology Centre, Copenhagen. Personal communication.

Mosbach, K., Ramström, O., 1996. The emerging technique of molecular imprinting and its future impact on biotechnology; Bio/Technology 14, p. 163-170

Patel, P.D., 2000. Biosensors for measurement of analytes implicated in food safety: A review. Trends in Analytical Chemistry, 21, p. 96-116

Tan, T., Schmitt, V., Isz, S., 2001. Electronic tongue: A new dimension in sensory analysis. Food Technology; 55(10), p. 44-50

Data management: A bottleneck or the ultimate success factor?

F.P.J.M. Leijse
ITC Validation Consultants, Haarlem, the Netherlands

Abstract

Rapid detection of biological and chemical contaminants in food, feed and environments can be done efficiently and effectively by deploying automated (computerised) systems. Assuring structural integrity of data, which is processed by such systems, is a pivotal task and requires proper management. International standards (i.e., ISO 17025) and EU legislation (i.e. General Food Law) require management of processes (traceability) and structural integrity of (electronic) data. An appropriate way to meet those requirements is by validating automated systems.

Keywords: traceability, safety, integrity, risk, validation, computerised, systems, automated

1. Introduction

The professional food and foodstuffs industries have always worked in a controlled environment. The focus of the regulators will always be on health and safety. Over the years regulations have not only increased in number, but also in the areas in which they are applicable. New methods of examining the quality of the "food product" are constantly being developed; however, probably the biggest change is the use of computerised (automated) systems. Computerised systems have become an integral part of laboratory processes. The data input and results that pass through the processes are as much part of core business as the food product itself. EU legislation and international standards now require proof that not only the data, but also all the (relevant) processes are properly controlled. Risk management on automated business processes and demonstration of structural integrity of (electronic) data are now requirements. Validation of automated systems - i.e. to demonstrate that one is in control- is the appropriate way to meet these requirements.

2. Data management

Data management is all about managing electronic information. Over the last decades, computerised systems have become an integral part of laboratory processes. Due to the ever increasing volume of assays - hence data - laboratories will be increasingly dependent on automated systems. Therefore, by deploying computerised systems such as CDS (Chromatography Data Systems) and LIMS (Laboratory Information Management System) it has become a necessity to manage the data produced by such systems.

In the next decade the laboratory world is undergoing a transition. There are currently hybrid laboratory environments where data records - produced by automated systems - are kept electronically and kept on paper (hard copy). These environments will gradually change to environments where almost all data are processed and archived electronically.

Laboratory instruments such as UV/VIS spectrometry analysers, GC/MS and HPLC generate information (raw data). Processed data generated by CDS become result data. Output created by laboratories is usually in the form of reports or certificates, which are printed on paper and/or shown on a display (PC-Client). These laboratory reports and certificates are based on electronic information (data) generated by an automated system and as such also require control and management. The same is true for methods that are developed on laboratory instruments to perform certain assays. Methods are after all a set of instructions in the form of an electronic data file that is written so that a laboratory instrument will perform an assay and report the results in a pre-determined and controlled way.

Given the above, proper data management will be a critical success factor. The other huge advantage of electronic data is that it can be processed in a fully controlled and controllable manner. Functions currently implemented in automated systems enable us to review, annotate, approve, e-sign (accept with an electronic signature), store and archive electronic data. This enables data to be traceable and to provide proof of its integrity. This is meant by data management.

3. Compliance to standards and legislation

In short, international standards like ISO 17025 and the EU General Food Law require management of processes (traceability) and structural integrity of (electronic) data. The process to meet such requirements is known as validation.

The definition of validation has evolved although the basics are unchanged. The US Food and Drug Administration (FDA) define validation as "Establishing documented evidence which provides a high degree of assurance that a specific process will consistently produce a product meeting its' predetermined specification and quality attributes" (see www.fda.gov). This FDA definition for processes was modified by the PDA (Parenteral Drug Association, www.pda.org) for computerised systems: "Establishing documented evidence which provides a high degree of assurance that a specific computer-related system will consistently produce a product meeting its predetermined specifications".

In ISO 17025 § 5.4.5.1 the definition of validation is: "The confirmation by examination and the provision of objective evidence that the particular requirements for a specific intended use are fulfilled".

The key elements in the definitions are:
- documented evidence;
- high degree of assurance; and
- predetermined specification.

In short, validation demonstrates that an - automated - process is fit for its purpose.

4. Method validation and (automated) system validation

Method validation is recognised to be of critical importance to a laboratory. It is common practice to validate methods in laboratories around the globe. The main reasons to validate methods are:
- to show fitness for purpose; and
- to build confidence in results.

Method validation is about providing objective evidence that a given method is fit for its purpose. Being fit for purpose one can say that one has a high degree of assurance that a given method delivers a pre-defined result.

Interestingly to note is that the reasons to validate methods equally apply to computerised systems. It is our vision that validation of computerised systems in laboratories will be commonplace over the next years.

5. Demonstrating compliance

Today readers are probably quite accustomed to inspectors asking questions about method validation and uncertainty of measurement. In the near future one can expect inspectors and auditors to be looking for system compliance. Therefore different questions might be asked that most readers might not be familiar with. Here are some of the most infamous questions being asked today:

- What does the system do?
- What does the system consist of?
- How do you know that the system does what it is supposed to do?
- Who uses it?
- How do know that you are in control?

It is interesting to note that the scope of questioning has broadened focus. The current shift towards automation in laboratories - similar to what has been happening in the pharmaceutical domains - will cause a similar shift in the line of questioning by inspectorates. The questions asked will increasingly focus on the automated system that has produced the results.

6. Why and when to validate a process or a system?

As we have seen, the two main reasons to validate a system are to build confidence in results and to show fitness for purpose. Basically there are three validation approaches for organisations to gain that confidence:

- Prospective validation
  Develop a validation strategy and build a document base before a decision is made to purchase an automated system.

- Concurrent validation
  Develop a validation strategy and build a document base while implementing an automated system.
- Retrospective validation
  Develop a validation strategy and build a document base when an automated system has been in use for a period of time.

There are pros' and cons' for each scenario and of course things depend on local circumstances. However, it emerges from both literature and field experience that a prospective validation is by far the best approach.

One of the most appealing reasons why prospective validation is the best approach is based on the cost aspects of fault propagation in the project. In any project errors will be made, this is only human. When detected, faults and errors can be corrected. The key to success is the moment faults are detected. When a fault is detected early on in the project it can be corrected with minimal cost and effort. At a later stage the cost and effort required to correct errors increases dramatically. It is estimated that the costs to correct errors when a system is already implemented and in place, are a factor 200 as compared to correcting errors in the concept/design phase.

Another reason for prospective validation is that it puts an organisation into a position of system control and controlling the produced data throughout its lifecycle. Validating a computer system at a later stage - or validating a system retrospectively - is not only more difficult and more expensive, but also denies the possibility of being in control of the data and the structural integrity of the outcome.

7. A bottleneck or the ultimate success factor?

In the regulated (pharmaceutical) industries one can observe a gradient scale of how companies view and deal with validation of IT systems. On one end of the scale there are companies perceiving validation of IT systems as a hinder to their business process and thus validation is carried out only to please the inspectors. Inspections are also perceived as an obstacle or hindrance, a necessary evil! The only objective is to obtain approval or a license, the ticket to start or continue research activities or production.

At the other end of the scale there are an increasing number of organisations that use computerised system validation as a method to be in control of their systems and to provide the management with the evidence of that control. Compliance is regarded as one of the most important business goals. Audits are not only seen as a mean to obtain the accredited status, but also to provide valuable information. The observations in the ensuing audit report serve as a basis and a justification for the next project; not only to redress the omissions, but more important to improve and to ensure the continuing 'quality' of the entire system.

8. Conclusion

It is now clear that electronic data has become critical corporate information and therefore must be regarded as an asset. Organisations must develop corporate (data) security management and cultivate a Computer System Validation capability. This is necessary to comply with international standards and regulations and it is imperative to be able to fulfil business missions.

Nowadays IT systems are the basis of all business processes and are in fact an integral part of them. To manage these processes the company must be able to fully rely on its IT systems' architecture. IT systems' architecture does not only include IT Systems (applications and hardware platforms), but also the infrastructures.

Developing a computer system Quality Assurance (QA) capability will be of equal importance to laboratories as the need to rapid process assays. The core business of laboratories needs to be secured for the future by deploying rapid methods. However, the future is a gamble without securing the management of processes. Computer system QA capability enables laboratories to manage the entire life cycle of IT systems. As a result the TCO (Total Cost of Ownership) of such systems can be minimised, while their beneficial deployment can be maximised. This renders computerised systems validation an ultimate success factor.

References

Accenture - White paper on computer systems validation, 2001. http://www.accenture.com/xdoc/en/industries/hls/pharma/csv.pdf

ISO 17025:1999. General Requirements for the Competence of Calibration and Testing Laboratories.

Regulation (EC) No. 178/2002 of 28 January 2002, laying down the general principles and requirements of food law, establishing the European Food Safety Authority and laying down procedures in matters of food safety.

Recommended literature

The Good Automated Manufacturing Practice (GAMP) Guide for Validation of Automated Systems in Pharmaceutical Manufacture - GAMP4, Available through ISPE (International Society for Pharmaceutical Engineering) www.ispe.org.

Computer Systems Validation: Quality Assurance, Risk Management and Regulatory Compliance for Pharmaceutical and Healthcare Companies, edited by Guy Wingate, 2003. Interpharm/CRC, ISBN 0849318718.

Key for tomorrow: Nanotechnology in food analysis

I.J. Bruce, T. Sen and A. del Campo
University of Kent, Canterbury, Kent, CT2 7NJ, United Kingdom and University of Greenwich, School of Science, London, SE10 9LS, United Kingdom

Abstract

Nanotechnology is an emerging engineering discipline based upon nanoscience which is a fusion of materials science, chemistry, biology, physics and engineering. It intends to deliver radical new products and processes at nanoscale dimensions through the characterisation and manipulation of matter at the level of atoms, molecules and macromolecular assemblies. Its conceptual roots can be traced back to Richard Feynman (1959) and currently the technology is being applied to areas as diverse as biomedicine, computing and engineering. In the context of food, nanotechnology is already engaged in both manufacturing and testing. In manufacturing nanotechnology is involved in formulation and packaging and in testing in identity preservation, characterisation, salubrity and contamination monitoring.

Keywords: nanotechnology, nanoscience, NST, food, testing, analysis

1. Introduction

What is nanoscience/nanotechnology (NST)? Nanoscience and nanotechnology as terms derive from the unit of length, the nanometer, and in their broadest definitions intend science and technology that is associated with the assembly, manipulation and control of matter on length scales of between 1 to 100 nanometers. For comparative purposes we can consider that an 'average' atom is a fraction a nanometer in size, whilst a small molecule is of the order of 1 nanometer and an average protein is 10 nanometers in diameter. On a larger scale, a bacterial cell is between 2 to 5 micrometers in length (several thousand nanometers) and an animal cell is between 5 to 20 micrometers (tens of thousands of nanometers) in size depending on type.

Nanoscience already exists and was what was once referred to as molecular science. It can be considered as the convergence of physics, chemistry, materials science and biology dealing with the characterisation and manipulation of matter on length scales between the molecular and the micron size.

Nanotechnology on the other hand is an emerging technology. It can almost be considered as an engineering discipline applying methods from nanoscience to create products.

One description of nanotechnology is that it represents a technology, which permits "the creation of functional materials, devices and systems through control of matter on the nanometer length scale (1 to 100 nanometers). It involves the exploitation of novel phenomena and properties (physical, chemical, biological, mechanical, electrical...) of materials at that length scale".

Nanoscience and nanotechnology are therefore multidisciplinary.

2. Some considerations concerning NST

What is so special about NST? One thing that makes it different is that at very small scales the laws of physics do not operate in the same ways as for larger dimensions. This imposes limitations in the way that we can do things but may also help in the design and creation of things on a small scale. In turn this could mean that nanoscale devices and processes work on very different principles from those that function on the macroscopic level. Finally, this will and already has been observed to have consequences for nanofabrication (see later).

Some key differences about the way physics works at the nanoscale include:

- Quantum effects. On small scale matter behaves in ways that respect the laws of quantum mechanics rather than Newtonian mechanics, which operate in the macroscopic world.
- Brownian motion. Small particles (submicron) in suspension are subject to continuous bombardment by the molecules around them causing them to move around and flex in a random and uncontrollable way.
- Surface forces. Surfaces and interfaces play an important role in the ways things can interact. As things become smaller there is a natural tendency for them to stick together. We do not completely understand why this happens but this sticking together gives us problems with keeping things

apart and as well as leading to high frictional forces when things have to move against each other.

These various factors also have an important effect on our ability to fabricate truly efficient nanostructures.

In fact a major limitation in nanofabrication is linked to our ability to manipulate individual components of the desired product into place, fix them there so that they do not move or change, and the speed with which we can do so; all in a controlled fashion. In this context the concept of self-assembly is becoming more and more important. The more we understand and can control the property of matter and its composites on the nanoscale the more ability we have to control its self or auto assembly into truly tailored, efficient end products. In that type of process the ability to create the final outcome/product is literally encoded into its component parts.

Self-assembly is an example of a 'bottom-up' process. A benefit of this type of approach is that it should be scalable and it represents a parallel process, limited only by the number of molecules present. This is in contrast to macro-fabrication where generally one thing is made at a time; a serial process.

3. Current areas and interests involving nanotechnology

Disruptive technologies are those that can be defined as those that displace older technologies and which permit radically new products and processes. Disruptive technologies are ones, which can also establish whole new market areas and products not previously feasible or possible. NST is disruptive!

A characteristic of truly disruptive technologies its that they can have very many different applications and Figure 1 indicates the possible application areas which either already involve nanoparticle technology or could benefit from it. Figure 2 on the other hand lists the current major developmental themes in nanotechnology. From this it is clear that there are few areas which either have not already been, or will not be, touched by this technology.

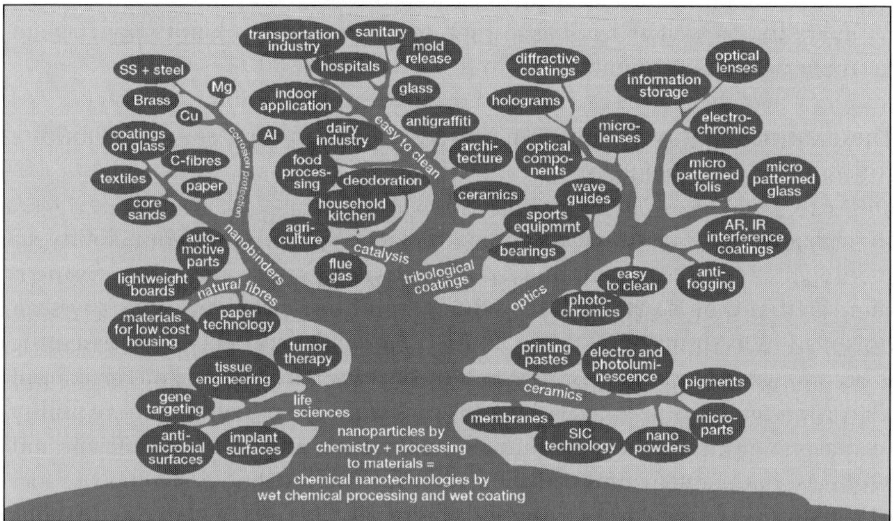

Figure 1. Potential application areas for nanoparticle technology (Institute of New Materials, University of Saarbrucken, Saarbrucken, Germany).

Current themes in nanoscale science and technology

Materials science
- New forms of carbon
- Nanocomposites
- Quantum dots and wires made by colloid chemistry
- Nanostructured materials by self assembly
- Toxicity

Biomedical sciences
- Drug delivery
- Tissue engineering
- Laboratory-on-a-chip

Electronics and optoelectronics
- Electronics and optoelectronics
- Semiconductor optoelectronics
- Photonics
- Memory and data storage
- Plastic electronics
- Molecular electronics
- New general concepts in computing

Figure.2. Current themes in nanotechnology.

4. Why bother with nanotechnology and testing (molecular diagnostics)?

The ability to discriminate and define things in a quicker, more sensitive, easier, cheaper and more reliable and reproducible way depends in turn upon our ability to apply materials separately and together in diagnostic (test) processes. This necessarily means that the better our understanding of their structural and chemical characteristics so that we can formulate them into effective products, then the more successful they will be in their applications. It also means we can better tailor the materials and their properties to specific applications. This is also coupled with a process of miniaturisation. Nanotechnology addresses all these issues.

Separately the benefits to the organisations and individuals involved in the nanotechnology industrial sector are potentially enormous. When coupled to the benefits likely to accrue to the population as a whole in terms of health and well being it appears that nanotechnology represents a truly win-win situation (see later).

5. Economics (from the nanotechnology industry and service provider prospective)

If one considers the least speculative areas of nanotechnology in the context of diagnostics, the development of nano-support materials with enhanced properties, improved surface activation chemistries and incremental improvement to existing thermal cyclers and nucleic acid modification chemistries then it is possible to make reasonable estimates as to the likely future scope for growth and size of impact on the market. These can be said to address large market areas of the world economy, *viz. 'in vitro diagnostics'* and *'health care'*. In the former case an analysis of current and future market potential is indicated in Figure 3.

The total world market for molecular diagnostics MD_X was approximately 1.5 billion euro in 2000 and is projected to grow to 4.7 billion euro by 2005 and 16,800 billion euro by 2010. The 'in vitro diagnostics' market segment of the total health care market can be sectioned into major subgroups and one of them DNA testing, includes forensics, infectious disease, tissue typing, food testing, paternity and prenatal screening, plus veterinary and agricultural application (see Table 1).

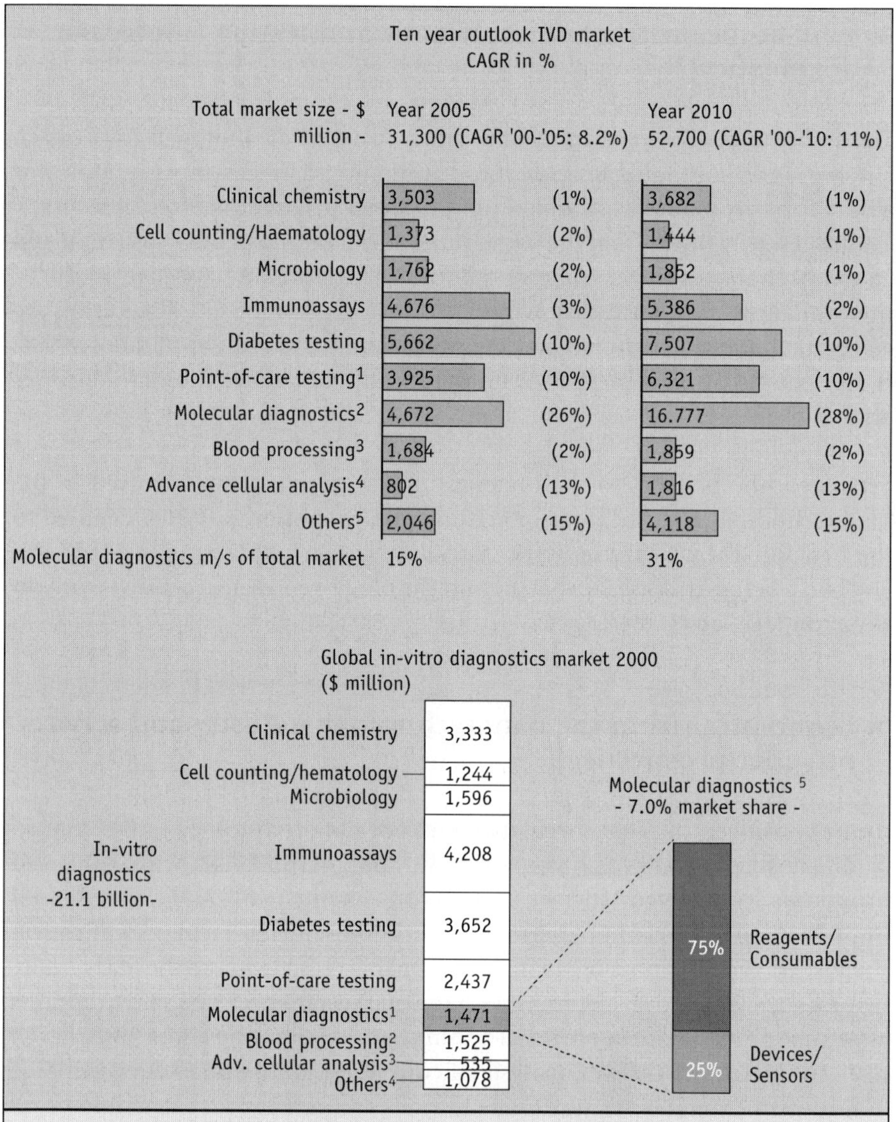

Figure 3. Global 'in vitro diagnostics' market now and in the future (A.T. Kearney, Inc., 'In-Vitro Diagnostics Update', Presentation at Philips Research Laboratories, Aachen, Germany, January 2003).

Table 1. Worldwide market for DNA testing by segment.

Type of test	Revenues 2000 ($ million)	(%)
Infectious disease	956	65
Genetic/predisposition	279	19
Oncology	44	3
Tissue typing	44	3
Identity/forensics	29	2
Other	118	8
Total	1,470	100

In the second area noted above, health care, the total world market is currently estimated as a total of 3,000 billion euro 45% of which in located in the US. The medical equipment, services and supplies section of this market encompass a total of approx. 250 billion euro and Figure 4 illustrates the most relevant segments of this part to the global health care market.

Separately, the world market revenue for nucleic acid analytical instrumentation was estimated at $1 billion in 2001 of which thermal cycler based instrumentation accounted for 23% (Frost and Sullivan, www.frost.com). The laboratory instrument market is also widely reported to be growing at 17% per annum. DNA sequencers accounted for 51% of this market, a significant proportion of which are used in DNA based molecular diagnostics.

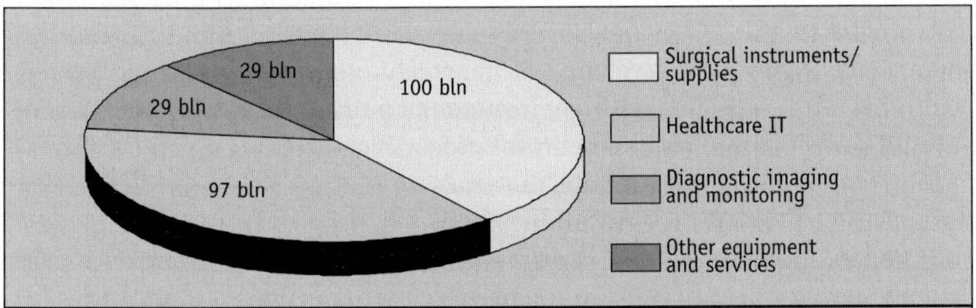

Figure 4. Important segments of the global 250 billion euro 'medical equipment, services and supplies' market (Frost and Sullivan, 2001).

Finally, the provision of reagents for the synthesis of oligonucleotides is a highly competitive market. One major and several smaller players are likely to share an estimated market volume of approximately $180 million in 2003 with an expected growth rate of 10 to 12% per annum. It is an objective of the project to improve oligonucleotide synthesis technology through research and development on the next generation monomers with advanced features over the state of the art, and on novel solid supports modifiers with competitive advantages.

Further longer term commercial benefits are envisaged in the biochip array area, which includes protein chips and lab-on-a-chip devices. The entire biochip market was $42 million in 1998, but is predicted to rise to some $632 million by 2005 (Front Line Strategic Management Consulting, Inc.). DNA chips form the major part of this market (94%). The radical technological breakthroughs predicted in preparation of novel magnetisable suspension, nanoporous and flat surface arrays will have application in the entire biochip sector, offering the same improvements as are expected for the oligonucleotide arrays: automation, high sample throughput, reduced handling, lower cost, improved detection sensitivity/accuracy and enhanced range of applications.

6. Health

Nanotechnology outputs address health via disease and disease causing agents detection, characterization and management, preservation of food quality and animal welfare. This will lead to better access to and improved methods for health care, maintenance monitoring and treatment and consequently will benefit and improve the quality of life of the population as a whole. Lower costs and better efficiency will also be an important contribution. The processes and materials that will be developed via nanotechnology will also lead to indirect health benefits associated with environmental gains in terms of reduction of harmful emissions and requirements for processing wastes (eg organic solvents) or long term storage of hazardous materials.

The same technologies will also contribute indirectly to the healthcare of citizens through development of improved and more cost-effective methodologies for identification of inherited genetic diseases and tissue typing for organ transplant operations, using the platform technologies that will be developed. The use of DNA-based tests in particular will also demonstrate clear advantages of genetic technologies to the wider population, some quarters of which are either antipathetic or hostile to the development of genetic technologies. Finally, it will also improve public access to new drug treatments through reduction in time to market, and cost, by the improvements envisaged in microarray technology and combinatorial chemistry which all major pharmaceutical companies are either now considering or already using to streamline their drug discovery and development processes (Front Line Strategic Management Consulting, 1998).

7. How can nanotechnology help in food analysis for purposes of identity testing (characterisation), contamination monitoring and to ensure salubrity?

Primary reasons for analysis:
- To confirm identity.
- Opportunistic confirmation.
- For identity preservation.
- To test for contamination with organisms, entities that could be harmful, e.g., bacteria and viruses.
- To test for presence of substances which could be harmful, e.g., toxins.
- To test for presence of substances which are banned or illegal, e.g., hormones and antibiotics.
- To test for salubrity, e.g., presence and level of particular components.
- To test for adulteration.

Primary needs in terms for analysis:
- Well designed support materials (particles and flat surfaces).
- Appropriate surface chemistries for ligand grafting.
- Specific and effective ligands for surface grafting.
- Simple, cheap and robust protocols for use which can be easily automated.

8. Some basic considerations in analysis (matching materials to applications)

8.1 Biomolecular separations

Biomolecular separation and discrimination represents the basis for most current day molecular diagnostic processes as biomolecules are frequently used as indicators of state of well being (or the converse), in environmental monitoring, determination of food hygiene and quality, as indicators of the efficiency of drug therapy and in forensic identification and profiling of individuals.

Unfortunately, current understanding at a physico-chemical level of the properties of most of the materials used to mediate these processes is often limited; this is particularly the case when they are used in conjunction as composites. For example, the functionalities associated with pure silica, which impart its usefulness in purification procedures, are yet to be fully explained in any other way than they frequently involve weak bonding forces and are relatively non-specific. The surface of silica under aqueous conditions, the basis for most biomolecular separative processes, is naturally heterogeneous, hydrophilic, weakly acidic and represents various silanol groups and siloxane bridges. The principal types of identified surface groups can be represented as isolated silanols, germinal silanols, vicinal and siloxane bridges. The terminal SiOH of a chain of vicinal silanols is also known as a terminal silanol. A fully hydroxylated silica surface will contain a mixture of isolated, germinal, and vicinal silanol groups, as well as physically adsorbed water molecules. These water molecules can be hydrogen bonded to the surface hydroxyl groups or they can occur within the structure of the colloidal particles as structurally bound water. Internal silanol groups can also occur, buried within the structure of the silica particle. These cause distortions in the silica surface, facilitating the presence of germinal silanol groups, and in some cases are considered to be structurally bound water. Figure 5 illustrates a range of possible architectures, which can be adopted by silica. Each different form possesses its own specific characteristics with respect to applications and uses but all represent essentially the same material at the chemical level, so far as we can tell.

Given this level of complexity, it is not unsurprising that even in the case of silica, one of the most widely used materials in the context of bioseparations, we cannot give a full description, nor are we able to fully control the manner

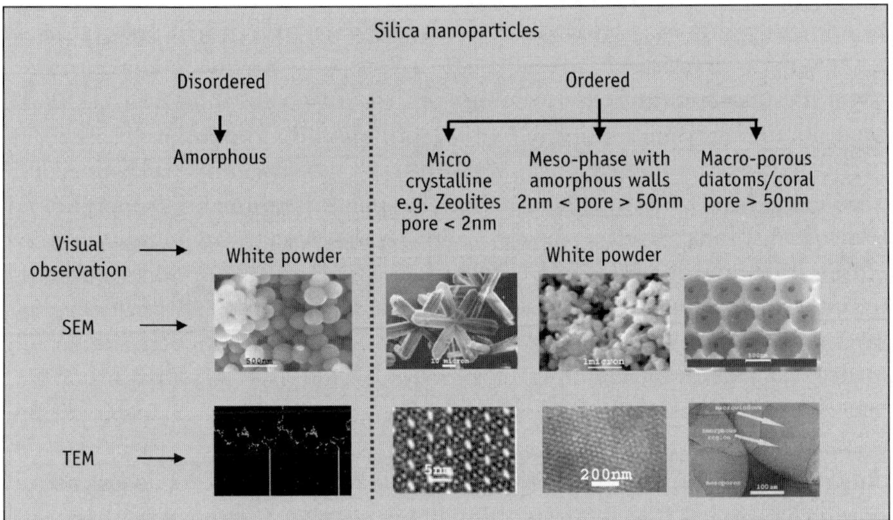

Figure 5. Various silica and silica nanoparticle architectures.

in which it will interact with a particular molecular species, or in a more likely context for a bioseparative process, a range of biomolecular species.

Figure 6 illustrates just how materials need to be matched to applications and why one material may not necessarily be good for use in all processes.

Further issues that are important when functionality and performance of support materials used in bioseparative processes are considered concern the attachment of useful ligands to their surfaces (see next section) and surface physical properties. In the latter context intrinsic properties such as porosity, reticulation, density of accessible grafting functions, surface charge/density, ratio of hydrophobicity/hydrophilicity, ability to H-bond, and ligand non-specific binding need to be considered in tailoring a support for a particular purpose. In the former context current surface chemistries are generally lacking or limiting that permit us tofunctionalise materials in a way which is fully controllable in terms of ligand density, organisation (spatial distribution), orientation and which take account of chirality.

Many of the nanomaterials which are currently available (and which possess significant potential) and are being applied in one or another nanobiotechnology

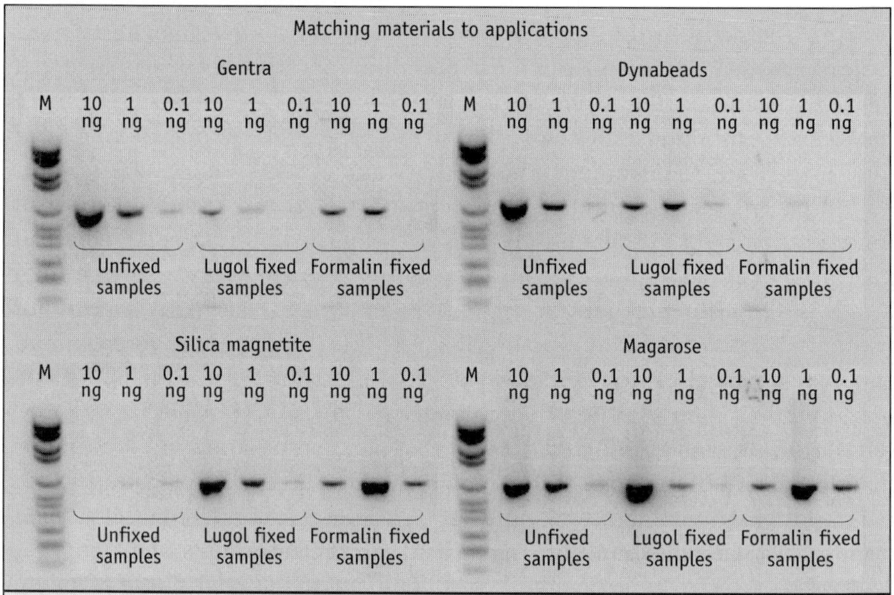

Figure 6. Extraction and amplification of toxic marine algal DNA using various commercial supports and ones prepared in our laboratory.

area, have yet to be fully characterised in any of these terms including mesoporous molecular sieves, multi-walled carbon nanotubes (MWCNTs) and paramagnetic silica composites. For example MWCNTs can contain residues from the plasma deposition process involved in their synthesis, imparting paramagnetic properties, whilst silica-magnetite composites can vary widely in their ability to absorb molecular species depending upon the way in which silica has been deposed to a magnetite core.

Materials which can be specified and closely defined in terms of their properties will be increasingly more useful in the future for applications in nanobiotechnology. The most important principles involved will be (i) the physical and chemical properties and architectures of the matrix materials themselves, (ii) the regularity and reproducibility with which desirable ligands can be associated with such materials, and (iii) the ease, reproducibility and efficiency with which they can be applied in the process of interest.

Consequently, with respect to analysis and testing and given the diversity of applications it would be useful to generate a 'library' of tailored media (nanomaterials) with well-characterised properties for application in bioseparative processes/molecular diagnostics, and which are fully biocompatible and toxicologically evaluated. These should include novel paramagnetic nanoparticles/constructs that will have separative properties specifically tailored for particular nucleic acid and protein molecules, and which can be used in automated high-throughput isolation and detection systems (particle/suspension arrays) and as drug delivery vehicles either alone, or in conjunction with biological structures. Additionally mesoporous molecular sieves (silica), substrate attached carbon nanotubes of various architectures and polymeric, functionalised nanorods produced using mesoporous molecular sieves as templates and bidisperse silica. Also MWCNTs that are capable of being produced in a variety of controlled architectures and substrate attached. These could have significant potential for use in the fabrication of flat surface microarrays by improving surface area for probe immobilisation, and the dynamics of the hybridisation process by reduction of steric hindrance effects, ligand crowding etc (see later).

8.2 Surface chemistries and functionalisation/grafting

Membranes, glass, silicon wafers, polysaccharides and silica most commonly use chemistries for their modification which do not permit high levels of control over where and how much grafted material can be associated with their surfaces. This consequently results in problems with respect to density, distribution and orientation of the associated ligands and therefore the surface's efficiency to take part in a bioseparative/diagnostic process. Also it is very difficult to differentially functionalise a surface with more than one type of ligand in a controlled way. These are particular problems in the molecular diagnostics arena where control and reproducibility are paramount considerations. As an example, silanisation which generates surface coatings that are inherently non specific in terms of their ability to be functionalised with useful groups is also, paradoxically, one of the most commonly adopted approaches to surface modification of glass and silicon wafers for their subsequent fabrication into microarrays and nano supports used in bioseparations. Considering diagnostics and arrays, very high-density arrays are the end objective of most microarray manufacturers especially in the context of genomics or proteomic type studies and using silanisation in their fabrication, is a limiting factor. Arrays (or supports used in bioseparations) fabricated using silanisation as a basis for immobilisation suffer from steric interaction effects between adjacent immobilised groups which limit the number of subsequent groups which can be attached. Also adjacent oligonucleotides which are surface attached can interact, effectively removing them from the population able to interact/hybridise with complimentary sequences. This latter occurs as a result of the fact that reagents commonly used in the first derivatisation step for immobilisation are non-rigid and flexible molecules. In the case of microarrays this means a decrease in the overall sensitivity of the chip and in the case of nanomaterials used in bioseparative processes a decrease in the amount of desired end product recovered from the purification process. Although there is much research in the area of surface chemistries and physical approaches to surface deposition of ligand, for example in the case of microarrays, microphotolithography, combinatorial generation of surface attached oligonucleotides and 'on-chip'/'off-chip' and electrospray deposition, there are still significant limitations as a consequence of the above. Figure 7 illustrates a range of commercially available silanes and how they can be used in surface activation and functionalisation.

Other issues when considering immobilisation of ligands to surfaces also involve the physical method of application of ligand to substrate and the type of surface involved. For example in microarray fabrication spot inhomogeneity (coffee-ring effects within separate dots) can occur during robotic printing of chemically activated slides based on silicated monolayers which can adversely affect array performance. Also nylon membranes function more efficiently for proteins/antibodies/enzymes immobilisation rather than silicated monolayers, sulfide-based monolayers, or acrylamide pads.

It is obvious that chemistries that allow more controllable, reproducible immobilisation of ligands to surfaces are not just desirable but essential to tailoring of existing and new nanomaterials for specific applications. Matching materials and the chemistries used for their surface functionalisation is likely to be a critical factor in the performance and characteristics of any end product.

Surface activation chemistries that will overcome these defects in immobilisation, and permit better control over organisation and higher densities of immobilised ligands, reduced steric hindrance effects, increased versatility with respect to groups immobilised, improved dynamics in terms of interaction with molecular species of interest and consequently permit

Alkoxysilanes chemistry: possible surface functionalities

Silane precursor	Further surface modification	Final functionality
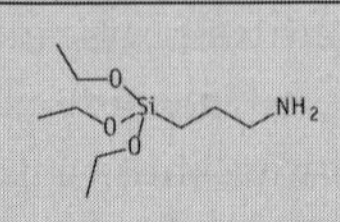		$-NH_2$
	Coupling with Gluteraldehyde	-COH
	Coupling with biotin molecule	Biotin
	Coupling with Steptavidin	Strept avidin
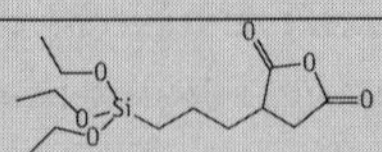	Hydrolysis of the anhydride	-COOH
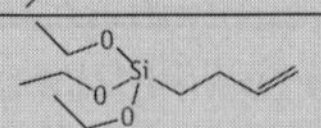	Oxidation of the double bond	
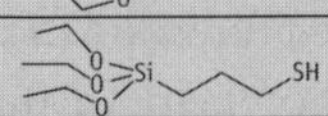		-SH
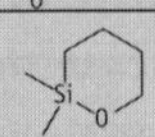		-OH

Surface chemistry: Overview of a typical modification process

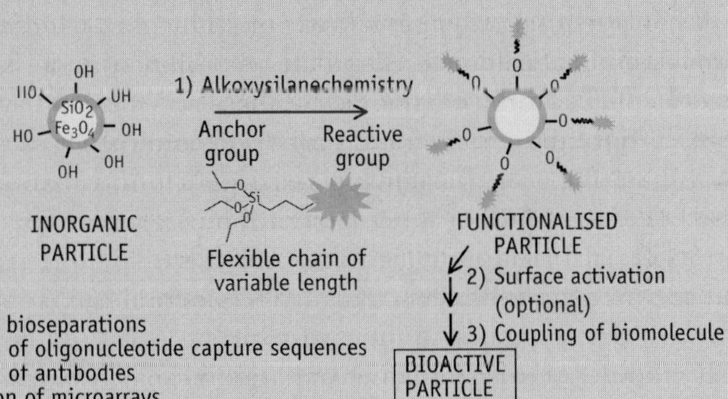

Applications:
(i) directly in generic bioseparations
(ii) for immobilisation of oligonucleotide capture sequences
(iii) for immobilisation of antibodies
(iv) for the construction of microarrays

Figure 7. Silane chemistry and its application in surface activation and functionalisation.

better sensitivities, yields and end product quality with respect to the species to be separated or discriminated. These also need to be tailored to specific applications (see later).

Chemistries to be developed for use in surface engineering should address chirality, ligand attachment and control of probe orientation. For example, matched or mismatched interactions between chiral, chemically activated surfaces and chiral macromolecules such as proteins, antibodies, enzymes, oligopeptides, and oligosaccharides can be used to direct, manipulate and control the non-random covalent attachment of ligands onto surfaces. Surface films constructed using chiral polypyrrole and polycarbazoles films could be one way of making more regular, chiral and controllable substrates for grafting. End activation of carbon nanotubes to -COOH and biotinylation to direct the macro molecular assembly of such structures is additional area for exploitation in nanoconstuct and surface modification technology. Finally, the synthesis and functionalisation of nanorods holds potential in analytical applications. Combinatorial surface engineering of nanomaterials, paramagnetic or otherwise, using a given palette of monomers which can be put together in variable binary, ternary or even quaternary mixtures could lead to access to an almost unlimited vaiety of novel nanomaterials. Resulting chiral polymers deposited around/on nanomaterials could contain COOH groups for covalent grafting of ligands, which can be close to, or spaced from the asymmetric centres of the polymer frameworks. Such films could also be doped so that the adlayers could be electrically conducting. This would enable the electrical wiring of amperometric enzymes such as polyphenol oxidase, glucose oxidase, and horseradish peroxidase onto their surfaces and the possibility of novel biosensor constructs. Finally, combinatorial engineering of these polymeric adlayers, combined with possible sintering effects could enable the simultaneous covalent grafting of more than one type of ligand creating multifunctionalized nanomaterials. A different type of surface functionalisation is required for the modification of nanomaterials' surfaces to facilitate their use as support materials in relatively non-specific generic bioseparations, to reduce multistep to single step processes with concomitant gains in time, ease but no loss in associated quality of the final product. Amongst other chemistries recent novel technology, relating to surface film coatings consisting of modified or unmodified hydrophobic perfluorpolymers and hydrophilic polyanilin have been demonstrated which offer significant and radical benefits over existing materials for nucleic acid and protein purification reducing the process to a single step, faster approach, which can be used on large scale.

8.3 The selection and immobilisation of ligands

In making surfaces and materials useful in applications the selection and immobilisation of ligands to the materials is of prime consideration Often this commences with *in silico* and literature searches for nucleic acid and protein sequences which after their synthesis and purification can be used in the construction of novel, better and improved protein and DNA arrays (see later). Principally but not exclusively ligands for surface grafting in food analysis could be antibodies and oligonucleotides capture sequences (see Figure 8)

8.4 Microarrays

Characterisation and detection of PCR products and protein/protein/other molecular interactions based upon antibody/antigen (immuno assays) are much-used mechanisms in molecular diagnostics. This characterisation is more and more frequently being performed via the use of microarrays. However, current microarrays suffer from a range of limitations. In the first case most commonly used methods are imprecise and the detection of sequence variants, including SNPs which is important in many applications, is not possible without complex and time consuming experimental procedures. In the second place there are serious limitations as to the number of parallel assays that can be conducted. New array formats responding to these issues to be delivered by nanotechnology based upon nanoporous flat surfaces and particle arrays, could bring significant advantages to existing array fabrication strategies and materials.

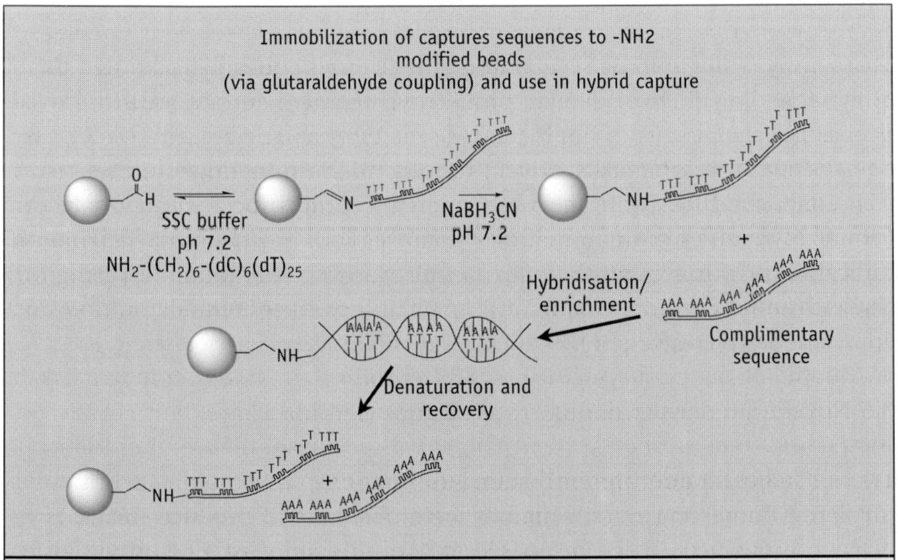

Figure 8. Oligonucleotide grafting to nanoparticle surfaces.

Novel high density encoded magnetisable particle/suspension arrays could be developed using variants of paramagnetic nanoparticles activated with appropriate ligands. Specific oligonucleotides could be immobilised to individual batches of nanoparticles permitting high degrees of surface activation whilst overcoming many of the problems associated with present day flat surface array construction, spotting and activation. Novel diene/dienophile chemistries and dipyrrole/carbazole polymers could be used in these immobilisations as well as standard classical approaches involving silanisation. Particle coding could be by incorporation of a fluorescent marker that would permit unambiguous, sequence specific, on-bead fluorescence detection of hybridised target DNA.

A possible alternative to the method described above for particle coding could be the use of quantum dots (QDs). Quantum dots are nanocrystalline materials with size-dependent photon emission properties, zinc sulfide-cadmium selenide clusters, that are intrinsically fluorescent, and whose emission maxima vary with particle size. In theory, a selection of different sized QDs in combination with different amounts of QDs can generate a total possible number of coding options of (m x [n-1]) where m is the number of different emission maxima (different sized QD crystals) and n is the number of different fixed wave length intensities (different amounts of QDs).

These arrays will combine several desirable features currently lacking in other array technologies: high density probe coverage/high surface area/improved fluorescent characteristics (for increased sensitivity and decreased S-to-N); ease of handling in magnetically driven, robotic, high throughput systems, using well-established hardware (reduced operator contact, potential closed tube format, lower risk of contamination); unambiguous 'yes-no' signal generation indicating presence of target sequence, no washing step required (using MF fluorochromes in probe sequence); encodability for potential multiplexing applications (presence of QD codes).

8.5 Novel high-density nanoporous oligonucleotide arrays

Two substrates have very recently been innovated that when used in hybrisations involving fluorescence-based signals were observed to produce increases in detection sensitivity when compared to common silicated glass monolayers. Colloidal silica films of variable thicknesses have been deposited onto glass slides resulting in a two-fold increase in hybridization signal and dendrimeric 2D-organized monolayers (64 NH_2 groups per dendrimer) have been attached to silicated glasses resulting in hybridization signals which were between five and ten fold greater than that from silicated glass monolayers. In this case fluorescence signal increase corresponded directly with the number of 2D-disposed NH_2 groups per dendrimer potentially available for binding DNA probes.

DNA and protein/antibody microarrays based upon nanoporous surfaces that will possess greater sensitivity and definition than those in already in existence could be researched and developed. Unique 3D-matrix (nanocolloidal based) surfaces could be constructed to which DNA and protein, antibody or antibody fragments could be immobilised in the fabrication of microarrays using classical robotic printing methodology. Microarrays can be produced on these surfaces using chiral, chemically activated polypyrrole/polycarbazole films grown on the nanocolloids that will permit non-random covalent grafting of protein/antibody ligands using "matched-mismatched" properties. Optically pure electrosynthesized carboxy-polypyrrole, or polycarbazole-covered SnO_2 Sb/F-doped SnO_2 and TiO_2 nanocolloids could be employed. Surface roughness could be varied as a factor of nanocolloid size within the range 100 and 1000 times (from 5 to 100 nm) and layer thickness could also be varied to improve surface area (50 nm to 10 µm). Nanoporous, transparent microarrays, possessing surfaces of controlled pore distribution (5 to 100 nm / 50 to 1,000 Å) should bind more ligand probes per square unit area than their non nanoporous counterparts. Another way to improve surface area for grafting in microarray construction could be to covalently graft starburst dendrimers such as highly-branched G1 to G10 *poly*NH_2-starbust poly(amidoamine) and poly(propyleneimine) dendrimers (PAMAM and PPI).

Dendrimerized carboxy-polypyrrole- or polycarbazole-nanocomposite colloids sintered onto glass slides, as vertically and horizontally homogenous transparent adlayers, could both simultaneously address existing flat surface array limitations, i.e., available surface area for probe immobilization and maximum achievable probe density and dot homogeneity. A further advantage of this material and the surface engineering approach taken to its construction will arise from its modular nature. This would allow better, more complete and reproducible optimization of its structure and function and therefore reproducibility in array fabrication.

Finally, carbon nanotubes based arrays could be constructed by chemical vapor deposition of carbon nanotubes to flat surfaces in arrays, which could radically increase surface area available for immobilisation of useful ligands (see Figure 9).

9. Public acceptance of new science and technology

Any new technology needs to be fully validated and determined to be safe and efficient from the public point of view so as to aid in its general acceptance and uptake. This is particularly so in the case of a 'new science' such as nanotechnology. However in the context of food testing its application in a diagnostic fashion rather than as a component of a finished product, and the psychological benefit of 'better' testing regimes, means that its uptake should actually be promoted.

It is also important to note that nanotechnology may be a new subject to the general public but in reality represents a synthesis of disciplines already existent i.e., materials science, chemistry, surface chemistry and biology. It is particularly important that potential resistance to the uptake of the benefits of nanotechnology on the basis of misinformation concerning its novelty and potential hazards is set within that context.

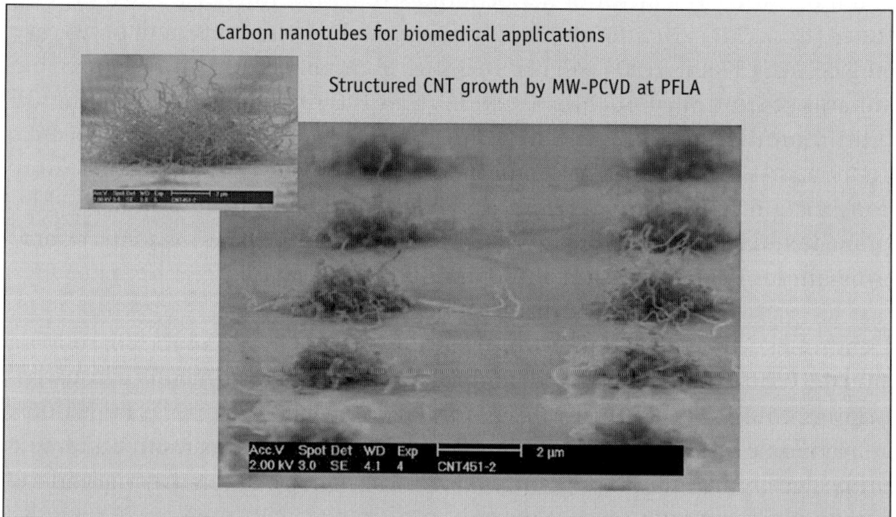

Figure 9. Carbon nanotubes based flat surface microarrays.

10. Conclusion

Nanotechnology is an enabling technology impacting health, computing and electronics, public safety, the environment and manufacturing. etc. The field is multidisciplinary but starts with chemistry and materials science.

Principal challenges are:
- novel synthesis strategies;
- characterisation of nanoconstructs;
- large scale fabrication;
- applications development.

Opportunities and rewards are great (in our sector estimated to be €29 billion by the end of the project).

Useful and interesting reading

Drexler, K.E., 1986. Engines of creation. The coming era of nanotechnology. Anchor Books Doubleday Publishing, New York.

Drexler, K.E., 2001. Machine-phase nanotechnology. Scientific American, September issue. (http://www.sciam.com)

Feynman, R.P., 1959. There's plenty of room at the bottom. http://www.zyvex.com/nanotech/feynman.html

Wood., S., R. Jones and A. Geldart, 2003. The social and economic challenges of nanotechnology. (http://www.esrc.ac.uk/esrccontent/downloaddocs/nanotechnology.pdf)

Websites

The Institute of Nanotechnology (IoN), http://www.nano.org.uk
Nanoforum (a European Union sponsored Thematic Network), http://www.nanoforum.org
Foresight Institute, http://www.foresight.org

Index